T0214344

Lecture Notes in Computer Science 11315

Commenced Publication in 1973
Founding and Former Series Editors:
Gerhard Goos, Juris Hartmanis, and Jan van Leeuwen

More information about this series at http://www.springer.com/series/7409

Hujun Yin · David Camacho
Paulo Novais · Antonio J. Tallón-Ballesteros (Eds.)

Intelligent Data Engineering and Automated Learning – IDEAL 2018

19th International Conference
Madrid, Spain, November 21–23, 2018
Proceedings, Part II

Springer

Editors
Hujun Yin 🆔
University of Manchester
Manchester, UK

David Camacho 🆔
Autonomous University of Madrid
Madrid, Spain

Paulo Novais 🆔
Campus of Gualtar
University of Minho
Braga, Portugal

Antonio J. Tallón-Ballesteros
University of Seville
Seville, Spain

ISSN 0302-9743   ISSN 1611-3349   (electronic)
Lecture Notes in Computer Science
ISBN 978-3-030-03495-5   ISBN 978-3-030-03496-2   (eBook)
https://doi.org/10.1007/978-3-030-03496-2

Library of Congress Control Number: 2018960396

LNCS Sublibrary: SL3 – Information Systems and Applications, incl. Internet/Web, and HCI

This Springer imprint is published by the registered company Springer Nature Switzerland AG
The registered company address is: Gewerbestrasse 11, 6330 Cham, Switzerland

Preface

This year saw the 19th edition of the International Conference on Intelligent Data Engineering and Automated Learning (IDEAL), which has been playing an increasingly leading role in the era of big data and deep learning. As an established international forum, it serves the scientific communities and provides a platform for active, new, and leading researchers in the world to exchange the latest results and disseminate new findings. The IDEAL conference has continued to stimulate the communities and to encourage young researchers for cutting-edge solutions and state-of-the-art techniques on real-world problems in this digital age. The IDEAL conference attracts international experts, researchers, academics, practitioners, and industrialists from machine learning, computational intelligence, novel computing paradigms, data mining, knowledge management, biology, neuroscience, bio-inspired systems and agents, distributed systems, and robotics. It also continues to evolve to embrace emerging topics and trends.

This year IDEAL was held in one of most beautiful historic cities in Europe, Madrid. In total 204 submissions were received and subsequently underwent the rigorous peer-review process by the Program Committee members and experts. Only the papers judged to be of the highest quality were accepted and are included in the proceedings. This volume contains 125 papers (88 for the main rack and 37 for workshops and special sessions) accepted and presented at IDEAL 2018, held during November 21–23, 2018, in Madrid, Spain. These papers provided a timely sample of the latest advances in data engineering and automated learning, from methodologies, frameworks, and techniques to applications. In addition to various topics such as evolutionary algorithms, deep learning neural networks, probabilistic modeling, particle swarm intelligence, big data analytics, and applications in image recognition, regression, classification, clustering, medical and biological modeling and prediction, text processing and social media analysis. IDEAL 2018 also enjoyed outstanding keynotes from leaders in the field, Vincenzo Loia, Xin Yao, Alexander Gammerman, as well as stimulating tutorials from Xin-She Yang, Alejandro Martin-Garcia, Raul Lara-Cabrera, and David Camacho.

The 19th edition of the IDEAL conference was hosted by the Polytechnic School at Universidad Autónoma de Madrid (UAM), Spain. With more than 30,000 students, and 2,500 professors and researchers and a staff of over 1,000, the UAM offers a comprehensive range of studies in its eight faculties (including the Polytechnic School). UAM is also proud of its strong research commitment that is reinforced by its six university hospitals and the ten join institutes with CSIC, Spain's National Research Council.

We would like to thank all the people who devoted so much time and effort to the successful running of the conference, in particular the members of the Program Committee and reviewers, organizers of workshops and special sessions, as well as the authors who contributed to the conference. We are also very grateful to the hard work by the local organizing team at Universidad Autónoma de Madrid, especially Victor Rodríguez, for the local arrangements, as well as the help from Yao Peng at the University of Manchester for checking through all the camera-ready files. The continued support and collaboration from Springer LNCS are also greatly appreciated.

September 2018

Hujun Yin
David Camacho
Paulo Novais
Antonio J. Tallón-Ballesteros

Organization

Honorary Chairs

Hojjat Adeli	Ohio State University, USA
Francisco Herrera	Granada University, Spain

General Chairs

David Camacho	Universidad Autónoma de Madrid, Spain
Hujun Yin	University of Manchester, UK
Emilio Corchado	University of Salamanca, Spain

Programme Co-chairs

Carlos Cotta	Universidad de Málaga, Spain
Antonio J. Tallón-Ballesteros	University of Seville, Spain
Paulo Novais	Universidade do Minho, Portugal

International Advisory Committee

Lei Xu (Chair)	Chinese University of Hong Kong and Shanghai Jiaotong University, China
Yaser Abu-Mostafa	CALTECH, USA
Shun-ichi Amari	RIKEN, Japan
Michael Dempster	University of Cambridge, UK
José R. Dorronsoro	Autonomous University of Madrid, Spain
Nick Jennings	University of Southampton, UK
Soo-Young Lee	KAIST, South Korea
Erkki Oja	Helsinki University of Technology, Finland
Latit M. Patnaik	Indian Institute of Science, India
Burkhard Rost	Columbia University, USA
Xin Yao	Southern University of Science and Technology, China and University of Birmingham, UK

Steering Committee

Hujun Yin (Chair)	University of Manchester, UK
Laiwan Chan (Chair)	Chinese University of Hong Kong, Hong Kong, SAR China
Guilherme Barreto	Federal University of Ceará, Brazil
Yiu-ming Cheung	Hong Kong Baptist University, Hong Kong, SAR China

Emilio Corchado	University of Salamanca, Spain
Jose A. Costa	Federal University of Rio Grande do Norte, Brazil
Marc van Hulle	K. U. Leuven, Belgium
Samuel Kaski	Aalto University, Finland
John Keane	University of Manchester, UK
Jimmy Lee	Chinese University of Hong Kong, Hong Kong, SAR China
Malik Magdon-Ismail	Rensselaer Polytechnic Inst., USA
Peter Tino	University of Birmingham, UK
Zheng Rong Yang	University of Exeter, UK
Ning Zhong	Maebashi Institute of Technology, Japan

Publicity Co-chairs/Liaisons

Jose A. Costa	Federal University of Rio Grande do Norte, Brazil
Bin Li	University of Science and Technology of China, China
Yimin Wen	Guilin University of Electronic Technology, China

Local Arrangements Chairs

Antonio González Pardo	Raúl Lara Cabrera
Cristian Ramírez Atencia	Raquel Menéndez Ferreira
Víctor Rodríguez Fernández	F. Javier Torregrosa López
Alejandro Martín García	Ángel Panizo Lledot
Alfonso Ortega de la Puente	Marina de la Cruz

Programme Committee

Paulo Adeodata	Zoran Bosnic
Imtiaj Ahmed	Vicent Botti
Jesus Alcala-Fdez	Edyta Brzychczy
Richardo Aler	Andrea Burattin
Davide Anguita	Robert Burduk
Ángel Arcos-Vargas	José Luis Calvo Rolle
Romis Attux	Heloisa Camargo
Martin Atzmueller	Josep Carmona
Javier Bajo Pérez	Mercedes Carnero
Mahmoud Barhamgi	Carlos Carrascosa
Bruno Baruque	Andre Carvalho
Carmelo Bastos Filho	Pedro Castillo
José Manuel Benitez	Luís Cavique
Szymon Bobek	Darryl Charles
Lordes Borrajo	Francisco Chavez

Raquel Menéndez Ferreira
José M. Molina
Mati Mottus
Valery Naranjo
Susana Nascimento
Tim Nattkemper
Antonio Neme
Ngoc-Thanh Nguyen
Yusuke Nojima
Fernando Nuñez
Eva Onaindia
Jose Palma
Ángel Panizo Lledot
Juan Pavón
Yao Peng
Carlos Pereira
Sarajane M. Peres
Costin Pribeanu
Paulo Quaresma
Juan Rada-Vilela
Cristian Ramírez-Atencia
Izabela Rejer
Victor Rodriguez Fernandez
Zoila Ruiz
Luis Rus-Pegalajar
Yago Saez

Jaime Salvador
Jose Santos
Matilde Santos
Dragan Simic
Anabela Simões
Marcin Szpyrka
Jesús Sánchez-Oro
Ying Tan
Ricardo Tanscheit
Renato Tinós
Stefania Tomasiello
Pawel Trajdos
Stefan Trausan-Matu
Carlos M. Travieso-González
Milan Tuba
Turki Turki
Eiji Uchino
José Valente de Oliveira
José R. Villar
Lipo Wang
Tzai-Der Wang
Dongqing Wei
Michal Wozniak
Xin-She Yang
Weili Zhang

Additional Reviewers

Mahmoud Barhamgi
Gema Bello
Carlos Camacho
Carlos Casanova
Laura Cornejo
Manuel Dorado-Moreno
Verónica Duarte
Antonio Durán-Rosal
Felix Fuentes
Dušan Gajić
Brunno Goldstein
David Guijo

César Hervás
Antonio López Herrera
José Ricardo López-Robles
José Antonio Moral Muñoz
Eneko Osaba
Zhisong Pan
Pablo Rozas-Larraondo
Sancho Salcedo
Sónia Sousa
Radu-Daniel Vatavu
Fion Wong
Hui Xue

Workshop on RiskTrack: Analyzing Radicalization in Online Social Networks

Organizers

Javier Torregrosa   Universidad Autónoma de Madrid, Spain
Raúl Lara-Cabrera   Universidad Autónoma de Madrid, Spain
Antonio González Pardo   Universidad Autónoma de Madrid, Spain
Mahmoud Barhamgi   Université Claude Bernard Lyon 1, France

Workshop on Methods for Interpretation of Industrial Event Logs

Organizers

Grzegorz J. Nalepa   AGH University of Science and Technology,
                             Poland
David Camacho   Universidad Autónoma de Madrid, Spain
Edyta Brzychczy   AGH University of Science and Technology,
                             Poland
Roberto Confalonieri   Smart Data Factory, Free University of
                             Bozen-Bolzano, Italy
Martin Atzmueller   Tilburg University, The Netherlands

Workshop on the Interplay Between Human–Computer Interaction and Data Science

Organizers

Cristian Mihăescu   University of Craiova, Romania
Ilkka Kosunen   University of Tallinn, Estonia
Ivan Luković   University of Novi Sad, Serbia

Special Session on Intelligent Techniques for the Analysis of Scientific Articles and Patents

Organizers

Manuel J. Cobo   University of Granada, Spain
Pietro Ducange   eCampus University, Italy

Antonio Gabriel López-Herrera   University of Granada, Spain
Enrique Herrera-Viedma   University of Granada, Spain

Special Session on Machine Learning for Renewable Energy Applications

Organizers

Sancho Salcedo Sanz   Universidad de Alcalá, Spain
Pedro Antonio Gutiérrez   University of Cordoba, Spain

Special Session on Evolutionary Computing Methods for Data Mining: Theory and Applications

Organizers

Eneko Osaba   TECNALIA Research and Innovaton, Spain
Javier Del Ser   University of the Basque Country, Spain
Sancho Salcedo-Sanz   University of Alcalá, Spain
Antonio D. Masegosa   University of Deusto, Spain

Special Session on Data Selection in Machine Learning

Organizers

Antonio J. Tallón-Ballesteros   University of Seville, Spain
Ireneusz Czarnowski   Gdynia Maritime University, Poland
Simon James Fong   University of Macau, SAR China
Raymond Kwok-Kay Wong   University of New South Wales, Australia

Special Session on Feature Learning and Transformation in Deep Neural Networks

Organizers

Richard Hankins   University of Manchester, UK
Yao Peng   University of Manchester, UK
Qing Tian   Nanjing University of Information Science
    and Technology, China
Hujun Yin   University of Manchester, UK

Special Session on New Models of Bio-inspired Computation for Massive Complex Environments

Organizers

Antonio González Pardo   Universidad Autónoma de Madrid, Spain
Pedro Castillo   Universidad de Granada, Spain
Antonio J. Fernández Leiva Universidad de Málaga, Spain
Francisco J. Rodríguez   Universidad de Extremadura, Spain

Contents – Part II

Special Session on Intelligent Techniques for the Analysis of Scientific Articles and Patents

Special Session on Machine Learning for Renewable Energy Applications

**Special Session on Evolutionary Computing Methods for Data Mining:
Theory and Applications**

Special Session on Data Selection in Machine Learning

**Special Session on New Models of Bio-inspired Computation for Massive
Complex Environments**

Contents – Part I

Workshop on RiskTrack: Analyzing Radicalization in Online Social Networks

Ontology Uses for Radicalisation Detection on Social Networks

Mahmoud Barhamgi[1]([✉]), Raúl Lara-Cabrera[2], Djamal Benslimane[1], and David Camacho[2]

[1] Université Claude Bernard Lyon 1, LIRIS lab, Lyon, France
mahmoud.barhamgi@univ-lyon1.fr
[2] Universidad Autonoma Madrid, Madrid, Spain
{raul.lara,david.camacho}@uam.es

Abstract. Social networks (SNs) are currently the main medium through which terrorist organisations reach out to vulnerable people with the objective of radicalizing and recruiting them to commit violent acts of terrorism. Fortunately, radicalization on social networks has warning signals and indicators that can be detected at the early stages of the radicalization process. In this paper, we explore the use of the semantic web and domain ontologies to automatically mine the radicalisation indicators from messages and posts on social networks.

Keywords: Semantic web · Ontology · Radicalization · Semantics
Social networks

1 Introduction

Social networks have become one of the key mediums through which people communicate, interact, share contents, seek information and socialize. According to recent studies published by Smart Insight Statistics, the number of active users on social networks has reached 2.8 billion users, accounting for one-third of the world population. Unfortunately, terrorist groups and organisations have also understood the immense potential of social networks for reaching out to people around the world and as a consequence, they now rely heavily on such networks to propagate their propagandas and ideologies, radicalise vulnerable individuals and recruit them to commit violent acts of terror.

Social networks can additionally play an important role in the fight against radicalisation and terrorism. In particular, they can be seen as an immense data source that can be analysed to discover valuable information about terrorist organisations, including their recruitment procedures and networks, terrorist attacks as well as the activities and movements of their disciples. They can be also analysed to identify individuals and populations who are vulnerable to radicalisation in order to carry out preventive policies and actions (e.g., psychological and medical treatments for individuals, targeted education plans for communities) before those populations fall into the radicalisation trap.

Social network data analysis raises important scientific and technical challenges. Some of the key challenges involve the need to handle a huge volume of data, the high

© Springer Nature Switzerland AG 2018
H. Yin et al. (Eds.): IDEAL 2018, LNCS 11315, pp. 3–8, 2018.
https://doi.org/10.1007/978-3-030-03496-2_1

dynamicity of data (as contents of social networks continue to evolve with the continuous interactions with users), and the large value of noise present in social network data which affects the quality of data analysis. These challenges emphasis the need for automating the data analysis to the most to reduce the human intervention required from data analysts.

One of the vital research avenues for pushing further the limits of existing data mining techniques is the use of semantics and domain knowledge [1], which has resulted into the Semantic Data Mining (SDM) [2]. The SDM refers to the data mining tasks that systematically incorporate domain knowledge, especially formal semantics, into the process of data mining [2]. The utility of domain knowledge for data mining tasks has been demonstrated by the research community. Fayyad et al. [1] pointed out that domain knowledge can be exploited in all data mining tasks including, data transformation, feature reduction, algorithm selection, post-processing, data interpretation. For these purposes, domain knowledge should first be captured and represented using models that can be processed and understood by machines. Formal ontologies and associated inference mechanisms [2] can be used to specify and model domain knowledge. An ontology is a formal explicit description of concepts in a domain of discourse along with their properties and interrelationships. Domain concepts are often referred to as ontology classes. An ontology along with the instances of its concepts is often called a knowledge base. The Semantic Web research community has defined over the last decade several standard ontology specification languages such as the Ontology Web Language (OWL), RDF and RDFS, as well as effective tools that can be exploited to create, manage and reason on ontologies. These standard languages can be exploited to represent and model domain knowledge.

In this paper, we explore the use of domain knowledge and semantics for mining social data networks. We use online violent radicalization and terrorism as the application domain. We explore the use of ontologies to improve the radicalisation detection process on social media.

2   Ontology Uses for Radicalisation Detection

In this section, we present and explore the different uses of a semantic knowledge-base to improve the process of identifying violent radicalised individuals on social networks. Ontologies can be useful in two major phases: (i) Data analysis and (ii) Data exploration. We detail in the following how ontologies can be exploited to enrich these two phases.

2.1   Ontologies in the Data Analysis Phase

The objective of the data analysis phase is to compute the considered radicalisation indicators to determine whether an individual is radicalised or not. This computation is usually carried out for a population of individuals (e.g., the members of an online group, the inhabitants of a city, district). The indicators themselves are defined by domain experts (e.g., experts in psychology, criminology). This phase is often supervised by a *"Data Analyst"* who could choose the indicators they desire among the ones defined by experts. In this phase, different data analysis algorithms could be applied to compute the indicators including data clustering, community detection and sentiment analysis. The output of this phase is the indicators computed for the complete set of the considered population.

Most of the data analysis algorithms in the data analysis phase compute the radicalisation indicators by counting the occurrences of some specific keywords that relate to the indicator considered. Data analysts are, therefore, required to supply the applicable keywords for each indicator. For example, an indicator such as the *"identification to a specific terrorist group"*, for example the so-called *"Islamic State of Iraq and Syria"*, may be computed by counting the frequency of its different acronyms and abbreviations such as *"Daesh"*, *"ISIS"*, *"ISL"*, *"Daech"*, to name just a few. Relying merely on keywords has the following major limitations:

- *Missing relevant names, acronyms and abbreviations of an entity:* The analyst is required to figure out the different names and acronyms for an entity. Missing some of these names and acronyms leads to erroneous value for the computed indicator. For example, the Islamic State of Iraq and Syria has numerous names and abbreviations (e.g., *ISIS, ISIL, DAECH, DAESH, AL-KILAFA, Dawlatu-AL-Islam, the state of Islam*, etc.). Supplying the complete list of names and abbreviations may become very difficult, if not impossible, for a human to realize.
- *Missing relevant keywords:* A radicalised individual may not be referring directly to ISIS in his or her social messages that could still reflect his or her ISIS identification. For example, if a phrase such as *"I declare my entire loyalty to Al-Baghdadi"* was analysed by a data analysis algorithm with a limited scope, then the individual may not be considered as identifying himself to ISIS, while he should be considered as such, as *Al-Baghdadi* is the leader of ISIS.
- *Missing related or similar keywords:* The analyst may miss the keywords that are similar to a given keyword but do not refer to the same concept or entity that is referred to by the considered keyword. For example, the analyst maybe interested in measuring the identification of an individual to a given terrorist group such as ISIS. However, terrorist groups are numerous (e.g., ISIS, Hezbollah, Al-Qaeda, Al-Nusra, etc.) and by focusing on ISIS, the analyst may miss the individuals who identify themselves with Hezbollah and Al-Qaeda.

Formal domain ontologies can be exploited to address those limitations. Specifically, by modelling and representing domain knowledge with an ontology, the data analysis phase could be extended as follows:

- *Analyse-by-Concept:* Domain concepts (represented as ontological concepts) could be exploited by the data analysis algorithms instead of relying solely on keywords to analyse the social content. For example, the concept *"Islamic-State-of-Iraq-and-Sham"* can be defined formally as an instance of the concept *"Terrorist-Group"* in a domain ontology, and associated with its different names and abbreviations (please refer to Fig. 1). Using ontological concepts (instead of keywords) would relieve data analysts from the need to cite all names, acronyms and abbreviations of an entity or concept, as those would be already identified and incorporated into the ontology by domain experts.

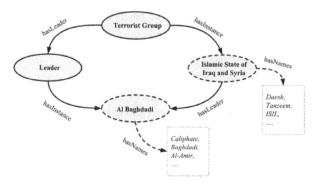

Fig. 1. Part of a domain ontology

- *Inclusion of related concepts:* Semantic relationships among concepts can be exploited to extend the data analysis to involve all concepts that relate to an indicator. For example, the relationship *"hasLeader"* that exists between the concepts *"Terrorist-Organization"* and *"Leader"* can exploited to infer that the identification to ISIS can also be inferred from messages that show an identification to *Al-Baghdadi*, the leader of ISIS.
- *Inclusion of similar concepts:* Sibling relationships can be represented by the instantiation mechanism of ontologies. For example, all of the entities *"Islamic-State-of-Iraq-and-Sham"*, *"Hezbollah"*, *"Al-Qaeda"*, *"Al-Nusrah"* would be represented as instances of the concept *"Terrorist-Organization"*. The analyst can query the domain ontology for those similar concepts/instances and enhance further the outcome of his algorithms.

2.2 Ontologies in the Data Exploration Phase

In the data exploration phase, the analysed dataset (i.e., the initial dataset plus the computed indicators) could be queried by end-users in different ways and for different purposes, depending on their needs and profiles. Examples include:

- An officer in a law enforcement agency could be interested just in identifying the list of individuals who could constitute an imminent danger to their society;
- An expert in sociology could be interested in exploring the correlation between religious radicalisation and the socio-economical situation of a population, or the correlation between an indicator and a specific class of individuals (involving some specific personal traits), etc.
- An educator or a city planner could be interested in exploring the correlation between an indicator (e.g., frustration, discrimination for being Muslim) and the geographic distribution of a population.

As with the data analysis phase, querying computed indicators relying solely on keywords has several limitations. We explain those limitations based on some query examples. Figure 2 shows some possible queries, along with their answers, that could be issued over a radicalised individual dataset in the data exploration phase. The figure shows also a sample of radicalised individuals along with their computed indicators.

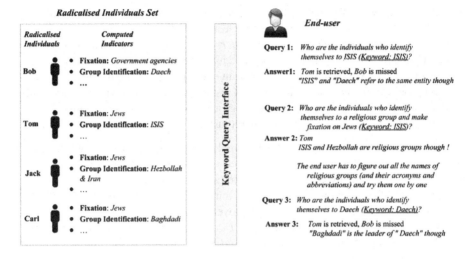

Part-A: The limitations of querying with mere keywords

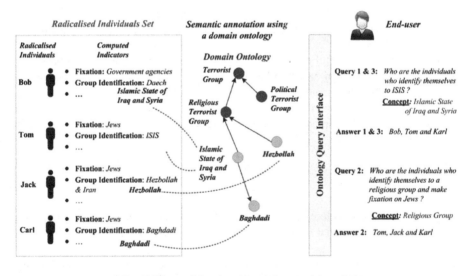

Part-B: The use of domain ontology to improve data analysis

Fig. 2. Part-A: the limitations of querying with mere keywords; Part-B: the use of ontology to improve data analysis

- *Missing relevant results:* Query-1 employs the keyword "*ISIS*" to search for the individuals who identify themselves to ISIS. Without the use of a domain ontology that could formally define the concept of ISIS (i.e., *Islamic State of Iraq and Syria*), the answer to that query would include only *Tom*. That is, even though *Bob* identifies himself to ISIS, he might not be considered, as the keyword "*Daech*" was used in his group identification indicator instead of "*ISIS*". Similarly, both *Query-1* and *Query*-3 will miss the individual *"Carl"* as he would be annotated with the keyword *"Baghdadi"* who, in turn, is the leader of *"Daech"*.

- *Missing similar and related results:* *Query-2* searches for the individuals who identify themselves to a religious group and make fixation on jaws. *Query-2* uses the keyword *"ISIS"*. Without the use of a domain ontology that would define the concept of religious group and specify its different instances that are known so far, the end-user has to figure out the names of these groups along with abbreviations used for each specific group which could a difficult task for a human, if not impossible.

The data exploration phase could be largely improved by the use of a domain ontology. Domain ontologies can be exploited to address the aforementioned limitations as follows. The analysed dataset (i.e., the radicalized individuals along with their computed radicalisation indicators) could be annotated, as shown in Fig. 2, with the ontological concepts that could be also employed in queries.

For example, the keywords *"ISIS"* and *"Daech"* can both be associated with the ontological concept *"Islamic-State-of-Iraq-and-Syria"* which is an instance of the concept *"Religious-Group"*. The instance *"Islamic-State-of-Iraq-and-Syria"* can be used to directly annotate *"Bob"* and *"Tom"*. Similarly, the instance *"Baghdadi"* can be used to annotate the information of *"Carl"*. The ontology states that *"Islamic-State-of-Iraq-and-Syria"* has *"Baghdadi"* as a leader. Therefore, *"Carl"* would appear in *Query-1*'s results. The same also applies to the term *"Hezbollah"* which is an instance of the concept *"Religious Group"*.

Annotation is a powerful mechanism that allows end-users to query data based on their semantics, rather than based on keywords. It is worth to mention that annotation with semantic information can be applied on *individuals* (for example, to express the fact that an individual is exhibiting a given radicalization indicator) and *messages* (to express the fact that the content of a message relates to an indicator and should be used to compute the indicator's value).

3   Conclusion

In this paper, we have explored the use of semantics and domain ontologies to improve the process of mining social networks for the radicalisation indicators.

Acknowledgment. This work is under the European Regional Development Fund FEDER, and Justice Programme of the European Union (2014–2020) 723180 – RiskTrack –JUST-2015-JCOO-AG/JUST-2015-JCOO-AG-1.

References

1. Fayyad, U., Piatetsky-Shapiro, G., Smyth, P.: From data mining to knowledge discovery in databases. AI Mag. **17**, 37 (1996)
2. Stephan, G., Pascal, H., Andreas, A.: Knowledge Representation and Ontologies. In: Studer, R., Grimm, S., Abecker, A. (eds.) Semantic Web Services. Springer, Heidelberg (2007)

Measuring Extremism: Validating an Alt-Right Twitter Accounts Dataset

Joshua Thorburn[1], Javier Torregrosa[2(✉)], and Ángel Panizo[3]

[1] RMIT University, Melbourne, Australia
[2] Biological and Health Psychology Department,
Universidad Autónoma de Madrid, 28049 Madrid, Spain
francisco.torregrosa@uam.es
[3] Computer Engineering Department, Universidad Autónoma de Madrid,
28049 Madrid, Spain

Abstract. Twitter is one of the most commonly used Online Social Networks in the world and it has consequently attracted considerable attention from different political groups attempting to gain influence. Among these groups is the alt-right; a modern far-right extremist movement that gained notoriety in the 2016 US presidential election and the infamous Charlottesville Unite the Right rally. This article details the process used to create a database on Twitter of users associated with this movement, allowing for empirical research into this extremist group to be undertaken. In short, Twitter accounts belonging to leaders and groups associated with the Unite the Right rally in Charlottesville were used to create this database. After compiling users who followed these leading alt-right figures, an initial sample was gathered (n = 549). This sample was then validated by two researchers. This involved using a scoring method created for this process, in order to eliminate any accounts that were not supportive of the alt-right. Finally, a total amount of 422 accounts were found to belong to followers of this extremist movement, with a total amount of 123.295 tweets.

Keywords: Extremism · Far-right · Twitter · Alt-right · White supremacy

1 Introduction

Very little academic research has been conducted analysing the alt-right movement, despite the far-right white supremacist groups gaining widespread cultural notoriety. The Oxford Dictionary defines the alt-right as "An ideological grouping associated with extreme conservative or reactionary viewpoints, characterized by a rejection of mainstream politics and by the use of online media to disseminate deliberately controversial content" [1]. The alt-right is primarily an online phenomenon, comprising of supporters from the US, but also from across the world [2, 3]. Initially starting off as a movement created by a small grouping of obscure far-right activists in 2008, the alt-right has extensively relied on social media to grow [4]. With the use of a "deceptively benign" name, ironic humour and the appropriation of the 'Pepe the Frog' internet meme, the movement has sought to avoid the negative stigma that white supremacism, fascism and neo-Nazism normally entail [4, 5].

© Springer Nature Switzerland AG 2018
H. Yin et al. (Eds.): IDEAL 2018, LNCS 11315, pp. 9–14, 2018.
https://doi.org/10.1007/978-3-030-03496-2_2

By disseminating its propaganda on social media websites including Twitter, 4chan, YouTube and Reddit, the alt-right has been able to reach a young audience [6]. The alt-right featured prominently on various social media sites during the 2016 US presidential election and gathered considerable press coverage. However, the Charlottesville Unite the Right rally on August 11–12, 2017, was a significant moment in the history of the alt-right which contributed to a factional split in the group [7, 8]. The rally, organised to protest the removal of a statue of Confederate leader Robert E. Lee, was marred by the actions of James Alex Fields Jr., who drove a car into a group of counter-protesters, killing one woman and injuring nineteen others [9]. The perpetrator was photographed earlier in the day wearing a shirt emblazoned with the logo of Vanguard America, a white supremacist group associated with the alt-right [9]. Following negative media attention from this event, many political and media figures disavowed themselves from the alt-right [10]. Consequently, this event was used as a basis to study the alt-right, as it has been the most prominent event in the movement's history.

Members of the alt-right commonly use Twitter. They have utilised Twitter's various functions to spread its ideology, including through the use of hashtag slogans ("#ItsOkToBeWhite", "#WhiteGenocide"), media (such as memes featuring "Pepe the Frog") and the group's own vernacular ("kek", "cuck", "red pill", etc.). Although accounts associated with jihadi extremism that are frequently suspended or deleted by Twitter, accounts belonging to members of the alt-right are still relatively prolific on Twitter [11]. The presence of the Alt-Right on Twitter, however, can lead to an opportunity to study this phenomenon. Due to the general lack of empirical research on the alt-right, this research aimed to create a dataset of accounts associated with the alt-right, in order to study and understand this modern political movement. To create this database, a team of researchers gathered and validated hundreds of Twitter accounts related to this far-right movement.

2 Methodology

The Twitter API tool was used to create the database of 422 accounts of alt-right supporters presented in this paper. The Unite the Right rally was used as a starting point in the process of creating it. Three different posters that were created for this event were found, with twelve different Twitter accounts that were linked to speakers and groups advertised to be attending the rally. In collaboration with a team of computer scientists, all Twitter accounts that followed any of these accounts were collected. Due to some of these accounts being followed by tens of thousands of people, our database was refined to only include accounts that followed six or more of these initial twelve accounts. This initial filtering process resulted in 2,533 users. It is worth noting that nine of these twelve initial accounts associated with the Unite the Right rally in Charlottesville have subsequently been deleted or suspended by Twitter in violation of the content policies.

Following this, 546 accounts were randomly extracted. After removing accounts with less than 20 tweets, this database consisted of 541 accounts with a total of 129,884 tweets. The following information was extracted from these accounts:

- User ID.
- Public name.
- Description
- Tweet ID.
- Text of the tweet.
- Time when the tweet was published.
- Language of the tweet.

After this process, an evaluation method was created to ensure that all accounts were a valid representation of the alt-right (see Table 1). This method involved two researchers reading tweets belonging to each account and scoring each account using the criteria that is simplified in Table 1. Accounts that reached a score of five or more were included, whilst those that did not were excluded. This threshold was decided after taking into account the several indicators of the accounts showing extremist ideas: the more they had, the more likely they would be radicals. However, exceptions could be made based on the researcher's opinion, for accounts that failed to reach the threshold score of five, but had less than 50 tweets available. Due to the clear linguistic differences of journalists and news publications, accounts of this nature were excluded by default, even if they were associated with the alt-right.

Following this process, 422 accounts were found to be clearly belonging to individuals who supported the alt-right. Common reasons for accounts being excluded were: the user failed to reach the threshold; accounts belonged to journalists; the accounts contained limited data; or the user was opposed to the alt-right. More information regarding accounts that were excluded can be found in the results section.

A summary of the scoring system can be seen below, along with some examples of the variables that were used for the analysis of the accounts.

Fig. 1. "Pepe the frog" meme.

Table 1. Criteria for the validation score.

Group	Variable	Score
Traditional indicators of the alt-right/far-right	Anti-immigrant sentiment expressed	1
	Explicitly racist views	
	Mention of discrimination against white people	
	Misogyny/anti-feminism ideas	
	Meme containing "pepe the frog", see Fig. 1	
	Support of Alt-right leaders	
Indicators exclusively associated with the alt-right/far-right	Pro-nazi/fascist expressions	2
	Support for violent action/terrorism/retribution	
	Use of language commonly used by alt-right members	
Explicitly identifies as part of a white supremacist movement	Use of white supremacist hashtags (related to the alt-right)	3
	Mention of a white genocide	

3 Results

Upon completion of the validation process, 422 accounts were confirmed to be a suitable representation of the alt-right on Twitter. 127 of the original 549 accounts were excluded from the database during this validation process.

The following is an explanation of the reasons why accounts were excluded, with

- **Low score**: all users that did not reach the threshold originally established (5 points using all the information on their profiles). This includes supporters of President Trump and individuals who were right-wing politically, but did not reach the threshold to satisfy that they were likely supporters of the alt-right.
- **Detractor**: the user explicitly criticized the Alt-Right movement or the far-right ideology, or stated that they are left-wing politically.
- **Journalist/media**: the user said that they were a journalist. Or, alternatively, the account only shared news stories without any personal commentary, or was a media publication. This included journalists and publications that were supportive of the alt-right.
- **Suspended account**: the account was banned or suspended by Twitter, or was deleted by the user.
- **Private account/low data**: the account did not present enough information (less than 20 tweets or a huge number of links) to make an accurate assessment possible, or the user's data was blocked and only visible for approved followers of the user.

A summary of the discarded accounts is presented on Table 2.

Table 2. Summary of the articles excluded.

Exclusion cue	n
Low score	57
Journalist/media	36
Detractor	12
Suspended account	12
Private account/low data	10
Total	**127**

After this screening process, and once the accounts not fulfilling the criteria previously presented were erased, 422 accounts remained, with a total amount of 123.295 tweets.

4 Discussion

The present research aimed to create a dataset based on accounts of followers of the alt-right movement. This process resulted in the creation of a dataset comprised of 123.295 tweets from 422 accounts belonging to individuals associated with the alt-right.

One advantage of this process is that it can be replicated to obtain more accounts. As explained in the methodology, the 2.533 accounts were found that followed six or more of the original twelve accounts associated with the Unite the Right rally, from which 549 accounts were randomly extracted. Therefore, the remaining accounts could also be validated to create a larger dataset.

The linguistic patterns of the alt-right supporters can be examined using this database, and compared with other politically extreme group. Furthermore, the data here could potentially be used to train predictive analysis software to detect signs of an individual becoming radicalised, as demonstrated by Fernández, Asif and Alani [12]. Similarly, Cohen, Johansson, Kaati and Mork [13] worked on the detection of linguistic cues related to violence on OSNs. This approach could also be used for the dataset presented here. Finally, it is also important to remember that there are other variables (such as descriptions or the time of creation of the tweet) that could also be utilised for research purposes. Thus, the database created for this study could have many different applications in the academic study of radicalisation.

In conclusion, the main contribution of this paper is the creation of a dataset of alt-right followers. This database could help advance further research in the field of radicalisation, especially in online environments. It also provides the opportunity for a greater understanding of the language usage of the alt-right and its supporters.

References

1. Peters, M.A.: Education in a post-truth world. In: Peters, M.A., Rider, S., Hyvönen, M., Besley, T. (eds.) post-truth. Fake News. Springer, Singapore (2018). https://doi.org/10.1007/978-981-10-8013-5_12
2. Southern Poverty Law Center: Alt-Right. https://www.splcenter.org/fighting-hate/extremist-files/ideology/alt-right. Accessed 29 May 2018
3. Hine, G.E., et al.: Kek, Cucks, and God Emperor Trump: A Measurement Study of 4chan's Politically Incorrect Forum and Its Effects on the Web. arXiv preprint arXiv:1610.03452 (2016)
4. Wendling, M.: Alt-right: From 4chan to the White House. Pluto Press, London (2018)
5. Kovaleski, S.F., Turkewitz, J., Goldstein, J., Barry, D.: An Alt-Right Makeover Shrouds the Swastikas. New York Times, 10 December 2016
6. Nagle, A.: Kill All Normies: Online Culture Wars From 4chan and Tumblr to Trump and the Alt-Right. John Hunt Publishing, New Alresford (2017)
7. Strickland, P.: Alt-Right Weakened but not Dead After Charlottesville. Al Jazeera, 3 October 2017
8. Marantz, A.: The Alt-Right Branding War has Torn the Movement in Two. The New Yorker, New York (2017)
9. Heim, J., Silverman, E., Shapiro, T.R., Brown, E.: One dead as car strikes crowds amid protests of white nationalist gathering in Charlottesville, two police die in helicopter crash. The Washington Post, 13 August 2017
10. Anti-defamation League.: From Alt Right to Alt Lite: Naming the Hate. https://www.adl.org/resources/backgrounders/from-alt-right-to-alt-lite-naming-the-hate. Accessed 29 May 2018
11. Olivia, S.: Alt-right retaliates against Twitter ban by creating fake black accounts. The Guardian, 17 November 2016
12. Fernandez, M., Asif, M., Alani, H.: Understanding the roots of radicalisation on Twitter. In: Proceedings of the 10th ACM Conference on Web Science, pp. 1–10. ACM, May 2018
13. Cohen, K., Johansson, F., Kaati, L., Mork, J.C.: Detecting linguistic markers for radical violence in social media. Terrorism Polit. Violence **26**(1), 246–256 (2014)

RiskTrack: Assessing the Risk of Jihadi Radicalization on Twitter Using Linguistic Factors

Javier Torregrosa[1(✉)] and Ángel Panizo[2]

[1] Biological and Health Psychology Department,
Universidad Autónoma de Madrid, 28049 Madrid, Spain
francisco.torregrosa@uam.es
[2] Computer Engineering Department, Universidad Autónoma de Madrid,
28049 Madrid, Spain

Abstract. RiskTrack is a project supported by the European Union, with the aim of helping security forces, intelligence services and prosecutors to assess the risk of Jihadi radicalization of an individual (or a group of people). To determine the risk of radicalization of an individual, it uses information extracted from its Twitter account. Specifically, the tool uses a combination of linguistic factors to establish a risk value, in order to help the analyst with the decision making. This article aims to describe the linguistic features used on the first prototype of the RiskTrack tool. These factors, along with the way of calculating them and their contribution to the final risk value, will be presented in this paper. Also, some comments about the tool and the next updates will be suggested at the end of this paper.

Keywords: Radicalization · Online social networks · Twitter · Terrorism Jihadism

1 Introduction

The increase of the Internet as a tool for radicalization (and, in some cases, eventually terrorism) is currently a topic increasingly studied by the scholars. The Internet works facilitating the sharing of radical content, publicizing their acts, or even allowing connections between recruiters and potential recruits [3]. In recent years, this increment has been maximized with the use of Online Social Networks (OSNs) to spread radical ideology [1, 2]. The relative anonymity of the OSNs allow, therefore, to use them on their way to access radicalized content, or the people that share it. Even though some of the platforms (like Facebook or Twitter) have tried to ban users related with radical groups, this is quickly solved by users creating new accounts.

The detection of these individuals on the OSNs represents, therefore, a prior objective in the counter-radicalization fight. To do so, it is necessary to create tools able to detect signs of radicalization on the different platforms, always adapting these tools to the specificities of the virtual context they are working on. There are already a few psychological tools created to assess the likelihood of getting radicalized, such as

© Springer Nature Switzerland AG 2018
H. Yin et al. (Eds.): IDEAL 2018, LNCS 11315, pp. 15–20, 2018.
https://doi.org/10.1007/978-3-030-03496-2_3

VERA-2 [4], ERG22+ [5], TRAP-18 [6], or the more suitable tool for online social media assessment, CYBERA [7]. All this tools, however, are conducted by an expert extracting information of the person of interest. But this represents a slow process, requires prior training, and also makes nearly impossible to make a final report without being in contact with the person assessed.

The present article aims to present the first prototype of RiskTrack, a software tool created to assess risk of Jihadi radicalization using the tweets of a user from Twitter. This tool, based on research conducted on risk factors for jihadi radicalization [8], facilitates the measurement over large quantities of information, giving back a risk value based on the punctuation of different factors (all of them related to Jihadi speech). This paper will give an overview of how the linguistic factors were chosen and used to compute the final value, along with how the final score is created using the groups to optimize the weight of the factors.

2 Linguistic Factors as Risk Factors

The use of radical rhetoric represents a risk factor associated to jihadi radicalization [8]. Jihadi groups assume a specific vocabulary, which is then used as a distinctive way of communication between the members of the group. Therefore, two kind of linguistic factors can be distinguished in the measurement. First, the linguistic factors that identifies a group as radical versus a non-radical one (for example, the common usage of first and third person plural words). Second, the linguistic factors that identifies the radical group with a specific ideology (in the case of jihadi groups, the usage of more words related to Islamic terminology, names of groups like ISIS, Al Qaeda, etc.).

With this differentiation in mind, an assessment was computed using a dataset related to Daesh supporters, published previously on Kaggle's webpage. From the initial 112 accounts, and after discarding those users which were not evidently supporting the radical group, 106 radical accounts were finally used (others studies have worked previously with this dataset [9]) 107 "control" accounts from Twitter were also used for the comparison. The total number of tweets used were 14758 and 15394 tweets respectively. For that assessment, the Linguistic Inquiry Word Count software (LIWC) was used [10]. The next linguistic factors were detected presenting significant differences between the groups:

- Frustration: words related with frustration. Like using swear words and words with negative connotations, such as: 'shit' or 'terrible'.
- Discrimination words: words showing perceived discrimination, like 'racism'.
- Jihadism: words supporting jihadist violence, such as: 'mujahideen' or 'mujahid'.
- Western: differentiation of West as a culture "attacking" them, like 'impure' or 'kiffar'.
- Islamic terminology: related to words of the Islamic religion and culture. Like 'Shunnah' or 'Shaheed'.
- Religion: contrary to the Islamic terminology, this factor is more focused on the Christian religion, but includes extra words related to other religions. Like 'afterlife' or 'altar'.

- First person singular pronouns: words as "I, me, mine"
- First person plural pronouns: words as "We, us"
- Second person pronouns: words as "You"
- Third person plural pronouns: words as "They, them, their".
- Power: domination words as "superior, bully".
- Death: words as "Kill, dead".
- Anger: anger based words as "Rage"
- Tone: positive emotions or negative emotions.
- Sixltr: words with six or more letters (long words).
- Weapons: name of different weapons or guns.
- Terrorist groups: name of terrorist groups, like 'ISIS' or 'Al Qaeda'.

The distribution of the mentioned different linguistic factors, separated by group, is presented on Fig. 1 (radical group) and 2. The values shown in the figures have been normalized in order to compute the global risk factor of an user. To normalize this factors, first we have calculated the percentiles of frequencies of each linguistic factor, mentioned above, for all the accounts in the radical dataset. Then we normalize the linguistics factors depending on the percentile an user falls.

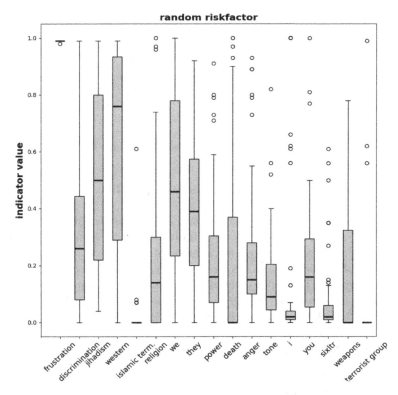

Fig. 1. Radical group scores distribution.

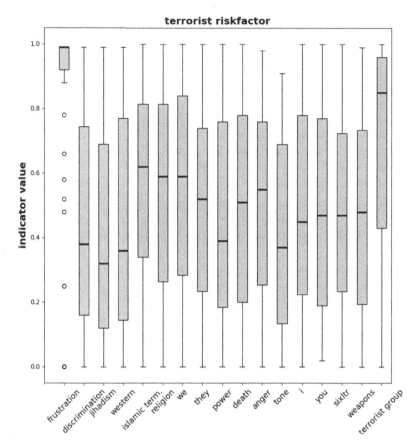

Fig. 2. Control group scores distribution.

3   Distribution of the Linguistic Factors Weight

All the linguistic factors presented above show that there are clear differences between both the radical and the control. The tool needs to estimate a final risk value, in order to help the analyst deciding the risk of radicalization of the assessed account. Therefore, the weights of all the linguistic factors were optimized to show the biggest differences between both groups. To do this optimization we have used the method '*Sequential Least SQuares Programming (SLSQP)*' of the *scipy* library. the Fig. 3 shows the distribution of the risk value for each group, while Fig. 4 represents the distribution of weights in the calculation of the risk value.

The risk value can be interpreted as it follows:

- [0–0.25) = low risk.
- [0.25–0.5) = medium risk.
- [0.5–0.75) = high risk.
- [0.75–1] = very high risk.

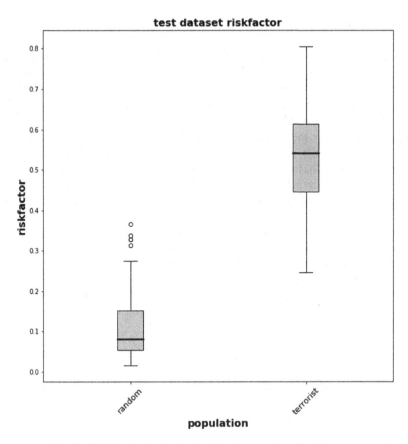

Fig. 3. Distribution in the final risk value scoring of each group.

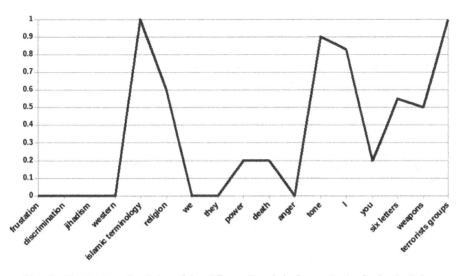

Fig. 4. Distribution of weights of the different linguistic factors in the final calculation

4 Conclusion

The present paper aimed to show how the linguistic factors are used by the RiskTrack tool to measure risk of radicalization on Twitter. The process used to optimize the tool to select the most useful factors has been overviewed, explaining which of them showed significant differences. Finally, the linguistic factors weight was calculated using both groups as ground truth.

As a first prototype, there are some limitations that shall be improved in next steps of the development. First, the linguistic factors shall be optimized using bigger radical samples as ground-truth, to avoid bias. Also, more dictionaries and an ontology could be used, to improve the detection of terminology which can be related with radicalization and is not covered by the current tool. Finally, as it can be seen in Fig. 3, there is a extreme value on the distribution of the radical group. When checking his tweets, this individual only shares links, even though they contain Jihadi propaganda. The creation of a new risk value (as a gray label) could be useful to show the analyst that the measurement can't be done, but that he shall check manually the account.

The development of this kind of tools will enhance the fight against radicalization and terrorism, giving the analysts extra tools combining the efforts of fields like criminology, psychology, linguistics, and computer engineering.

References

1. Edwards, C., Gribbon, L.: Pathways to violent extremism in the digital era. RUSI J. **158**(5), 40–47 (2013)
2. Rowe, M., Saif, H.: Mining pro-ISIS radicalisation signals from social media users. In: Proceedings of the Tenth International AAAI Conference on Web and Social Media (ICWSM 2016), pp. 329–338 (2016)
3. Wright, M.: Technology and terrorism: how the Internet facilitates radicalization. Forensic Exam. **17**(4), 14–20 (2008)
4. Elaine Pressman, D., Flockton, J.: Calibrating risk for violent political extremists and terrorists: the VERA 2 structured assessment. Br. J. Forensic Pract. **14**(4), 237–251 (2012)
5. Lloyd, M., Dean, C.: The development of structured guidelines for assessing risk in extremist offenders. J. Threat. Assess. Manag. **2**(1), 40 (2015)
6. Meloy, J.R., Genzman, J.: The clinical threat assessment of the lone-actor terrorist. Psychiatr. Clin. **39**(4), 649–662 (2016)
7. Pressman, D.E., Ivan, C.: Internet use and violent extremism: a cyber-VERA risk assessment protocol. In: Combating Violent Extremism and Radicalization in the Digital Era, pp. 391–440 (2016)
8. Gilpérez-López, I., Torregrosa, J., Barhamgi, M., Camacho, D.: An initial study on radicalization risk factors: towards an assessment software tool. In: Database and Expert Systems Applications (DEXA), 2017 28th International Workshop, pp. 11–16. IEEE (2017)
9. Fernandez, M., Asif, M., Alani, H.: Understanding the roots of radicalisation on Twitter. In: Proceedings of the 10th ACM Conference on Web Science, pp. 1–10. ACM (2018)
10. Pennebaker, J.W., Booth, R.J., Francis, M.E.: Linguistic inquiry and word count: LIWC [Computer software]. Austin (2007)

On Detecting Online Radicalization Using Natural Language Processing

Mourad Oussalah[1(⊠)], F. Faroughian[2], and Panos Kostakos[1]

[1] Centre for Ubiquitous Computing, University of Oulu, Oulu, Finland
Mourad.Oussalah@oulu.fi
[2] Aston University, Aston, UK

Abstract. This paper suggests a new approach for radicalization detection using natural language processing techniques. Although, intuitively speaking, detection of radicalization from only language cues is not trivial and very debatable, the advances in computational linguistics together with the availability of large corpus that allows application of machine learning techniques opens us new horizons in the field. This paper advocates a two stage detection approach where in the first phase a radicalization score is obtained by analyzing mainly inherent characteristics of negative sentiment. In the second phase, a machine learning approach based on hybrid KNN-SVM and a variety of features, which include 1, 2 and 3-g, personality traits, emotions, as well as other linguistic and network related features were employed. The approach is validated using both Twitter and Tumblr dataset.

Keywords: Natural language processing · Radicalization · Machine learning

1 Introduction

The variety, easy access and popularity of social media user-friendly platforms have revolutionized the sharing of information and communications, facilitating an international web of virtual communities. Violent extremists and radical belief supporters have embraced this changing digital landscape with active presence in online discussion forums, creating numerous virtual communities that serve as basis for sympathizers and active users to discuss and promote their ideologies. As well as disseminating events and inspiration to gain new resources and the demonization of their enemies [2, 13, 16]. They have exploited the Internet's easy to use, quick, cheap and unregulated, relatively secure and anonymous platforms.

Within the extremist domain, online forums have also facilitated the 'leaderless resistance' movement, a decentralized and diffused tactic that has made it increasingly difficult for law enforcement officials to detect potentially violent extremists [5, 17].

It is becoming increasingly difficult – and near impossible – to manually search for violent extremists or users that may embarrass radicalization through Internet because of the overwhelming amount of information, and the inherent difficulty to distinguish self-curiosity, sympathizer and real doctrine supporter or genuine participation in violence acts.

© Springer Nature Switzerland AG 2018
H. Yin et al. (Eds.): IDEAL 2018, LNCS 11315, pp. 21–27, 2018.
https://doi.org/10.1007/978-3-030-03496-2_4

Uncovering signs of extremism online has been one of the most significant policy issues faced by law enforcement agencies and security officials worldwide [6, 16], and the current focus of government-funded research has been on the development of advanced information technologies to identify and counter the threat of violent extremism on the Internet [13].

In light of these important contributions in digital extremism, an important question has been set aside: how can we uncover the digital indicators of 'extremist behavior' online, particularly for the 'most extreme individuals' based on their online activity? To some extent, criminologists have begun to explore this critical point of departure via a customized web-crawler, extracting large bodies of text from websites featuring extremist material and then using text-based analysis tools to assess the content [3, 9, 11]. Similarly, some computational-based research has been conducted on extremist content on Islamic-based discussion forums [1, 6]. Salem et al. [14] proposed a multimedia and content-based analysis approach to detect Jihadi extremist videos and the characteristics to identify the message given in the video. Wang et al. [15] presented a graph-based semi-supervised learning technique to classify intent tweets. They combined keyword based tagging (referred as an intent keyword) and graph regularization method for classifying tweets into six categories. Both Brynielsson et al. [4] and Cohen et al. [6] hypothesized a number of ways to detect online traces of lone wolf terrorists, although, no practical platform has been demonstrated and evaluated. Davidson et al. [7] annotated some 24,000 Tweets for 'hate speech', 'offensive language but not hate', and 'neither'. They began with filtering Tweets using a hate speech lexicon from Hatebase.org, and selected a random sample for annotation. The authors pointed out that distinguishing hate speech from nonhate offensive language was a challenging task, as hate speech does not always contain offensive words while offensive language does not always express hate. O'Callaghan et. al. [12] described an approach to identify extreme right communities on multiple social networking websites using Twitter as a possible gateway to locate these communities in a wider network and track its dynamic. They performed a case study using two different datasets to investigate English and German language communities and implemented a heterogeneous network employing Twitter accounts, Facebook profiles and YouTube channels, hypothesizing that extreme right entities can be mapped by investigating possible interactions among these accounts.

In this paper, we propose a multi-facet based approach for identifying hate speech and extremism from both Twitter and Tumblr dataset. Building on previous research (e.g., see [8, 10]), we use various n-gram based features such as the presence of religious words, war-related terms and several hashtags that are commonly used in extremist posts. Furthermore, other high-level linguistic cues like sentiment, personality change, emotion and emoticons as well as network related features are employed in order to grasp the rich and complexity of hate/extremism like text.

2 Method

Our general approach for radicalization identification undergoes a two-step strategy. First, a radicalization score is obtained by exploring mainly the characteristics of negative sentiments. Second, a machine learning strategy is explored to separate radical post from non-radical one using a wider and diverse set of features involving both linguistic and network features together with previously estimated radicalization score.

2.1 Radicalization Score

Similarly to alternative works in [7], we first explore the sentiment analysis of user's posts. The rationale behind this reasoning is to hypothesize that an extremist user is characterized by the dominance of negative materials over a certain period of time, suggesting that such user espouses an extremist view. Typically sentiment score enables quantifying such trend. Indeed, sentiment analysis is a well-known data collection and analysis method that allows for the application of subjective labels and classifications, by assigning an individual' sentiment with a negative, positive or neutral polarity value. We employed the established Java-based software SentiStrength, which allows for a keyword-focused method of determining sentiment near a specified keyword.

In line with Scrivens et al. [19], the radical score accounts for:

- Average sentiment score percentile (AS), it is calculated by accounting for the average sentiment score for all posts in a given forum. The scores for each individual were converted into percentiles scores, and percentile scores were divided by 10 to obtain a score out of 10 points.
- Volume of negative posts (VN). This is calculated in two parts: (1) the number of negative posts for a given member, and (2) the proportion of posts for a given member that were negative. To calculate the number of negative posts for a given member, we counted the number of negative posts for a given member and converted these scores into percentiles scores.
- Severity of negative posts (SN). This is calculated in two steps: (1) the number of very negative posts for a given member and (2) the proportion of posts for a given member that were very negative. 'Very negative' was calculated by standardizing the count variable; all posts with a standardized value greater than three were considered to be 'very' negative.
- Duration of negative posts (DN). An author who posted extreme messages over an extensive period of time should be classified as more extreme than an author who posted equally extreme messages over a shorter period of time. It is calculated by determining the first and last dates on which individual members made negative posts.

The radical score is therefore calculated as an aggregation of the four previous elements, see Fig. 1. Unlike Scrivens et al. [19] where a simple arithmetic operation was employed, we advocate a non-linear combination of these attributes:

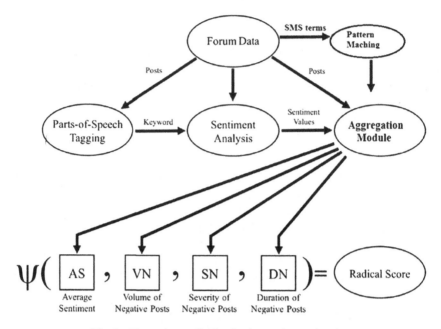

Fig. 1. Extremism radicalization/extremism estimation

$$Radical\, Score = \Psi(AS, VN, SN, DN) \tag{1}$$

Especially, Radical Score is intuitively increasing with respect to VN, SN and DN. Therefore, we argue that the combination operator Ψ linear but rather multiplicative where AS plays only a normalization like role yielding (for some constant factor K):

$$Radical\, Score = (K/AS^3)(VN.SN.DN) \tag{2}$$

2.2 Machine Learning Based Classification

The second phase in the radicalization identification is to use a one-class classification framework involving advanced machine learning techniques where SVM, K-NN, Random Forest were implemented. More specifically, the approach uses the following:

– Extensive preprocessing stage is employed at the beginning in order to filter out stop list words, unknown characters, links, and
– A hybrid SVM-KNN in the same spirit as Zhou et al. [18] is adopted.
– Three types of features are considered. The first one is related to the use of N-gram, especially, 1-g, 2-g and 3-g features were employed as primarily input to the classifier.
– The second type of features relate to personality traits (using the five personality model), emotion and writing cues. The implementation of personality trait identification is performed using the MRC Psycholinguistic database, Linguistic

Inquiry and Word Count (LIWC) feature and Random Forest classifier. Emotion recognition is performed using WordNet Affect, an extension of WordNet domains that concerns a subset of synsets suitable to represent affective concepts correlated with affective words together with Bayes classifier. Finally, writing cues were only considered from its basic content with respect to psychological process as quantified using LWIC features.

- The third set of features are related to various semantic and network related measures. This includes, the length of the post, emoticons, personal pronouns, interrogation and exclamation marks, offensive words, swear words, war words, religious words. We use both LIWC categorization as well as wordnet taxonomy in order to identify war and religious related words. Next, social network related features concern mainly the frequency of messages of the user, average number of posts by the user as well as centrality value whenever possible. Furthermore, the radicalization score computed in previous step is also employed as part of input to the two-class classifier (presence or absence of radicalization case) based on hybrid KNN-SVM.
- We utilize some of existing corpus gathered from DarkWeb project and repository (https://data.mendeley.com/datasets/hd3b6v659v/2) in order to enhance the training of the hybrid KNN-SVM classifier. The overall architecture of this classification scheme is highlighted in Fig. 2.

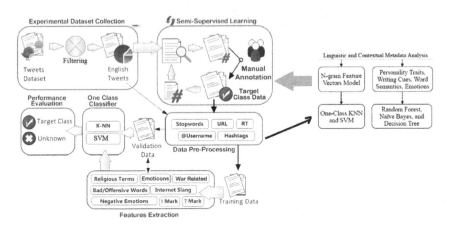

Fig. 2. One-class classification approach

3 Method

3.1 Dataset

Two types of dataset were employed: Twitter dataset and Tumblr dataset.

The initial attempt to collect related tweets is to crawl the hashtags that contains terms "#islamophobia", "#bombing", "#terrorist", "#extremist", "#radicalist". For each set of identified hashtags, Twitter Search API was employed to collect up to one hundred tweets per identified hashtag. A total of 12,202 tweets were collected. Eight thousands of these tweets were sent to Amazon Mechanical Turk in order to perform manual labelling. For each distinct user with a set of tweets, three independent annotators were employed to test whether the user is classified radical or not.

Similarly, we use close hashtags in order to collect data from Tumblr, especially, we employed keywords #islamophobia, #islam is evil, #supremacy, #blacklivesmatter, #white racism, #jihad, #isis and #white genocide. A total of 8000 posts were collected. We deliberately attempt to choose scenarios where a user is associated with several posts in order to provide tangible framework for application of our methodology. Likewise Twitter-dataset, close to 6000 of these posts are sent to Amazon Mechanical Turk in order to manually annotate the post whether the underlying user is considered radical/extremist or not. The results of this analysis are summarized in Table 1, which highlight the usefulness of the approach and its capabilities.

Table 1. Twitter and Tumblr classification scores for various feature sets

Dataset	Features	Classification score
Twitter	1, 2, 3-g	46%
	1, 2, 3-g + personality + emotion	57%
	All features + radicalization score	68%
Tumblr	1, 2, 3-g	54%
Tumblr	1, 2, 3-g + personality + emotion	63%
Tumblr	All features + radicalization score	72%

References

1. Abbasi, A., Chen, H.: Applying authorship analysis to extremist-group web forum messages. Intell. Syst. **20**(5), 67–75 (2005)
2. Bowman-Grieve, L.: Exploring "stormfront:" a virtual community of the radical right. Stud. Confl. Terror. **32**(11), 989–1007 (2009)
3. Bouchard, M., Joffres, K., Frank, R.: Preliminary analytical considerations in designing a terrorism and extremism online network extractor. In: Mago, V., Dabbaghian, V. (eds.) Computational Models of Complex Systems, pp. 171–184. Springer, New York (2014). https://doi.org/10.1007/978-3-319-01285-8_11
4. Brynielsson, J., Horndahl, A., Johansson, F., Kaati, L., Martenson, C., Svenson, P.: Analysis of weak signals for detecting lone wolf terrorist. In: Proceedings of the European Intelligence and Security Informatics Conference, Odense, Denmark, pp. 197–204 (2012)

5. Chen, H.: Dark Web: Exploring and Data Mining the Dark Side of the Web. Springer, New York (2012). https://doi.org/10.1007/978-1-4614-1557-2
6. Cohen, K., Johansson, F., Kaati, L., Mork, J.: Detecting linguistic markers for radical violence in social media. Terror. Polit. Violence **26**(1), 246–256 (2014)
7. Davidson, T., Warmsley, D., Macy, M., Weber, I.: Automated hate speech detection and the problem of offensive language. In: Proceedings of ICWSM (2017)
8. Foong, J.J., Oussalah, M.: Cyberbullying system detection and analysis. In: European Conference in Intelligence Security Informatics, Athens (2017)
9. Frank, R., Bouchard, M., Davies, G., Mei, J.: Spreading the message digitally: a look into extremist content on the internet. In: Smith, R.G., Cheung, R.C.-C., Lau, L.Y.-C. (eds.) Cybercrime Risks and Responses: Eastern and Western Perspectives, pp. 130–145. Palgrave Macmillian, London (2015). https://doi.org/10.1057/9781137474162_9
10. Kostakos, P., Oussalah, M.: Meta-terrorism: identifying linguistic patterns in public discourse after an attack. In: SNAST 2018 Web Conference (2018)
11. Mei, J., Frank, R.: Sentiment crawling: extremist content collection through a sentiment analysis guided web-crawler. In: Proceedings of the International Symposium on Foundations of Open Source Intelligence and Security Informatics, Paris, France, pp. 1024–1027 (2015)
12. O'Callaghan, D., et al.: Uncovering the wider structure of extreme right communities spanning popular online networks (2013). https://arxiv.org/pdf/1302.1726.pdf
13. Sageman, M.: Leaderless jihad: Terror networks in the twenty-first century. University of Pennsylvania Press, Philadelphia (2008)
14. Salem, A., Reid, E., Chen, H.: Multimedia content coding and analyzis: unraveling the content of Jihadi extremist groups's video. Stud. Confl. Terror. **31**(7), 605–626 (2008)
15. Wang, J., Cong, G., Zhao, W.X., Li, X.: Mining user intents in Twitter: a semi-supervised approach to inferring intent categories for tweets. In: AAAI (2015)
16. Weimann, G.: The psychology of mass-mediated terrorism. Am. Behav. Sci. **52**(1), 69–86 (2008)
17. Weimann, G.: Terror on Facebook, Twitter, and YouTube. Brown J. World Affairs **16**(2), 45–54 (2010)
18. Zhou, H., Wang, J., Wu, J., Zhang, L., Lei, P., Chen, X.: Application of the hybrid SVM-KNN model for credit scoring. In: 2013 Ninth International Conference on Computational Intelligence and Security (2013). https://doi.org/10.1109/cis.2013.43
19. Scrivens, R., Davies, G., Frank, R.: Searching for signs of extremism on the web: an introduction to sentiment-based identification of radical authors. Behav. Sci. Terror. Polit. Aggress. **10**(1), 39–59 (2017)

Workshop on Methods for
Interpretation of Industrial Event Logs

Automated, Nomenclature Based Data Point Selection for Industrial Event Log Generation

Wolfgang Koehler[✉] and Yanguo Jing

School of Computing, Electronics and Mathematics,
Faculty of Engineering, Environment and Computing, Coventry University,
Priory Street, Coventry CV1 5FB, UK
koehlerw@coventry.ac.uk

Abstract. Within the automotive industry today, data collection, for legacy manufacturing equipment, largely relies on the data being pushed from the machine's PLCs to an upper system. Not only does this require programmers' efforts to collect and provide the data, but it is also prone to errors or even intentional manipulation. External monitoring, is available through Open Platform Communication (OPC), but it is time consuming to set up and requires expert knowledge of the system as well. A nomenclature based methodology has been devised for the external monitoring of unknown controls systems, adhering to a minimum set of rules regarding the naming and typing of the data points of interest, which can be deployed within minutes without human intervention. The validity of the concept will be demonstrated through implementation within an automotive body shop and the quality of the created log will be evaluated. The impact of such a fine grained monitoring effort on the communication infrastructure will also be measured within the manufacturing facility. It is concluded that, based on the methodology provided in this paper, it is possible to derive OPC groups and items from a PLC program without human intervention in order to obtain a detailed event log.

1   Introduction

Advances in industrial automation often go with the increased complexity of the manufacturing equipment. Today an automotive manufacturing cell, consisting of a number of work stations, is controlled by a PLC which interfaces with dozens of process specific controllers for robots, welding and sealing systems. Many of these process specific controllers already have a built in function, that collects process related data. This data may or may not be accessible to the end user and is not the subject of this paper.

Stations, controlled by a PLC, often have dozens of actuators and sensors. In addition, ever increasing health and safety stipulations require that such equipment is contained, which makes the simple observation of the equipment's

© Springer Nature Switzerland AG 2018
H. Yin et al. (Eds.): IDEAL 2018, LNCS 11315, pp. 31–40, 2018.
https://doi.org/10.1007/978-3-030-03496-2_5

sequence difficult or impossible. Lean manufacturing efforts require, that equipment status information is made available in a central location. In order to achieve this, many manufacturing facilities have implemented so called 'plant floor systems'. They collect a predetermined set of alarm messages in order to display them on maintenance screens and also preserve them for statistical analysis. In addition these systems receive triggers for some predetermined events, such as the machine being blocked or starved, which are predominantly used for statistics and dashboards as well. The main issue with both of those data streams is, that the data has to be made available by the programmer of the equipment in a defined format. Often however the data generation is either neglected all together, is error prone or even subject to intentional manipulation.

Besides the 'plant floor system' described above, there are also software packages available, that allow the user to manually select any of the data points for monitoring, that are accessible through OPC. This approach is not only time consuming but also requires in-depth knowledge of the system in order to choose the tags suitable for the intended analysis. To prove this point one of the leading providers of such software was invited to implement it on a system, unknown to him, with one PLC. The vendor spent roughly 150 h until he was able to start the data collection.

2   Industry 4.0 Vision

Based on above observation, of the current situation, a vision has been developed suitable for the Industry 4.0 paradigm shown in Fig. 1. It's aim is to collect process related data from the PLCs and process equipment (number 2) which can be analysed to derive recommendations for process improvements as well as predictions of imminent failure (numbers 3 & 4). These findings are provided for the maintenance department to act upon. Once completion is reported the system is used to verify the work performed. The figure shows the current state marked as number 1 and highlighted in green. The goals set for research are marked with the numbers 2, 3 and 4. This paper strictly focuses on researching methodologies to automatically monitor and log events controlled by Rockwell PLCs (marked as number 2).

3   Creation and Evaluation of Industrial Event Log

3.1   Data Collection Procedure

In the realm of automotive manufacturing equipment the process is equal to the sequence of operation. Therefore, time stamps are required for every actuator and its corresponding sensors within a station. Robots however can be working in more than one station, which could potentially cause unidentified gaps within this sequence. Gaps would also be shown if the robot performed an automated maintenance task. This can only be avoided if the robots' working segments are logged as well.

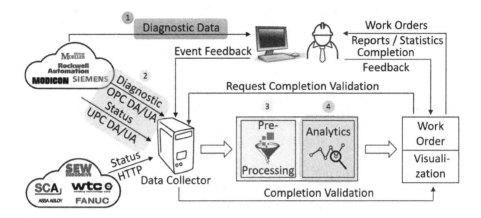

Fig. 1. Industry 4.0 vision (Color figure online)

In order to group the events into cases, an identifier needs to be recorded together with the events. In a first attempt, recording the parts sequence number was chosen. Since however not all manufacturing steps use a sequence number, ultimately the part present status was recorded instead.

For future predictive analysis the data needs to be correlated with events. That could be achieved either by tagging the data manually a posteriori, or by logging the events (alarms) concurrently.

Data collection could be achieved with multiple local, decentralised data collectors, run on the maintenance work stations, that periodically submit the collection results to a centralised location. Here the main advantage is that the traffic between the data collector and the PLC stays in the local sub net. The lower data volume also allows for a less expensive database version to be used and at the same time the error rate might decrease as well. In addition a single point of failure doesn't halt the whole system. Also the current version of the PLC software is readily available on the maintenance work station so that updates to the collection algorithm are possible.

On the other hand such approach requires multiple licenses of OPC server software along with increased effort to manage and maintain the system. The alignment of the data from different sources can become, due to time stamp inaccuracies, challenging.

The second option is a centralised setup, where the collection algorithm, the OPC server and the database are located on only one computer. For this scenario the above advantages and disadvantages can just be reversed. For manageability, the centralised setup was chosen, for the proof of this concept. However it is believed that a decentralized setup would be more beneficial for a permanent solution.

Table 1 shows, how the PLC program aligns with the actual equipment for this project. In addition it shows what keywords can be found within the PLC logic's text file and what regular expressions can be used to locate them.

Table 1. Relationship between equipment and software

Actual equipment	Software equivalent	Tags within L5K	Regular Expression (RegEx)
Cell	Controller	CONTROLLER	ˆCONTROLLER \s\w+\s\(
Station	Program	PROGRAM	ˆ(\t+|\s+)PROGRAM\s\w+\s\(
Advancing motion	(Advance) routine	ROUTINE	ROUTINE S
Returning motion	(Return) routine	ROUTINE	ROUTINE S
Action feedback	Rung	.comp (UDT za_Action)	(OTL\(|OTE\()[A-Z,a-z,0-9]+\.Comp\)
Sensor	Contact	XIC(...)	XIC(

Before discussing the parsing itself, it should be understood that the Rockwell PLCs work with two levels of tags. There are controller tags, which can be accessed from every program and there are program tags, which can only be accessed by the program in which they were defined. Addressing the tags through OPC also requires different nomenclatures. The first step in this parsing effort therefore needs to locate the controller tags and store them in an array for later referencing.

Program tags are identified by the name of the program in which they have been defined. This in turn means keeping track of the current program name while parsing the logic and associating it with the data point of interest, as long as that has not been found to be a controller tag.

Once the controller tags have been identified parsing can simply be done line by line from the beginning to the end. Since all the data to be gathered belong to sequence routines the parsing can be speeded up by simply skipping all lines that are not located within a section which starts with 'ROUTINE S' (where the 'S' stands for sequence) and 'END_ROUTINE'. Once a program rung that ends with a '.Comp' tag is encountered, that tags name is extracted and the extension is replaced with '.Out'. For this project, that is the name of the tag that initiates a motion. The resulting sensor inputs can also be found within the same rung, programmed as normally open (XIC) contacts. All the sensor addresses will be given by extracting all the tags that have been programmed with the XIC instruction.

If no such apparent relation between actuator and sensors is to be found within the PLC code, it would also be imaginable to discover the relations a posteriori with the help of some reasoning algorithms. Alternatively the software could 'learn' the relation through a 'teach in' process, where the machine is manually run once through its sequence step by step.

The current robot segment number also can be found within a defined controller tag. While parsing the program, all that needs to be extracted, are the names of the robots associated to that station.

For this project the actual alarm messages are included within the software as rung comments, with an alarm number, which represents the last three digits of the actual alarm. The leading one or two digits of the alarm are a program related offset. In order to decode an occurring alarm the alarm messages need to be extracted as well as the offsets associated with the program currently being parsed.

To summarise, there are three basic requirements for nomenclature based data point selection. The most important is a common tag naming structure or data type for the output initiating the actuators (e.g. Clamps1Close.Out; za_Action). The same applies to the naming structure and data type for the sensors (e.g. C01.PX1; zp_Cylinder). Ideally there is also an identifiable rung that sums up the sensors associated with the actuators to allow deriving the connection between the inputs and outputs.

3.2   Evaluation

The methodology described above was implemented at an automotive manufacturing plant in Thueringen/Germany using VB.net and the Advosol OPC DA .NET framework. It has been installed on a single workstation PC, along with an OEM version of RSLinx, as OPC server and a developers version of SQL Server. Currently 193 stations, controlled by 26 Rockwell PLCs, are being monitored, while logging ~1.000.000 events a day.

One of the main concerns, within IT, was the increased network load due to the data collection efforts. The system was implemented in a plant which has a 100 Mbit/s Ethernet system for the production floor level. While monitoring the network traffic, it was found that communication to the PLCs increased by 3.8%; from 371.164 packets/s to 385.171 packets/s. Since this was a minor increase of overall network traffic, the work was cleared to proceed.

The controls group on the other hand was concerned that the constant monitoring of the PLC through OPC would put an additional burden on the overhead time slice. Therefore some code was created that could measure the overhead time slice within the PLC and it was found that, although the monitoring caused an increase of 3.75%, from 7.7% to 8%, it was still well within the pre-set boundaries.

In order to determine, if the motion duration recorded by the system was realistic, a logging algorithm within the PLC was also created. The results of ~500 events were collected and compared and it was found that this system deviated from 0 ms to up to +100 ms from the times recorded within the PLC. This can largely be attributed to the OPC server's minimum sampling rate of 50 ms and the fact, that RSLinx, contrary to the OPC specification, does not provide a time stamp with the data. Since this circumstance can not be influenced, this error will be considered during the data analysis phase.

This paper argues that all motions that occur, while there is a part in the station, belong to the same case. As a consequence the set of motions occurring prior to this must be the load step, which also belongs to that case. This leaves us with the transition events, which happen while there is no part in the station and also while the station is not being loaded.

The transition events can be split up into three categories. The tooling 'reset' events which move units, particular to the previous style, out of the way; and the tooling 'set' events which prepare the station to receive the next style of parts. In addition there are common reset motions which are independent of a style. For example, an ejector, used to remove the part from the station, always needs to be returned prior to the next part being loaded.

The common reset motions can be identified by pinpointing transition motions, which occur prior or after all of the observed styles. The style dependant reset and set events remain.

This research yielded that the raw data collected is above 96% complete. Therefore the conclusion can be drawn that any event, that occurs at least 90% of the time, following a certain style, must be a reset event as long as the type mixture is such that a certain transition combination does not happen more than 90% of the time. Continuing with the same logic, any event that occurs at least 90% previous to the same style, must be a set event.

In order to evaluate the quality of the event log, the quality matrix, proposed by Bose et al. [1], was chosen and reduced to the measures applicable to industrial manufacturing equipment.

As previously described, it was at first chosen to record the parts sequence number together with the motions in an attempt to use that ID as a case identifier. This however was reflected in the two quality measures missing relationship, where there was no sequence ID recorded and incorrect relationships, which manifested as a wrong sequence ID being recorded. Both became measurable after implementing a case clustering approach based on the part being present in the station.

Another issue was found to be missing time stamps. This research chose that all motions be identified by a start and a corresponding complete time stamp. Therefore some of the missing time stamps could be identified if there was a start time stamp but no corresponding complete time stamp or vice versa. Based on the clustered cases it could also be determined how many events should be in a case. Deviations from that count were used to pinpoint missing events/time stamps.

The next category of quality issues were incorrect events. These were found to be caused by motions being triggered more than once, due to the interruption of clear conditions for that motion. In addition, events were found that came to completion, but were triggered once again within a short amount of time, which was caused by the motion reaching it's end position and bouncing back.

Last but not least incorrect time stamps were identified. As previously mentioned, time variances of approximately 100 ms were expected, due to the logging through OPC. It was concluded that the time stamps must have been recorded incorrectly if the events duration deviated more than $+/-100$ ms from it's mean duration.

A subsequent analysis of four selected stations within the body shop yielded the results shown in Table 2.

The deployment of this methodology to the 26, unknown, PLCs took approximately 30 min (compared to 150 h for 1 PLC as described previously). From this, 26 min were spent converting the PLC programs to text based files, which had to be done with keyboard macros, since the Rockwell software does not provide a command line instruction to do so. The remaining 4 min were needed to parse the text files and set up the OPC groups and items for monitoring.

Table 2. Quality assessment

description	station 1		station 2		station 3		station 4	
	count	%	count	%	count	%	count	%
events recorded	59724	100	89682	100	104064	100	76004	100
missing relationship	1198	2	2424	2.7	1224	1.18	1229	1.62
missing time stamps	242	0.41	2438	2.72	35	0.03	176	0.23
incorrect events	88	0.15	1137	1.27	203	0.2	2046	2.69
incorrect relationships	216	0.36	21	0.02	0	0	61	0.08
incorrect time stamps	127	0.21	315	0.35	251	0.24	165	0.22

Note: Areas in gray mark quality problems that could be compensated to 100% during pre-processing

4 Related Works

Modern fieldbus systems, like IO Link as described by Heynicke et al. [2], allow for very detailed status information to be retrieved directly from the sensor through a network connection. The data available are not limited to binary on/off indicators but also include parameters for temperature, sensing range and sensor maintenance requirements. Creating a log of all the data available promises to provide the basis for predictive maintenance systems. Unfortunately, as Hoffmann et al. [3] also point out, the majority of automation systems, installed within manufacturing facilities today, are not based on such technology creating the need for alternative data logging approaches.

The most common interface protocol for automation systems is OPC. Within OPC, the first version called OPC-DA. As described by Veryha et al. [4], should be differentiated from its predecessor OPC-UA, as explained by Reboredo et al. [5] as well as Schleipen et al. [6]. Hoffmann et al. [3] in their paper provide a wrapper that allows OPC-DA systems to be integrated into production networks along with OPC-UA systems. Oksanen et al. [7] on the other hand provide a framework for accessing OPC based data of mobile systems through the internet. Haubeck et al. [8] propose to monitor PLC in- and outputs through OPC-UA. They conclude however: "In order to obtain additional value of the data, the signals must be enriched with semantics to become automatically interpretable".

Feldmann et al. [9] discuss in their paper "feature-based monitoring using the information contained in the process interface and in the logic control structure" without going into details of how to extract the features from the logic control structure and how to access the data within the controller. Also [10–13] propose in their papers systems for monitoring PLC based production system. Their focus however is mostly on real time fault detection based on specialised frameworks and algorithms. The data points to be monitored are often manually chosen by a domain expert.

5   Discussion

As described above, the final goal is to use the event log for creation of a process model for industrial manufacturing equipment. This model then is to be used to derive recommendations and predictions. There are multiple additional research steps needed to achieve that goal.

The first step is pre-processing and data cleaning. Here the main challenge will be clustering the events into cases which conform to process mining requirements. In addition, quality issues will be identified and addressed. The discovered cases will be grouped, based on their sequence of events, into trace classes and validated for completeness.

Once all quality issues are known, possible repair algorithms will be defined. These will include the calculation of missing time stamps as well as replacement of missing events and cases.

After successful pre-processing, cleaning and repair of the log, the detection of potential for equipment and sequence improvements will come into focus. This will be based on, for industrial processes common, Gantt charts with the objective to identify gaps and setup problems.

The final step is the application of Process Mining techniques with the goal of guiding the operators to follow the so called 'happy path' in order to optimise the equipments throughput. In addition the process model will be used to predict running times of the different stations within a manufacturing cell and to subsequently schedule maintenance tasks for predicted idle times.

For the prediction of imminent equipment failures it is hypothesised that most equipment will follow a degradation curve which can be correlated with ever increasing mean duration times for the different motions. If the log contains all the records, including the actual failure, it will be possible to apply machine learning algorithms to predict similar future events. Due to the need for long term equipment monitoring this objective might however become it's own research topic.

There are also alternative use cases for the event log data. In an unrelated project it was shown, that near real time processing of the event log allows for a diagnostic system which is independent of the equipment's controls logic. For yet another project the parsing algorithm and the mean motion data were used for a basic cell level simulation system.

6   Conclusion and Future Works

It is concluded that a nomenclature based approach, as described in this paper, is possible as long the software to be parsed complies with a minimum set of naming and/or typing requirements for the data points to be monitored. It was demonstrated that parsing can then be done following this structure while identifying the desired data points with the help of regular expressions or based on their assigned data type. Not only were the resulting OPC groups and items compiled within a much shorter time frame, but it was also ensured that, due to automation, the result was much more accurate compared to manual data point selection. In addition, measurements were provided proving that such an in-depth monitoring effort can be achieved with little impact on the plants communication infrastructure.

The next step of this research focuses on evaluating the quality of the log obtained and the steps that need to be taken to make the log conform to the requirements put forward by Process Mining. These include, but are not limited to, case clustering, trace class identification/validation and log repair.

References

1. Bose, R.J.C., Mans, R.S., van der Aalst, W.M.: Wanna improve process mining results? Its high time we consider data quality issues seriously. In: Proceedings of the 2013 IEEE Symposium Series on Computational Intelligence and Data Mining, CIDM 2013, pp. 127–134 (2013). https://doi.org/10.1109/CIDM.2013.6597227
2. Heynicke, R., et al.: IO-link wireless enhanced sensors and actuators for industry 4.0 networks. In: Proceedings-AMA Conferences 2017 with SENSOR and IRS2, pp. 134–138 (2017). https://doi.org/10.5162/sensor2017/A8.1
3. Hoffmann, M., Büscher, C., Meisen, T., Jeschke, S.: Continuous integration of field level production data into top-level information systems using the OPC interface standard. Procedia CIRP **41**, 496–501 (2016). https://doi.org/10.1016/j.procir.2015.12.059
4. Veryha, Y.: Going beyond performance limitations of OPC DA implementation. In: 10th IEEE Conference on Emerging Technologies and Factory Automation, p. 4 (2005). https://doi.org/10.1109/ETFA.2005.1612501
5. Reboredo, P., Keinert, M.: Integration of discrete manufacturing field devices data and services based on OPC UA. In: 39th Annual Conference on Industrial Electronics Society, IECON 2013, pp. 4476–4481 (2013). https://doi.org/10.1109/IECON.2013.6699856
6. Schleipen, M.: OPC UA supporting the automated engineering of production monitoring and control systems. In: 2008 IEEE International Conference on Emerging Technologies and Factory Automation, pp. 640–647 (2008). https://doi.org/10.1109/ETFA.2008.4638464
7. Oksanen, T., Piirainen, P., Seilonen, I.: Remote access of ISO 11783 process data by using OPC unified architecture technology. Comput. Electron. Agric. **117**, 141–148 (2015). https://doi.org/10.1016/j.compag.2015.08.002
8. Haubeck, C., et al.: Interaction of model-driven engineering and signal-based online monitoring of production systems. In: 40th Annual Conference of the IEEE Industrial Electronics Society (2014). https://doi.org/10.1109/IECON.2014.7048868
9. Feldmann, K., Colombo, A.W.: Monitoring of flexible production systems using high-level Petri net specifications. Control. Eng. Pract. **7**(12), 1449–1466 (1999). https://doi.org/10.1016/S0967-0661(99)00107-0
10. Lee, J., Bagheri, B., Kao, H.A.: A cyber-physical systems architecture for industry 4.0-based manufacturing systems. Manuf. Lett. **3**, 18–23 (2015)
11. Palluat, N., Racoceanu, D., Zerhouni, N.: A neuro-fuzzy monitoring system: application to flexible production systems. Comput. Ind. **57**(6), 528–538 (2006). https://doi.org/10.1016/j.compind.2006.02.013

12. Phaithoonbuathong, P., Monfared, R., Kirkham, T., Harrison, R., West, A.: Web services-based automation for the control and monitoring of production systems. Int. J. Comput. Integr. Manuf. **23**(2), 126–145 (2010). https://doi.org/10.1080/09511920903440313
13. Ouelhadj, D., Hanachi, C., Bouzouia, B.: Multi-agent architecture for distributed monitoring in flexible manufacturing systems (FMS). In: Proceedings of the IEEE International Conference on Robotics and Automation, ICRA 2000, vol. 3, pp. 2416–2421 (2000). https://doi.org/10.1109/ROBOT.2000.846389

Monitoring Equipment Operation Through Model and Event Discovery

Sławomir Nowaczyk$^{(\boxtimes)}$, Anita Sant'Anna, Ece Calikus, and Yuantao Fan

CAISR, Halmstad University, Halmstad, Sweden
{slawomir.nowaczyk,anita.santanna,ece.calikus,yuantao.fan}@hh.se

Abstract. Monitoring the operation of complex systems in real-time is becoming both required and enabled by current IoT solutions. Predicting faults and optimising productivity requires autonomous methods that work without extensive human supervision. One way to automatically detect deviating operation is to identify groups of peers, or similar systems, and evaluate how well each individual conforms with the group.

We propose a monitoring approach that can construct knowledge more autonomously and relies on human experts to a lesser degree: without requiring the designer to think of all possible faults beforehand; able to do the best possible with signals that are already available, without the need for dedicated new sensors; scaling up to "one more system and component" and multiple variants; and finally, one that will adapt to changes over time and remain relevant throughout the lifetime of the system.

1 Introduction

In the current "Internet of Things" era, machines, vehicles, goods, household equipment, clothes and all sorts of items are equipped with embedded sensors, computers and communication devices. Those new developments require, and at the same time enable, monitoring the operation of complex systems in real-time. The ability to predict faults, diagnose malfunctions, minimise costs and optimise productivity requires new cost-effective autonomous methods that work without extensive human supervision.

One way to automatically detect deviating operation is to identify groups of peers, or similar systems, and evaluate how well each individual conforms with the rest of the pack. This "wisdom of the crowd" approach focuses on understanding the similarities and differences present in the data, and require a suitable representation or model – one that captures events of interest within sensor data steams. Building such models requires a significant amount of expert work, which is justifiable in case of safety critical products, but does not scale for the future as more systems need to be monitored. A different approach is needed for complex, mass-produced systems where the profit margins are slim.

We propose a monitoring approach that can construct knowledge more autonomously and relies on human experts to a lesser degree: without requiring the designer to think of all possible faults beforehand; able to do the best

© Springer Nature Switzerland AG 2018
H. Yin et al. (Eds.): IDEAL 2018, LNCS 11315, pp. 41–53, 2018.
https://doi.org/10.1007/978-3-030-03496-2_6

possible with signals that are already available, without the need for dedicated new sensors; scaling up to "one more system and component" and multiple variants; and finally, one that will adapt to changes over time and remain relevant throughout the lifetime of the system.

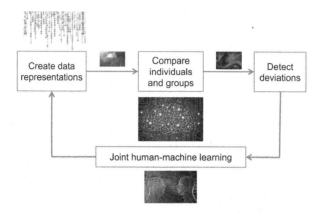

Fig. 1. Architecture for self-monitoring systems

Modern systems need to become more "aware" to construct knowledge as autonomously as possible from real life data and handle events that are unknown at the time of design. Today, human experts generally define the task that the system should perform and guarantee that the collected data reflects the operation of the system sufficiently well. This means that such designed systems cannot function when their context changes in unforeseen ways. The next step is to approach the construction of AI systems that can do life-long learning under less supervision, so they can handle "surprising" situations.

Humans plas an important role in all steps of the knowledge creation process, e.g., giving clues on interesting data representations, clustering events, matching external data, providing feedback on suggestions, etc. What is important, though, is for machine and human to create knowledge together – unlike traditional AI or ML, where humans provide expertise that the machine is expected to replicate. Domain experts are a crucial resource, since their expertise goes beyond the technical specification and includes business and societal aspects. Automatically derived solutions must interact with them through a priori knowledge, justifying the solutions given, as well as accept feedback and incorporate it into further processing. We refer to this as *joint human-machine learning*.

A necessary component for monitoring is the semi-supervised discovery of relevant relations between various signals, and the autonomous detection and recognition of events of interest. It is often impossible to detect problems by looking at the characteristics of a single signal. Models created based on the interrelations of connected signals are more indicative.

Our general framework, presented in Fig. 1, based on domain-specific input, provides group based monitoring solutions. This framework is capable of learning from streams of data and their relations and detecting deviations in an unsupervised setting. In addition, the framework interactively exploits available expert knowledge in a joint human-machine learning fashion.

In this paper we present initial implementations of our framework in two diverse domains: automotive and district heating. We investigate how models suggested by domain expertise can be combined with the unsupervised knowledge creation.

In the next section we discuss related work, followed by presenting the two domains of interest. In Sect. 4 we briefly describe the Consensus Self-Organizing Models (COSMO) framework, a flexible method that we have used for monitoring across multiple domains. We showcase how it can be used to combine the distinct perspectives (methods from data mining with background domain knowledge for conceptual analysis) in Sect. 5.

2   Related Work

Self-exploration and self-monitoring using streams of data are studied within the fields of Autonomic Computing, Autonomous Learning Systems (ALS), and Cognitive Fault Diagnostic Systems (CFDS), among others. We cannot provide here a full overview of all relevant fields, therefore we restrict ourselves to maintenance prediction using a model space.

Particularly relevant within ALS are [9] and [8], presenting a framework for using novelty detection to build an autonomous monitoring and diagnostics based on dynamically updated Gaussian mixture model fuzzy clusters. Linear models and correlations between signals were used in [5] to detect deviations. Vedas and MineFleet® systems [14–16] monitor correlations between vehicle on-board signals using a supervised paradigm to detect faulty behaviours, with focus on privacy-preserving methods for distributed data mining. [1,2] used linear relationship models, including time lagged signals, in their CFDS concept and provided a theoretical analysis motivating their approach. [22] showed how self-organised neural gas models could capture nonlinear relationships between signals for diagnostic purposes. [4] and [19] used reservoir computing models to represent the operation of a simulated water distribution network and detect faults through differences between model parameters. [18] and [17] have used groups of systems with similar usage profiles to define "normal" behaviour and shown that such "peer-clusters" of wind turbines can help to identify poorly performing ones. [23] used fleets of vehicles for detecting and isolating unexpected faults in the production stage. Recently, [21] provided categorisation of anomalies in automotive data, and stressed the importance of designing methods that handle both known and unknown fault types, together with validation on real data.

The ideas presented here were originally suggested in [13], with the initial feasibility study only done using simulation. The study of a real bus fleet was first presented in [3], followed by a comprehensive analysis in [20]. Specifically focusing on air compressor, [6] have recently evaluated the COSMO algorithm, while

[7] presents a comparison between automatically derived features and expert knowledge, as described in a series of patents by Fogelstrom [10, 11].

3   Description of the Domains

In this paper we showcase monitoring solutions in two diverse settings: in the automotive domain we analyse operation of a small bus fleet; and in the smart cities domain we analyse a network of district heating substations.

The automotive data used in this study was collected between 2011 and 2015 on a bus fleet with 19 buses in traffic around a city on the west coast of Sweden. Each bus was driven approximately 100.000 km per year and the data were collected during normal operation. Over one hundred on-board signals were sampled, at one Hertz, from the different CANs. All buses are on service contracts offered by the original equipment manufacturer, which also includes on-road service that is available around the clock, at an additional cost. For a bus operator the important metric is the "effective downtime", i.e., the amount of time that a bus is needed but not available for transportation. In this case, the bus fleet operator's goal was to have one "spare" bus per twenty buses, i.e., that the effective downtime should be at most 5% and the bus operator took very good care of the vehicles in order to meet this goal.

The district heating data used in this study consists of smart meter readings from over 50.000 buildings connected to a network within two cities in the southwest of Sweden. It includes hourly measurements of four important parameters: heat, flow, supply, and return temperatures on the primary side of the substations. In addition, information about the type or category of each customer is also available, for example multi-dwelling buildings, industry, health-care and social services, public administration, commercial buildings, etc. One of the most important goals for district heating operators is to decrease distribution temperatures. The current system operates at high supply and return temperatures, which leads to large heat losses in the network and inefficient use of renewable energy sources. Faults in customer heating systems and substations are an important factor contributing to the need for high supply temperature. Although, in many cases, faults do not affect customer comfort, they influence the performance of the network as a whole. Inspecting the behaviour of all buildings is prohibitively time-consuming, however, automated solutions are challenging due to the complex and dynamic nature of the district heating distribution system.

4   Method

This paper builds upon the Consensus Self-organising Models (COSMO) approach, i.e., on measuring the consensus (or the lack of it) among self-organised models. The idea is that representations of the data from a fleet of similar equipments are used to agree on "normality." The approach consists of three parts: looking for clues by finding the right models for data representation, evaluating consensus in parameter space to detect deviating equipment, and determining causes for those deviations for fault isolation and diagnosis. This paper is mainly concerned with the first step, therefore the latter two will be discussed in less detail.

Looking for Clues: monitoring requires that systems are able to collect information about their own state of operation, extracting clues that can be communicated to others. This can be done by embedded software agents that search for interesting relationships among the available signals. Such relationships can be encoded in many different ways, e.g., with histograms or probability density models describing signals; with linear correlations expressing correspondence across two or more signals, or signals with different time shifts; with principal components, autoencoders, self-organising feature maps and other clustering methods, etc.

The choice of model family can and should be influenced by domain knowledge, but a self-organising system should be able to take any sufficiently general one and make good use of it. A model is a parameterised representation of a stream of data consisting of one or more signals. There are endless possible model families, and hierarchies of models of increasing complexity. It is interesting to study methods to automatically select models that are useful for detecting deviations and communicating system status to human experts. In this paper we showcase several quite different examples, but this is by no means an exhaustive list.

Useful relationships are those that look far from random, since they contain clues about the current state of the equipment. If, for example, a histogram is far from being uniform, or a linear correlation is close to one, or the cluster distortion measure is low, then this is a sign that a particular model can be useful for describing the state of the system. At the same time, the variation in the models across time or fleet of equipments is also of value. Relationships that have a large variation will be difficult to use, due to difficulty in detecting a change. Whereas changes occurring in models that are usually stable are likely to indicate meaningful events.

A model considered "interesting" based on the above measures does not guarantee usefulness for fault detection. However, the "uninteresting" models are unlikely to contain valuable information. The goal of this stage is simply to weed out bad candidates, to make the subsequent steps more efficient.

Consensus in Parameter Space: In this step, all equipments compute the parameters for the most interesting models and send them to a central server. The server then checks whether the parameters from different systems are in consensus. If one or more units disagree with the rest, they are flagged as deviating and potentially faulty.

The z-score for any given system m at time t is computed (as in conformal anomaly detection) based on the distance to *most central pattern* (the row in distance matrix with minimum sum, denoted by c):

$$z(m,t) = \frac{|\{i = 1, ..., N : d_{i,c} > d_{m,c}\}|}{N}. \tag{1}$$

The z-score for a pattern m is the number of observations that are further away from the centre of the fleet distribution than m is. If m is operating normally, i.e., its model parameters are drawn from the same distribution as the rest of the group, the z-scores are uniformly distributed between zero and one. Any statistical test over a period of time can be used to decide whether it is the case or not, and the p-value from such a test measures the deviation level of m.

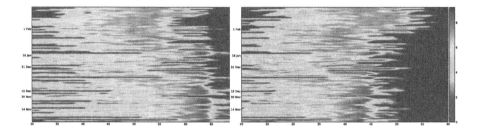

Fig. 2. Histograms of Engine Coolant Temperature signal between November and February (left: original raw data, right: difference from fleet average).

Fault Isolation and Diagnosis: When a deviation is observed in the parameter space, then this particular system is flagged as potentially faulty. The next step is to diagnose the reason for the deviation. One way is to compare against previous observations and associated repairs, using a supervised case-based reasoning approach. It requires a somewhat large corpus of labelled fault observations; however, for most modern cyberphysical systems such data are available in the form of maintenance databases or repair histories.

5   Results

The first result we would like to showcase comes from the automotive domain. In Fig. 2 we plot a sequence of histograms for the signal called *Engine Coolant Temperature* over a period of four months. Each horizontal line corresponds to a set of 20.000 data readings, presented as a colour map histogram with logarithmic scale. It is interesting to note that the actual amount of data we obtain from a bus varies according to usage in a given period. The left plot reveals a critical flaw of looking at signals in isolation: there is a clear trend in the data, however, to a human expert it does not indicate a wearing out component, but rather an influence of outside temperature. In "wisdom of the crowd" approach such trends can be compensated, for example by normalising against fleet average (shown in the right subplot of Fig. 2).

Another way to increase robustness is to look at combinations of different but related signals. A large number of "interesting" relationships exist and many of them are good predictors of faults. An example is relation between *Oil Temperature* and the aforementioned *Engine Coolant Temperature*, depicted in Fig. 3

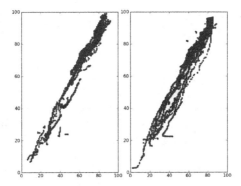

Fig. 3. Scatter plot of Oil Temperature against Engine Coolant Temperature (left: October, right: January)

(left sub-plot shows October 2011, while right sub-plot shows January 2012, each containing 40.000 readings). As can be expected due to the basic laws of thermodynamics, there is a strong linear relation between those two signals. The plots are definitely not identical (for example, both signals reach higher values in October), but the fundamental structure has not changed.

This fundamental structure should be captured in a model – faults that affect one of the subsystems but not the other would then introduce a systematic shift that would change parameters of that model. Of course, relations between signals can be arbitrarily complex, and finding good balance between purely data-driven exploration and taking advantage of available domain expertise is an interesting challenge.

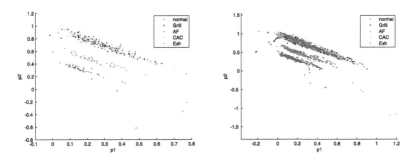

Fig. 4. Model parameters (left: LASSO method, right: RLS method) for normal data and in presence of four different injected faults.

We have performed an experiment on a Volvo truck with four different faults injected, as well as four runs under normal operating conditions. The exact details of faults are not important here, but they include clogged of *Air Filter* and *Grill*, leaking *Charge Air Cooler* and partially congested *Exhaust Pipe*.

The relations that exist among signals were discovered by modelling each signal using all other signals as regressors:

$$\Psi_k = \underset{\Psi \in \mathbb{R}^{s-1}}{\mathrm{argmin}} \left(\sum_{t=1}^{n} \left(y_k(t) - \Psi^\top \varphi_k(t) \right)^2 \right) \qquad (2)$$

where s is number of signals, Ψ_k is a vector of parameter estimates for the model of y_k and φ_k is the regressor for y_k (i.e. the set of all *other* signals). Figure 4 shows parameters of a model capturing the relation between *charge air cooler input pressure* and *input manifold temperature*, calculated using from two different methods, LASSO (Least Absolute Shrinkage and Selection Operator) and RLS (Recursive Least Squares). One of the injected faults, clogged *Air Filter* (AF), can be very clearly discovered based on this particular relation. Other relations are useful for other faults, of course.

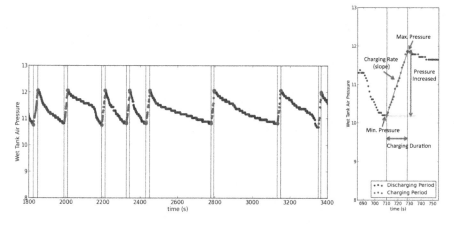

Fig. 5. *Wet Tank Air Pressure* signal (left: red marks charging periods and blue marks discharging, right: expert-defined parameters in a single charging cycle). (Color figure online)

Simple combinations of signals such as linear relations can be exhaustively enumerated and analysed by a monitoring system fully autonomously. In many cases, however, more complex models will outperform such simple ones. For example, Fig. 5 shows the *Wet Tank Air Pressure* signal for one vehicles during normal operation (left sub-plot), as well as parameters that have been identified by an expert as important (right sub-plot). Figure 6 shows the comparison of the deviation levels from both methods for one of the buses. Vertical lines correspond to repairs performed, and true positive, false negative, false positive and true negative samples are illustrated using different colours. Using such expert-defined parameters as models for COSMO, led to higher AUC in detecting failures (3% increase compared to using COSMO in an unsupervised manner, and 9% increase compared to methods described in [10,11] patents).

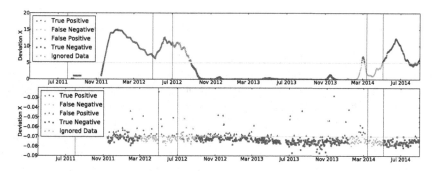

Fig. 6. Deviation level using COSMO (top) and expert method (bottom); Vertical lines correspond to repairs: red is compressor replacement that required towing, while blue indicate other faults related to the air system. (Color figure online)

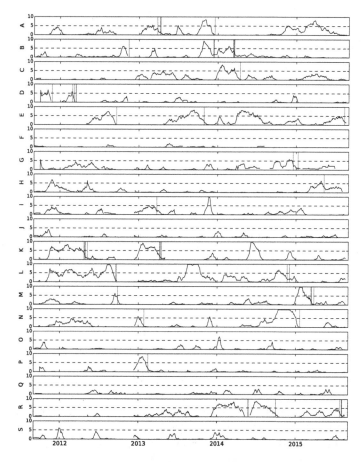

Fig. 7. Deviations levels based on *wheel speed sensor* signals for 19 buses. Vertical lines correspond to repairs after which deviations disappeared.

An important resource for evaluating monitoring systems is any data containing information about past repairs and maintenance operations during the lifetime of a vehicle. It allows monitoring system to not only inform the user that there is a problem with their vehicle, but also what had to be done to fix it last time similar thing happened. The usefulness of such information can be seen in Figs. 7 and 8. Figure 7 shows deviation levels for all 19 buses as indicated by models of the *wheel speed sensor* signal. We have identified a total of 33 "serious" deviations, and 51 workshop visits within 4 weeks after them (marked with red vertical lines). There are 150 operations that occur more than once within those 51 workshop visits – and in Fig. 8 we show operations that are likely to be related to those deviations, since they are more common during periods of interest than they are at other times.

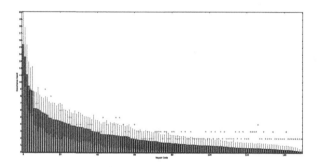

Fig. 8. Expected frequency of 150 repair codes (grey bars, with standard deviations in blue). Stars indicate the actual number during workshop visits from Fig. 7 (in green frequency more than 3σ above the mean). (Color figure online)

In the second domain of interest, district heating, there is also a large body of knowledge related to models of interest. One example are *heat load patterns* [12], which capture recurrent behavior of the buildings as weekly aggregated heat loads across four seasons of the year. In this case, weekly representation was chosen because it best captures social behaviour. Given that the data is collected once per hour, seven days a week, each heat load pattern consists of 168 values per season. By automatically analysing suitable models we have been able to identify several previously unknown heat load patterns, with examples shown in Fig. 9.

The final result we report in this paper is related to ranking abnormal district heating substations based on dispersion of heat power signatures. Heat power signature models the heat consumption of a building as a function of external temperature. Many previous studies have been analyzing heat power signatures to diagnose abnormal heat demand, however, those studies have usually focused on number of outliers, not dispersion. Figure 10 (left) shows an example power signature plot for one of the customers, and (right) the comparison in ranking accuracy where dispersion based method outperforms outlier based method.

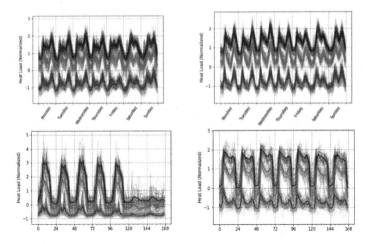

Fig. 9. Examples of typical heat load patterns (solid lines), with variations within cluster shown with transparency. Colours correspond to different year seasons. (Color figure online)

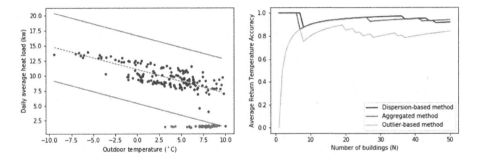

Fig. 10. Estimation of power signature using RANSAC method to measure dispersion; red points are outliers based on three standard deviations threshold (left). Comparison of the three anomaly ranking methods (right).

6 Conclusions

We propose a framework for monitoring that consists of four main steps: selecting the most suitable models based on the properties of the available data, available domain knowledge, the details of the task to be solved, and other constraints of the domain; discovering an appropriate group of peers for comparison; determining what is considered a deviation; and finally, the interactions with human experts, including techniques to present results, the scheme for tracing back the reasoning process, and how the expert's feedback should be taken into account.

In this paper we have presented several examples of how this approach can be used, and showcased the generality of it using two diverse domains: automotive and district heating.

References

1. Alippi, C., Roveri, M., Trovò, F.: A "Learning from Models" cognitive fault diagnosis system. In: Villa, A.E.P., Duch, W., Érdi, P., Masulli, F., Palm, G. (eds.) ICANN 2012. LNCS, vol. 7553, pp. 305–313. Springer, Heidelberg (2012). https://doi.org/10.1007/978-3-642-33266-1_38

2. Alippi, C., Roveri, M., Trovò, F.: A self-building and cluster-based cognitive fault diagnosis system for sensor networks. IEEE Trans. Neural Netw. Learn. Syst. **25**(6), 1021–1032 (2014)

3. Byttner, S., Nowaczyk, S., Prytz, R., Rögnvaldsson, T.: A field test with self-organized modeling for knowledge discovery in a fleet of city buses. In: IEEE ICMA, pp. 896–901 (2013)

4. Chen, H., Tiňo, P., Rodan, A., Yao, X.: Learning in the model space for cognitive fault diagnosis. IEEE TNNLS **25**(1), 124–136 (2014)

5. D'Silva, S.H.: Diagnostics based on the statistical correlation of sensors. Technical paper 2008-01-0129. Society of Automotive Engineers (SAE) (2008)

6. Fan, Y., Nowaczyk, S., Rögnvaldsson, T.: Evaluation of self-organized approach for predicting compressor faults in a city bus fleet. Procedia Comput. Sci. **53**, 447–456 (2015)

7. Fan, Y., Nowaczyk, S., Rögnvaldsson, T.: Incorporating expert knowledge into a self-organized approach for predicting compressor faults in a city bus fleet. Frontiers in Artificial Intelligence and Applications, vol. 278, pp. 58–67 (2015)

8. Filev, D.P., Chinnam, R.B., Tseng, F., Baruah, P.: An industrial strength novelty detection framework for autonomous equipment monitoring and diagnostics. IEEE Trans. Ind. Inform. **6**, 767–779 (2010)

9. Filev, D.P., Tseng, F.: Real time novelty detection modeling for machine health prognostics. In: North American Fuzzy Information Processing Society (2006)

10. Fogelstrom, K.A.: Air brake system characterization by self learning algorithm (2006)

11. Fogelstrom, K.A.: Prognostic and diagnostic system for air brakes (2007)

12. Gadd, H., Werner, S.: Fault detection in district heating substations. Appl. Energy **157**, 51–59 (2015)

13. Hansson, J., Svensson, M., Rögnvaldsson, T., Byttner, S.: Remote diagnosis modelling (2013)

14. Kargupta, H., et al.: VEDAS: a mobile and distributed data stream mining system for real-time vehicle monitoring. In: Fourth International Conference on Data Mining (2004)

15. Kargupta, H., et al.: MineFleet: the vehicle data stream mining system for ubiquitous environments. In: May, M., Saitta, L. (eds.) Ubiquitous Knowledge Discovery. LNCS (LNAI), vol. 6202, pp. 235–254. Springer, Heidelberg (2010). https://doi.org/10.1007/978-3-642-16392-0_14

16. Kargupta, H., Puttagunta, V., Klein, M., Sarkar, K.: On-board vehicle data stream monitoring using mine-fleet and fast resource constrained monitoring of correlation matrices. New Gener. Comput. **25**, 5–32 (2007)

17. Lapira, E.R.: Fault detection in a network of similar machines using clustering approach. Ph.D. thesis, University of Cincinnati (2012)

18. Lapira, E.R., Al-Atat, H., Lee, J.: Turbine-to-turbine prognostics technique for wind farms (2011)

19. Quevedo, J., et al.: Combining learning in model space fault diagnosis with data validation/reconstruction: application to the Barcelona water network. Eng. Appl. Artif. Intell. **30**, 18–29 (2014)

20. Rögnvaldsson, T., Nowaczyk, S., Byttner, S., Prytz, R., Svensson, M.: Self-monitoring for maintenance of vehicle fleets. Data Min. Knowl. Discov. **32**(2), 344–384 (2018)
21. Theissler, A.: Detecting known and unknown faults in automotive systems using ensemble-based anomaly detection. Knowl.-Based Syst. **123**, 163–173 (2017)
22. Vachkov, G.: Intelligent data analysis for performance evaluation and fault diagnosis in complex systems. In: IEEE ICFS, pp. 6322–6329 (2006)
23. Zhang, Y., Gantt Jr., G.W., Rychlinski, M.J., Edwards, R.M., Correia, J.J., Wolf, C.E.: Connected vehicle diagnostics and prognostics, concept, and initial practice. IEEE Trans. Reliab. **58**, 286–294 (2009)

Creation of an Event Log from a Low-Level Machinery Monitoring System for Process Mining Purposes

Edyta Brzychczy[(⊠)] and Agnieszka Trzcionkowska

AGH University of Science and Technology, Kraków, Poland
{brzych3, toga}@agh.edu.pl

Abstract. Industrial event logs, especially from low-level monitoring systems, very often have no suitable structure for process-oriented analysis techniques (i.e. process mining). Such a structure should contain three main elements for process analysis, namely: timestamp of activity, activity name and case id.

In this paper we present example data from a low-level machinery monitoring system used in underground mine, which can be used for the modelling and analysis of the mining process carried out in a longwall face. Raw data from the mentioned machinery monitoring system needs significant pre-processing due to the creation of a suitable event log for process mining purposes, because case id and activities are not given directly in the data.

In our previous works we presented a mixture of supervised and unsupervised data mining techniques as well as domain knowledge as methods for the activity/process stages discovery in the raw data. In this paper we focus on case id identification with an heuristic approach. We summarize our experiences in this area showing the problems of real industrial data sets.

Keywords: Event logs · Process mining · Low-level monitoring system
Underground mining · Longwall face

1 Introduction

The Internet of Things and Industry 4.0 in the mining industry have become a fact. The great step in underground industrial advancement which completed automatization in the field is the convergence of industrial systems with the power of advanced computing, analytics, low-cost sensing and new levels of connectivity. Smart sensor technologies and advanced technics of analysis play an important role in mining process monitoring and improvement [11].

Automation has enabled access to very detailed data characterizing the operation of machines and devices (stored in monitoring systems). Longwall automation and monitoring systems allows a closer look at the ongoing processes underground. A vast amount of data is generated that should be used and handled more efficiently in a modern mining operation [4]. Nowadays, the analytics of collected data is mainly based on data-oriented techniques: BI techniques for operational report creation as well as on more advanced analytic data mining and machine learning techniques for predictive

© Springer Nature Switzerland AG 2018
H. Yin et al. (Eds.): IDEAL 2018, LNCS 11315, pp. 54–63, 2018.
https://doi.org/10.1007/978-3-030-03496-2_7

maintenance purposes [8]. Thus, for acquiring new knowledge about ongoing processes underground, we proposed process-oriented analysis of the gathered data [2].

In the paper we present example data from a low-level machinery monitoring system used in underground mine, which can be used for the modelling and analysis of the mining process carried out in a longwall face.

Our work aims proposal of original extension of data analysis from low-level longwall machinery monitoring system with process mining techniques and according to the authors' best knowledge, it is the first attempt of process mining usage in the mining domain [6].

The basic challenge raising from the proposed analysis extension is the creation of a suitable event log for process mining purposes containing [1]: timestamp of activity, activity name and case id. This is not a trivial task since case id and activities (process stages) are not given directly in the raw data from the low-level longwall machinery monitoring system. Moreover, there is no procedure to identify case id in the raw data from the longwall monitoring system, since no one has applied process mining in the mining domain.

To address the mentioned challenges we prepared two procedures for data processing:

1. For activities' name definition we proposed the mixture of supervised and unsupervised data mining techniques as well as domain knowledge, presented in more detail in [2]. That proposal contains among others: data cleaning, clustering and classification for labelling the process stages in the raw data.
2. For case id identification, we propose the heuristic approach presented in this paper. Our solution is an example of how we can handle with raw data related to the cyclic process in a specific production domain without clear marking of the start and the beginning of the case.

Our both procedures are written in R, mainly using libraries: `dplyr`, `arules`, `cluster`, `forecast`, `CHAID` and `rpart`.

The paper is structured as follows: Sect. 2 includes mining process description. In Sect. 3 identification of case id in raw data is presented. An example of created event log is described in Sect. 4. Conclusions are presented in Sect. 5.

2   Process Description

The mining process can be defined as a collection of mining, logistics and transport operations. One of the most complex and difficult examples of its realisation is underground mining characterized by changeable geological and mining conditions as well as natural hazards not occurring on the surface. Very interesting is the nature of the mining process in the longwall system that is performed by machines and devices moving in a workspace and also in relation to each other.

Main longwall equipment includes (Fig. 1): a shearer (A), an armoured longwall conveyor (B) and mechanized supports (C).

Each of the mentioned machines realises its own operation process, consequently the mining process in a longwall face can be seen as collection of machinery processes.

Fig. 1. Longwall machinery (https://famur.com/upload/2016/09/FAMUR_01-1.jpg)

The mining process includes even up to a hundred processes (depending on the dimensions of a mining excavation and number of mechanized supports).

In the paper we focus on the operation process of main longwall machinery, namely the shearer. The operation of the shearer indicates the cycle of a whole mining process [12], therefore it is the most intuitive choice for case id in an event log. The theoretical shearer operation process is presented in Fig. 2.

In general a shearer operation cycle consists of several characteristic phases. Firstly, the shearer starts cutting from the driver unit side (1). The next phase is indentation where a shearer is cutting into the turning station direction for a distance of 30–40 m. Together with the movement of the shearer a longwall conveyor is shifting (2). The third step is cutting into the driver unit side – longwall cleaning (3). Along with the shearer the powered roof support is moving. In the next phase a shearer is cutting without loading for a distance of 30–40 m (4), after that it is cutting throughout the longwall till the turning station. Along with the movement of the shearer, the conveyor and powered roof support are moved (5, 6, 7).

The basic indicator of the shearer's movement is the value of the "Location in the longwall" variable. The ideal model of the shearer operation with activity names is presented in Fig. 3.

The real location of a shearer in a raw data is presented in Fig. 4.

Two main challenges in modelling the mining process based on real data are illustrated well on the picture: data quality and cycles variability. The first challenge has a major source in technical problems in data transfer from the machinery to the surface, especially in the case of power off events and data retrieving from the machinery local data containers. The second challenge is strictly related to the mining and geological conditions of process realisation.

It should be also mentioned that raw data contain various quantitative and qualitative (mostly binary) variables that in some way describe the process stages, not directly as activity names. It needs a lot of analytic efforts to build event logs on top of it (activity recognition, abstraction level choice etc.). Our contribution in this area is presented in the following sections.

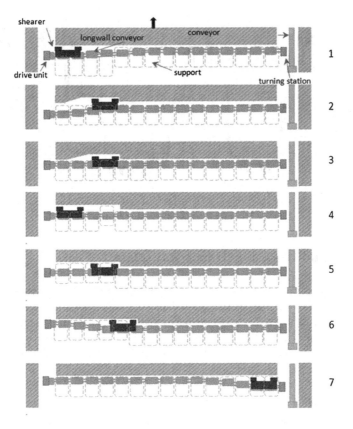

Fig. 2. Example cycle of shearer operation. Source: based on [10]

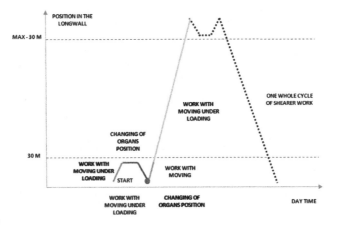

Fig. 3. Example cycles of shearer operation in time dimension

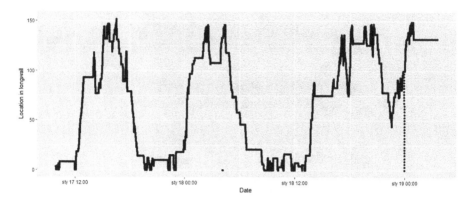

Fig. 4. Example cycles of the shearer operation

3   Identification of Case ID in a Raw Data

In this section we present issues related to case id identification for the purpose of event log creation based on the shearer operation data from the selected hard coal mine. Raw data related to the mentioned process include 2.5 million records from a monthly period obtained from one of the Polish mining companies.

In Table 1 the selected variables characterizing the shearer operation are presented.

Table 1. Selected variables characterizing the shearer operation

Variable	Type	Range
Location in the longwall	Numerical	0–200 [m]
Shearer speed	Numerical	0–20 [obr/min]
Arm left up/down	Binary	0/1
Arm right up/down	Binary	0/1
Move in the right	Binary	0/1
Move in the left	Binary	0/1
Current on the left organ	Numerical	0–613 [A]
Current on the right organ	Numerical	0–680 [A]
Current on the left tractor	Numerical	0–153 [A]
Current on the right tractor	Numerical	0–153 [A]
Security DMP left organ	Binary	0/1
Security DMP right organ	Binary	0/1
Security DMP left tractor	Binary	0/1
Security DMP right tractor	Binary	0/1

The identification of the shearer's work cycles (case id) was mainly based on the analysis of the attributes "Location in the longwall" (distance below 5 m and over 135 m) and "Shearer speed" (equal to 0 m/s).

The first approach of the cycle start and finish identification was based on the classic analysis of local minimum and maximum. This approach did not yield satisfactory results. The main problem was related to large local process variability.

Therefore, the heuristic approach with the following steps was proposed.

1. The shearer's position in the longwall face was split into three ranges (Fig. 5) according to the technological conditions and theoretical model of the cycle:

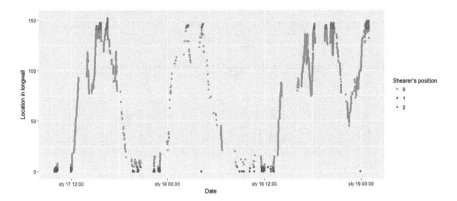

Fig. 5. Ranges of the shearer's position

- the beginning of the longwall face - distance below 5 m (was marked with 2),
- the end of the longwall face - over 135 m (was marked with 1),
- and in the middle of the longwall face (marked with 0).

2. In the range sets, below 5 m and over 135 m, the local minimum (1) and maximum (2) were detected accordingly.
3. Characteristic peaks (start and end of the cycle) were identified by the selection of sequences only with specific range (1) and (2) order (Fig. 6).

In the analysed dataset 75 cycles were identified (9 cycles are presented in Fig. 7)

In the most cases the proposed heuristic enabled the identification of the cycle start and end correctly. The errors in identification were caused mainly by data quality. The red arrows in the Fig. 7 points out one of the main issues: incorrect state of location. Thus, the presented approach is sensitive on data quality and further works will be focused on improving data cleaning at the early stages to avoid the mentioned issue.

In the case of lack of data in a shearer location variable extrapolation between the nearest two points existing in the data can be done. We know how the theoretical and technological cycle looks like, so extrapolation, based also on other variables values, could be verified.

The shearer cycle is crucial for all machinery working in the longwall face, because the rest of the machines and devices are just adjusting to the shearer position in the cycle. Therefore, for process modelling purposes, it is very important to find the way how in raw data a start and end of the cycle can be identified. Especially, when real cycles are varied very much from theoretical models.

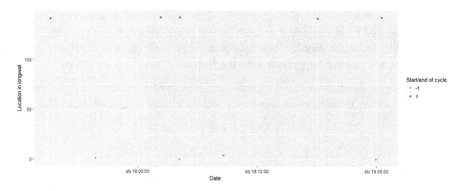

Fig. 6. Visualization of the beginning and the end of the cycles

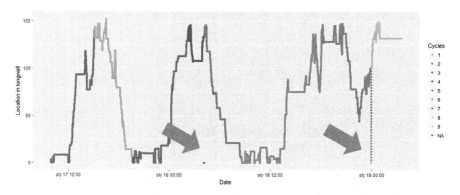

Fig. 7. Example of identified cycles

Although our approach is for a very specific domain, we think that it can be helpful for the creation of event logs for similar problems and processes.

4 Creation of an Event Log

The creation of an event log based on sensor data, beside a case id identification requires the recognition and identification of activities (process stages). In these cases supervised and unsupervised techniques of data mining can be applied [3, 5, 7, 9, 13].

Selected variables (Table 1) were used for distinguishing the unique states of the shearer operation, according to the procedure, described in [2]. The following stages were performed:

1. *Data preprocessing.* In this stage exploratory data analysis was conducted. Subsequently an analysis of correlation for the numerical variables and cross tables for the logical variables were performed to exclude the depended variables. Then the discretization of all continuous variables into a categorical variables was carried out. Furthermore, in the final data set, containing discretized and logical variables, duplicate rows were removed.

2. *Data clustering*. For the final data set dissimilarity matrix with Gower's distance was created. Then hierarchical clustering was carried out using the Ward's minimum variance method. Finally, selected clusters have been labelled with activity names, based on the statistical analysis results and an expert knowledge.
3. *Classification for labelling activity names in the raw data*. In this stage instances with a labeled activity name (process stage) have been used as a learning sample in the CHAID tree algorithm. For each label, according to the CHAID tree model, unique rules have been generated and, on this base, activity labelling in the raw data was done.

The identification of case id and activity definition enable the creation of an event log presented in Table 2. The process stages labelled on the example traces are shown in Fig. 8.

Table 2. Fragment of an event log (selected in the Fig. 8)

Case id/trace	Timestamp	Activity/process stage
1	17.01.2018 15:23:11	Mining – Type II
1	17.01.2018 15:24:14	Start mining
1	17.01.2018 15:25:20	Stope change – Type IX
1	17.01.2018 15:25:35	Start mining
1	17.01.2018 15:26:19	Stope change – Type IX
1	17.01.2018 15:26:21	Start mining
1	17.01.2018 15:27:15	Stope change – Type IX
1	17.01.2018 15:28:17	Start mining
1	17.01.2018 15:28:37	Shearer stoppage
1	17.01.2018 15:29:18	Start mining
1	17.01.2018 15:32:39	Shearer stoppage
1	17.01.2018 15:41:29	Start mining

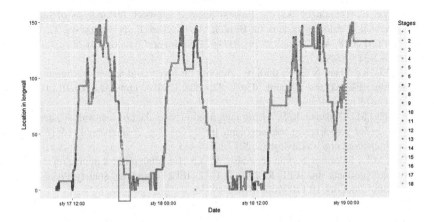

Fig. 8. Labelled process stages on example traces

A created event log enables the performance of process modelling with selected techniques and formalisms [1] and further works in this scope are carried out.

5   Conclusions

Current underground machinery monitoring systems can contain streaming data from hundreds of sensors of various types. The efficient processing of such an amount of data (Big Data) for process improvements is possible only with the specific techniques of advanced analysis from data mining and process mining fields.

Process mining techniques require a specific structure of an event log with activity names and case id, that very often are not present in raw industrial sensor data. The challenges related to activity recognition and case identification are strongly connected to the data quality and nature of an analyzed process. Therefore, cleaning and pre-processing activities are needed and adequate analytic approaches should be found.

In the paper we presented case id identification problems on a selected example from the longwall monitoring system in an underground mine. The classic approach in this case has not yielded correct results due to the high variability of the process and the existence of many local optima, thus the heuristic approach was developed.

We have contributed original solutions (procedures) for an event log creation from a low-level machinery monitoring system in underground mining for process mining purposes. Future challenges will be related to process modelling based on prepared event logs in the case of high process variability.

Acknowledgements. This paper presents the results of research conducted at AGH University of Science and Technology – contract no. 15.11.100.181.

References

1. van der Aalst, W.M.P.: Data science in action. In: van der Aalst, W.M.P. (ed.) Process Mining, pp. 3–23. Springer, Heidelberg (2016). https://doi.org/10.1007/978-3-662-49851-4_1
2. Brzychczy, E., Trzcionkowska, A.: Process-oriented approach for analysis of sensor data from longwall monitoring system. In: Burduk, A., Chlebus, E., Nowakowski, T., Tubis, A. (eds.) ISPEM 2018. AISC, vol. 835, pp. 611–621. Springer, Cham (2019). https://doi.org/10.1007/978-3-319-97490-3_58
3. Cook, D.J., Krishnan, N.C., Rashidi, P.: Activity discovery and activity recognition: a new partnership. IEEE Trans. Cybern. **43**(3), 820–828 (2013). https://doi.org/10.1109/tsmcb.2012.2216873
4. Erkayaoğlu, M., Dessureault, S.: Using integrated process data of longwall shearers in data warehouses for performance measurement. Int. J. Oil Gas Coal Technol. **16**(3), 298–310 (2017). https://doi.org/10.1504/ijogct.2017.10007433
5. van Eck, M.L., Sidorova, N., van der Aalst, W.M.P.: Enabling process mining on sensor data from smart products. In: IEEE RCIS, pp. 1–12. IEEE Computer Society Press, Brussels (2016). https://doi.org/10.1109/rcis.2016.7549355
6. Gonella, P., Castellano, M., Riccardi, P., Carbone, R.: Process mining: a database of applications. Technical report, HSPI SpA - Management Consulting (2017)

7. Guenther, C.W., van der Aalst, W.M.P.: Mining activity clusters from low-level event logs. BETA Working Paper Series, WP 165, Eindhoven University of Technology, Eindhoven (2006)

8. Korbicz, J., Koscielny, J.M., Kowalczuk, Z., Cholewa, W. (eds.): Fault Diagnosis: Models, Artificial Intelligence, Applications. Springer, Heidelberg (2004). https://doi.org/10.1007/978-3-642-18615-8

9. Mannhardt, F., de Leoni, M., Reijers, H.A., van der Aalst, W.M.P., Toussaint, P.J.: From low-level events to activities - a pattern-based approach. In: La Rosa, M., Loos, P., Pastor, O. (eds.) BPM 2016. LNCS, vol. 9850, pp. 125–141. Springer, Cham (2016). https://doi.org/10.1007/978-3-319-45348-4_8

10. Napieraj, A.: The method of probabilistic modelling for the time operations during the productive cycle in longwalls of the coal mines (in Polish). Wydawnictwa AGH, Cracow (2012)

11. Ralston, J.C., Reid, D.C., Dunn, M.T., Hainsworth, D.W.: Longwall automation: delivering enabling technology to achieve safer and more productive underground mining. Int. J. Mining Sci. Technol. 25(6), 865–876 (2015). https://doi.org/10.1016/j.ijmst.2015.09.001

12. Snopkowski, R., Napieraj, A., Sukiennik, M.: Method of the assessment of the influence of longwall effective working time onto obtained mining output. Archives Mining Sci. 61(4), 967–977 (2016). https://doi.org/10.1515/amsc-2016-0064

13. Tax, N., Sidorova, N., Haakma, R., van der Aalst, W.M.P.: Event abstraction for process mining using supervised learning techniques. In: Bi, Y., Kapoor, S., Bhatia, R. (eds.) IntelliSys 2016. LNNS, vol. 15, pp. 251–269. Springer, Cham (2018). https://doi.org/10.1007/978-3-319-56994-9_18

Causal Rules Detection in Streams of Unlabeled, Mixed Type Values with Finit Domains

Szymon Bobek$^{(\boxtimes)}$ (iD) and Kamil Jurek

AGH University of Science and Technology, al. Mickiewicza 30,
30-059 Krakow, Poland
szymon.bobek@agh.edu.pl

Abstract. Knowledge discovery from data streams in recent years become one of the most important research area in a domain of data science. This is mainly due to the rapid development of mobile devices, and Internet of things solutions which allow for obtaining petabytes of data within minutes. All of the modern approaches either use representation that is flat in time domain, or follow black-box model paradigm. This reduces the expressiveness of models and limits the intelligibility of the system. In this paper we present an algorithm for rule discovery that allows to capture temporal causalities between numeric and symbolic attributes.

Keywords: Rules · Knowledge discovery · Context-aware systems

1 Introduction and Motivation

Data mining methods gained huge popularity in recent decade. One of the reasons for that was a need for processing large volumes of data for knowledge extraction, automation of business processes, fault detection, etc. [10]. Machine learning algorithms are able to process petabytes of information to build models which later can be used on unseen data to generate smart and fast decisions. Humans are no longer capable of handling such a large volumes of data and thus need a support from artificial intelligence (AI), which became omnipresent in their daily lives. Industry, e-commerce and business were not the only beneficiaries of machine learning systems. Also regular users gained a huge support in their daily routines by ambient assistants, smart cognitive advisors and context-aware mobile intelligent systems [5] or ubiquitous computing systems [9].

In all of the areas where the human live is affected by the decisions of automated system, either in trivial situations like planning a day or in serious scenarios like diagnosing a disease, the **user trust** to the system is one of the most important factors. That trust is build upon intelligibility of the system, i.e. an ability of the system to being understood by the human [18].

The fact that humans left the data to machine learning algorithms is not because they did not want to understand what knowledge is hidden there, but

© Springer Nature Switzerland AG 2018
H. Yin et al. (Eds.): IDEAL 2018, LNCS 11315, pp. 64–74, 2018.
https://doi.org/10.1007/978-3-030-03496-2_8

because they could not handle it themselves. Therefore, we should not evince that opportunity to understand the data, but use machine learning algorithms to help us get that insight when needed. We argue that a good start to providing this can be by using explainable knowledge description methods.

One of the most human-readable and understandable method of representing knowledge are rules. Although there are mechanisms for rules mining such as: Apriori [1], FP-Growth [14], Rill [8], etc., most of them are as we call it *flat in time*, not handling well situations where the main goal is to capture dynamics of the system.

In certain situations it is desirable to have rules that are richer semantically than flat rules and are able to capture relations between changes of values of attributes of the system. Rules which can model such causalities can help to explain and forecast possible safety hazards, or other issues, in systems where changes of context and temporal factors matters most.

The example of a fragment of such system is presented in Fig. 1. The subjective responses to temperature is more understandable when considered in terms of changes, not simple associations of current temperature at given points in time. Changing environment from cold to mild may result in subjective feel of warm, while changing to the same mild environment from warm may result in feel of cold. The other changes, when they are gradual may result in no changes of other parameters of the system. Therefore, the same state without the respect of changes nor temporal factor may be a source of two disjoint responses, causing the system to produce wrong decisions. We solely focus on this type of temporal causality, where a change at some point of time of an attribute is assumed do be a trigger (cause) of a change for the other attribute.

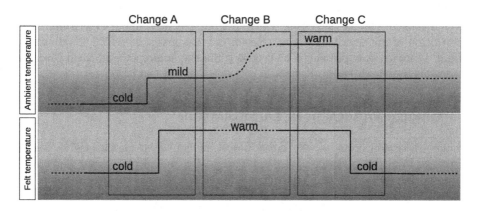

Fig. 1. Illustration of how change may result in different causal effects on the same parameter

The aim of our research was to provide a mechanism which will encode this temporal dependency between changes in a meaningful format that could be understand by the human. This approach fills the gap that existed between mechanisms for causality detection and algorithms for knowledge discovery and prediction from data streams and time series. The former, such as Bayesian networks, Markov Chains and Hidden Markov Models [17] are hardly explainable to the user.

The latter such as LSTM [12], or modified ARIMA [26] are more of a black box mechanisms with no explainability. Furthermore, they omit causal and temporal aspects of dependencies between attributes in data.

There were attempts to mine temporal patterns in data streams, such as these presented in [19], where authors propose hierarchical clustering to group similar patterns over time. Furthermore, there exists a huge number of sequential pattern mining methods implemented in SPMF framework[1] or SPAM framework[2] and others [16]. However, these methods mainly focus on association between attributes over time, rather than associations between changes of that attributes.

In this paper we propose a mechanism that detects changes in streams of values of mixed types (numeric, or nominal) and discovers relationships between these changes. We encode that knowledge into human readable format which can be later review, extended or modified by the user of by an expert. This aspect of designing modern AI systems is crucial in the face of the recent European Union General Data Protection Regulation, which sates that every user of an AI software should have the *right to ask for explanation* of an automated algorithmic decision that was made about them [13]. This feature should be crucial for AI systems that are designed for Industry 4.0, where large amount of data clearly exhibits temporal patterns and decision made based on that data should bas transparent as possible. This applies to areas such as logistics, mining industry, manufacturing and others where trust in black box algorithms is to risky due to large financial loss or human safety [23].

The rest of the paper is organized as follows. In Sect. 2 we discuss different mechanisms for change detection, which was implemented in our algorithm. The description of an algorithm and a rule format was given in Sect. 3. Evaluation of our approach was presented in Sect. 4 while the summary and future works directions was described in Sect. 5.

2   Change Detection in Data Streams

Change detection is one of the most important field in the area of knowledge discovery from data streams [10]. While originally, the aim of change detection is to capture so called *concept drift*, we exploited strengths of such methods to discover changes in values of predefined features with known and finite domains. In this work, we focused on rapid changes, that are defined as ones that take effect in relatively short period of time. Gradual changes, are in slow in time and expose more of the evolutionary than revolutionary nature. We left the former for future work.

---

[1] See: http://www.philippe-fournier-viger.com/spmf.

[2] See: http://www.cs.cornell.edu/database/himalaya/SequentialPatterns/seqPatterns_main.htm.

Change detectors can be divided into statistical change detectors and streaming window detectors Statistical change detectors monitor the evolution of some statistical indicators of the signal such as mean, standard deviation, variance and base on that decide if change occurred [10].

One of the basic approach is cumulative sum (CUSUM algorithm), first proposed by Page [20]. It is a sequential analysis technique that is typically used for monitoring change detection. It is a classical change detection algorithm that gives an alarm when the mean of the input data is significantly different from zero [4]. The CUSUM test is memoryless, and its accuracy and sensitivity to false alarms depends on the choice of parameters. As its name implies, CUSUM involves the calculation of a cumulative sum (which is what makes it sequential). When the value of calculated sum exceeds a certain threshold value, a change in value has been found.

Another change detection algorithm similar to CUSUM is the Geometric Moving Average Test, proposed in [24]. It introduces the forgetting factor used to give more or less weight to the last data arrived. The stopping rule is triggered when difference between last arrived data and average value is greater than a manually provided threshold value.

The Page-Hinkley test which is a sequential adaptation of the detection of an abrupt change of the average of a Gaussian signal. It allows efficient detection of changes in the normal behavior of a process which is established by a model. It is a variant of the CUSUM algorithm. This test considers a cumulated difference between the observed values and their average value till the current moment. When this difference is greater than given threshold the alarm is raised denoting change in the distribution.

Complementary to statistical change detectors are solutions based on streaming windows where we maintain statistics (such as mean) on two or more separate samples (or windows) of the data stream.

Z-Score algorithm is a change detector, that is based on sliding windows approach. Here two sets of statistics about data stream are maintained. The first one is computed based on entire stream using Welford's Method [25]. For the second set of statistics a moving window of the n most recent points is taken into account.

The calculated Z-score, tells the difference between local window and the global data stream. The stopping rule is triggered when the Z-score is greater than provided threshold value.

ADWIN (ADaptive WINdowing) is an algorithm proposed by Bifet in [2,3] is an adapting sliding window method suitable for data streams with sudden changes. The algorithm keeps in memory a sliding window W which length is adjusted dynamically to reflect changes in the data. When change seems to be occurring, as indicated by some statistical test, the window is shrunk to keep only data items that still seem to be valid. When data seems to be stationary, the window is enlarged to work on more data and reduce variance. This involves answering a statistical hypothesis: "Has the average μ remained constant in W with confidence δ"? The statistical test checks if the observed average in both sub-windows differs by more than threshold cut. The threshold is calculated using the Hoeffding bound [15].

Drift detection method (DDM) proposed by Gama et al. [11] controls the number of errors produced by the learning model during prediction. It compares the statistics of two windows: the first one contains all the data, and the second one contains only the data from the beginning until the number of errors increases. This method does not store these windows in memory. It keeps only statistics and a window of recent data. DDM uses a binomial distribution to describe the behavior of a random variable that gives the number of classification errors in a sample of size n. If the error rate of the learning algorithm increases significantly, it suggests changes in the distribution of samples, causing the change detection alarm.

The the next section we describe how the change detector mechanism was incorporated to our framework as a core element for causality detection.

3   Discovery of Rules in Streams with Unlabeled, Mixed Type Values

Figure 2 presents the workflow of a procedure for rules discovery in our framework.[3]

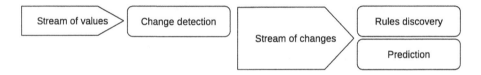

Fig. 2. Workflow of a process of knowledge discovery and prediction

At the input there is a stream of values for different attributes. The attributes have a well known type (nominal or numeric) and a finite domain. The stream can be noisy, i.e. changes may not be visible clearly. Figure 3 shows two streams of values generated for two different attributes with discrete numeric domains. In case of nominal attributes we use Helmert or Binary encoding [22] to convert categorical values into a form that allows for calculating statistics required by change detectors. We tested several other encoding mechanisms, but these two were giving the best results.

---

[3] The complete framework implementation is available on https://gitlab.geist.re/pro/CRDiS.

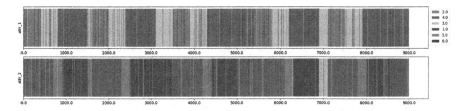

Fig. 3. Sample streams for two attributes. Different colors represent different values of the attribute, while the horizontal axis represents time. (Color figure online)

The stream is passed to change detectors, that mark changes of values at specific points of time. The changes discovered by different detectors were given in Fig. 4. These changes are then passed to rule generation mechanism, which is a single pass algorithm for rule mining. The algorithm creates rules with respect to target sequence, but due to its online nature and low computational complexity it can be executed in parallel to create all possible rules with respect to all target attributes.

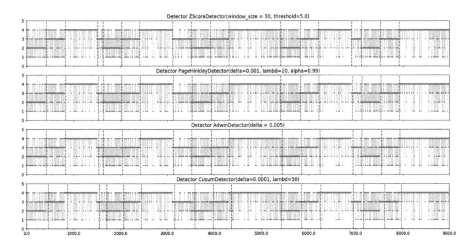

Fig. 4. Evaluation of different change detection mechanisms on a stream from Fig. 3

The rule format is **IF** $A(v_1)^{T_1}$ **THEN** $B(v_2)^{T_2}$. Here $A(v_1)^{T_1}$ and $B(v_2)^{T_2}$ are basic rule components, i.e., they are values of attributes that lasted for T_1 and T_2 respectively. In other words **IF** *attribute A had value of v_1 for T_1 time units* **THAN** *attribute B will have value of v_2 for T_1 time units.*

For each sequence a list of change points that where discovered is being remembered in a form of $change\_points[attr] = p_0, p_1, ..., p_i$, where p_i is the most recently found change in sequence of changes for attribute *attr*.

When change in target sequence is detected algorithm starts. Each sequence is analyzed in a window that starts in point p_{i-2} and ends in p_i of a target sequence. Length of the window, indicating how distant in time data are being analyzed can be adapted. Firstly, all possible left-hand-sides of the rule for currently processed sequence are generated in a sub-window stretching from p_{i-2} to p_{i-1}. For that window each change point that took place inside of it is being extracted. Using the information about changes that occurred, rule components are created.

Given a sequence S the support $Supp(A)$ of the rule component A is the number of occurrences of $A \in S$, The confidence $Conf(A \Rightarrow B)$ of the $A \Rightarrow B$ is the fraction of occurrences of A that are followed by a B.

$$Conf(A \Rightarrow B) = \frac{Supp(A, B)}{Supp(A)}$$

where $Supp(A, B)$ is the number of occurrences of A that are followed by a B. This differs from the usage of support for association rules.

The rule score is being calculated in the following way:

$$Score(A \Rightarrow B) = Supp(A, B) \times Conf(A, B) \times (|A| + |B|)$$

where $|A|$ and $|B|$ are the lengths(durations) of rule components A and B

Similarly right-hand-sides of the rule are being constructed but in sub-window from p_{i-1} to p_i and with values of target sequence. In the next step generated LHSs and RHSs are being combined. If the newly created rule was already known all necessary statistics (i.e. rule support, confidence, score) are being updated, otherwise the rule is being added to the discovered rules set and it's statistics are being calculated. Described steps are being repeated for each of analyzed sequences. For each sequence rules with highest score are being chosen and combined rule is being constructed. In combined rule antecedent part consists of LHS parts of rules generated for single sequences.

Currently two approaches fro handling subsumed rules are implemented. The basic one returns all the rules, sorted according to score. The other returns only most specific rules, i.e. rules that does not subsume any other rules.

4   Evaluation

The main goal of the evaluation was to show that the algorithm is able to correctly detect temporal patterns of causal relationship between attributes. The second step was a comparison of our algorithm with LSTM neural network [12].

For the evaluation purposes, the testing environment was implemented. It consists of a sequence generator that allows to define and generate sequences of values that represent given causal pattern and a simulator that allows for real-time simulation of the rule detection.

In Listing 1.1 a configuration for a sequence generator was given. It defines a value to hold for a given time spans and the noise factor that defines how stable the value is over that time span. A fragment of an output from the sequence generator for this configuration was presented in previous section in Fig. 3.

Listing 1.1. Sample configuration for a sequence generator

```
attr = 'attr_1';value=3;domain = [1,2,3,4]; from=-600; to=-200; probability=0.85
attr = 'attr_1';value=4;domain = [1,2,3,4]; from=-200; to=500; probability=0.95
attr = 'attr_2';value=4;domain = [1,2,3,4,5]; from=-1000; to=-300; probability=0.90
attr = 'attr_2';value=5;domain = [1,2,3,4,5]; from=-300; to=-100; probability=0.95
attr = 'attr_2';value=1;domain = [1,2,3,4,5]; from=-100; to=500; probability=0.85
attr = 'attr_3';value=1;domain = [1,2,3,4,5]; from=-1000; to=-100; probability=0.95
attr = 'attr_3';value=4;domain = [1,2,3,4,5]; from=-100; to=500; probability=0.95
attr = 'attr_4';value=1;domain = [1,2,3,4,5,6]; from=-1000; to=0; probability=0.90
attr = 'attr_4';value=6;domain = [1,2,3,4,5,6]; from=0; to=500; probability=0.85
```

The sequence is then passed to change detection mechanism. Table 1 shows the result from change detection done with different algorithms. The delay error represents the mean squared error between the real change timestamp and detected one. It is worth noting that depending on the noise level in the original stream one may consider different change detection mechanism to achieve better results.

Table 1. Evaluation of different change detection mechanisms on a sample stream with 19 change points.

	Z-score	Page-Hinkley	ADWIN	CumSUm	Noise
Changes detected	19/19	19/19	19/19	19/19	0.0
Delay error (MSE)	1.80	19.0	24.02	44.20	
Changes detected	19/19	19/19	19/19	19/19	0.05
Delay error (MSE)	11.20	22.19	28.06	49.92	
Changes detected	19/19	19/19	19/19	19/19	0.1
Delay error (MSE)	15.59	23.26	31.63	53.74	
Changes detected	18/19	19/19	19/19	19/19	0.15
Delay error (MSE)	1571.35	28.95	41.73	72.07	
Changes detected	18/19	19/19	19/19	19/19	0.20
Delay error (MSE)	1178.80	28.89	41.69	74.44	
Changes detected	15/19	19/19	19/19	19/19	0.25
Delay error (MSE)	1802.68	33.12	59.08	76.99	
Changes detected	13/19	21/19	19/19	20/19	0.30
Delay error (MSE)	2547.55	2088.89	70.01	1480.85	
Changes detected	11/19	22/19	18/19	19/19	0.35
Delay error (MSE)	2748.15	1812.31	1186.37	200.04	
Changes detected	10/19	22/19	18/19	19/19	0.40
Delay error (MSE)	3234.44	2140.62	1577.04	272.02	

The output of the change detection mechanism is presented in Listing 1.2. Every line is a compressed slice of 100 points of time.

Listing 1.2. Fragment of an output from change detection mechanism

```
      attr_1           attr_2           attr_3           attr_4
0     attr_1:2->2      attr_2:4->4      attr_3:1->1      attr_4:1->1
1     attr_1:2->2      attr_2:4->4      attr_3:1->1      attr_4:1->1
2     attr_1:2->2      attr_2:4->4      attr_3:1->1      attr_4:1->1
3     attr_1:2->3      attr_2:4->4      attr_3:1->1      attr_4:1->1
4     attr_1:3->3      attr_2:4->4      attr_3:1->1      attr_4:1->1
5     attr_1:3->3      attr_2:4->4      attr_3:1->1      attr_4:1->1
6     attr_1:3->3      attr_2:4->5      attr_3:1->1      attr_4:1->1
7     attr_1:3->4      attr_2:5->5      attr_3:1->1      attr_4:1->1
8     attr_1:4->4      attr_2:5->1      attr_3:1->4      attr_4:1->1
9     attr_1:4->4      attr_2:1->1      attr_3:4->4      attr_4:1->6
10    attr_1:4->4      attr_2:1->1      attr_3:4->4      attr_4:6->6
11    attr_1:4->4      attr_2:1->1      attr_3:4->4      attr_4:6->6
12    attr_1:4->4      attr_2:1->1      attr_3:4->4      attr_4:6->6
13    attr_1:4->4      attr_2:1->1      attr_3:4->4      attr_4:6->6
                                ...
```

Below, sample rules are presented that were discovered by the algorithm. It is worth noting that the rules reflect the configuration pattern passed to the sequence generator, and give in Listing 1.1.

```
[attr_1(3.0){400; 94%}] {supp: 43 conf: 0.5375 } AND
[attr_2(4.0){400; 79%}] {supp: 45 conf: 0.3515625 } AND
[attr_3(1.0){400; 94%}] {supp: 34 conf: 0.21794871794871795 } AND
==> attr_4(1.0){700; 89%}

[attr_1(3.0){300; 79%}, attr_1(4.0){200; 79%}]  AND
[attr_2(4.0){200; 87%}, attr_2(5.0){200; 93%}, attr_2(1.0){100; 93%}]  AND
[attr_3(1.0){400; 93%}, attr_3(4.0){100; 93%}]  AND
==> attr_4(6.0){400; 82%}
```

Such rules can be easily encoded in HMR+ format which is a native format for a HEARTDROID rule inference engine [7][4], thus allowing for instant execution of discovered knowledge.

We also evaluated our mechanism in comparison to LSTM neural network model, which is one of the most efficient mechanism for predictions based on temporal context. Figure 5 shows the output from both predictors. We used z-score change detector fro training set preparation. Mean squared error for LSTM was 3.30 and for our approach 2.20. This values may differ in favor for LSTM, depending on the input stream and change detector used. However, our experiments show that our approach is not worse than LSTM, giving at the same time far more intelligibility and interpretability of a model.

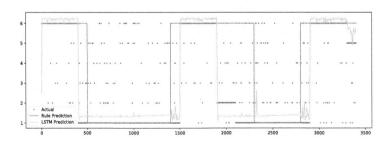

Fig. 5. Comparison of our approach to LSTM neural network predictions.

---

[4] See https://bitbucket.org/sbobek/heartdroid.

5   Summary and Future Works

In this paper we present an online algorithm for generating causal rules with temporal dependencies between changes of attributes' values. We argue that this approach can be very useful in the areas such as fault detection, logistics and others, where change in attribute value may cause changes in the future values of some attributes. We shown that our approach is not worse than LSTM neural network, giving at the same time far more insight into the model logic.

It is worth noting, that we intentionally left not tackled the problem of falling into a trap of "correlation does not imply causation" [21]. Our assumption was that given the intelligible and human-readable knowledge description in a form of rules, every (or most) false causalities can be filtered by an expert or a user. This issue though, can be a subject of further research.

As a future work we also plan to integrate the solution with HEARTDROID rule engine [7] and platform for mobile context-aware systems [6].

References

1. Agrawal, R., Mannila, H., Srikant, R., Toivonen, H., Verkamo, A.I.: Advances in knowledge discovery and data mining. chap. Fast Discovery of Association Rules, pp. 307–328. American Association for Artificial Intelligence, Menlo Park (1996). http://dl.acm.org/citation.cfm?id=257938.257975

2. Bifet, A., Gavaldà, R.: Kalman filters and adaptive windows for learning in data streams. In: Todorovski, L., Lavrač, N., Jantke, K.P. (eds.) DS 2006. LNCS (LNAI), vol. 4265, pp. 29–40. Springer, Heidelberg (2006). https://doi.org/10.1007/11893318_7

3. Bifet, A., GavaldÃǎ, R.: Learning from time-changing data with adaptive windowing, pp. 443–448 (2007). https://doi.org/10.1137/1.9781611972771.42

4. Bifet, A., Kirkby, R.: Data stream mining: a practical approach. Technical report. The University of Waikato (2009)

5. Bobek, S.: Methods for modeling self-adaptive mobile context-aware sytems. Ph.D. thesis, AGH University of Science and Technology (April 2016). (Supervisor: G.J. Nalepa)

6. Bobek, S., Nalepa, G.J.: Uncertain context data management in dynamic mobile environments. Futur. Gener. Comput. Syst. **66**(January), 110–124 (2017). https://doi.org/10.1016/j.future.2016.06.007

7. Bobek, S., Nalepa, G.J., Ślażyński, M.: HeaRTDroid - rule engine for mobile and context-aware expert systems. Expert Syst. (2018). https://doi.org/10.1111/exsy.12328

8. Deckert, M., Stefanowski, J.: RILL: algorithm for learning rules from streaming data with concept drift. In: Andreasen, T., Christiansen, H., Cubero, J.-C., Raś, Z.W. (eds.) ISMIS 2014. LNCS (LNAI), vol. 8502, pp. 20–29. Springer, Cham (2014). https://doi.org/10.1007/978-3-319-08326-1_3

9. Friedewald, M., Raabe, O.: Ubiquitous computing: an overview of technology impacts. Telemat. Inform. **28**(2), 55–65 (2011). http://www.sciencedirect.com/science/article/pii/S0736585310000547

10. Gama, J.: Knowledge Discovery from Data Streams, 1st edn. Chapman & Hall/CRC, Boca Raton (2010)

11. Gama, J., Medas, P., Castillo, G., Rodrigues, P.: Learning with drift detection. In: Bazzan, A.L.C., Labidi, S. (eds.) SBIA 2004. LNCS (LNAI), vol. 3171, pp. 286–295. Springer, Heidelberg (2004). https://doi.org/10.1007/978-3-540-28645-5_29

12. Gers, F.: Learning to forget: continual prediction with LSTM. In: IET Conference Proceedings, vol. 5, pp. 850–855, January 1999. http://digital-library.theiet.org/content/conferences/10.1049/cp_19991218

13. Goodman, B., Flaxman, S.: EU regulations on algorithmic decision-making and a "right to explanation" (2016). arxiv: 1606.08813. presented at 2016 ICML Workshop on Human Interpretability in Machine Learning (WHI 2016), New York, NY

14. Han, J., Pei, J., Yin, Y.: Mining frequent patterns without candidate generation. SIGMOD Rec. **29**(2), 1–12 (2000). https://doi.org/10.1145/335191.335372

15. Hoeffding, W.: Probability inequalities for sums of bounded random variables. J. Am. Stat. Assoc. **58**(301), 13–30 (1963). http://www.jstor.org/stable/2282952?

16. Inibhunu, C., McGregor, C.: Machine learning model for temporal pattern recognition. In: 2016 IEEE EMBS International Student Conference (ISC), pp. 1–4, May 2016

17. Koller, D., Friedman, N.: Probabilistic Graphical Models: Principles and Techniques. MIT Press, Cambridge (2009)

18. Lim, B.Y., Dey, A.K.: Investigating intelligibility for uncertain context-aware applications. In: Proceedings of the 13th International Conference on Ubiquitous Computing, UbiComp 2011, pp. 415–424. ACM, New York (2011). https://doi.org/10.1145/2030112.2030168

19. Magnusson, M.S.: Discovering hidden time patterns in behavior: T-patterns and their detection. Behav. Res. Methods, Instrum. Comput. **32**(1), 93–110 (2000). https://doi.org/10.3758/BF03200792

20. Page, E.S.: Continuous inspection schemes. Biometrika **41**(1/2), 100–115 (1954). https://doi.org/10.2307/2333009

21. Pearl, J.: Causal inference in statistics: an overview. Statist. Surv. **3**, 96–146 (2009). https://doi.org/10.1214/09-SS057

22. Potdar, K., Pardawala, T.S., Pai, C.D.: A comparative study of categorical variable encoding techniques for neural network classifiers. Int. J. Comput. Appl. **175**(4), 7–9 (2017). http://www.ijcaonline.org/archives/volume175/number4/28474-2017915495

23. ten Zeldam, S., de Jong, A., Loendersloot, R., Tinga, T.: Automated failure diagnosis in aviation maintenance using explainable artificial intelligence (XAI). In: Kulkarni, C., Tinga, T. (eds.) Proceedings of the European Conference of the PHM Society, vol. 4. PHM Society (2018)

24. Roberts, S.W.: Control chart tests based on geometric moving averages. Technometrics **1**, 239–250 (1959)

25. Welford, B.P.: Note on a method for calculating corrected sums of squares and products. Technometrics **4**(3), 419–420 (1962)

26. Zhang, G.: Time series forecasting using a hybrid arima and neural network model. Neurocomputing **50**, 159–175 (2003). http://www.sciencedirect.com/science/article/pii/S0925231201007020

On the Opportunities for Using Mobile Devices for Activity Monitoring and Understanding in Mining Applications

Grzegorz J. Nalepa[✉], Edyta Brzychczy, and Szymon Bobek

AGH University of Science and Technology, Cracow, Poland
{gjn,brzych3,szymon.bobek}@agh.edu.pl

Abstract. Over the last decades, number of embedded and portable computer systems for monitoring of activities of miners and underground environmental conditions that have been developed has increased. However, their potential in terms of computing power and analytic capabilities is still underestimated. In this paper we elaborate on the recent examples of the use of wearable devices in mining industry. We identify challenges for high level monitoring of mining personnel with the use of mobile and wearable devices. To address some of them, we propose solutions based on our recent works, including context-aware data acquisition framework, physiological data acquisition from wearables, methods for incomplete and imprecise data handling, intelligent data processing and reasoning module, hybrid localization using semantic maps, and adaptive power management. We provide a basic use case to demonstrate the usefulness of this approach.

Keywords: Mobile devices · Activity monitoring · Data understanding

1 Introduction

The mining industry is one of high risk. Two main types of risk can be specified. Systematic risk, related to functioning of the enterprise in local and global environment (markets, law, economy), and specific risk, related to internal conditions of enterprise. Main sources of specific risk in a mining enterprise comprise risks occurring in various industrial branches (i.e. financial, social, credit). However, the most characteristic risks for this type of industry are natural hazards. They bring dangerous events very often, causing serious accidents and injuries considering miners life and health. To prevent them and to support rescue actions in case of dangerous events, monitoring of activities of miners and underground environmental conditions should be constantly carried out.

Clearly there are technical solutions to deliver advanced monitoring. In last decades, number of embedded and portable computer systems for monitoring

Supported by the AGH University grant.

H. Yin et al. (Eds.): IDEAL 2018, LNCS 11315, pp. 75–83, 2018.
https://doi.org/10.1007/978-3-030-03496-2_9

have been developed. More recently, monitoring of workers as well as working conditions can be provided by mobile devices and wearables. Their great potential for these purposes can be seen in various industrial applications [5,19,22]. However, their potential in terms of computing power and analytic capabilities is still underestimated. What makes their use in mines challenging are the specific working conditions. Off-the-shelf mobile devices are mostly not suitable to underground operation, with constrained connectivity, dusty operating conditions, etc. As such, they require proper modifications. What is even more important, typical software solutions delivered with such devices might not meet the requirements of the mining industry.

In this paper we elaborate on the above mentioned challenges, in order to emphasize certain opportunities and to address them. We start with an overview of the recent examples of the use of wearable devices in the mining industry in Sect. 2. We then discuss challenges for high level monitoring of mining personnel with the use of mobile and wearable devices in Sect. 3. To address some of them, in Sect. 4 we propose a solution based on our recent works. We summarize the paper as well as present the directions for future works in the final Sect. 5.

2   Wearables in the Mining Industry

Two principal mining methods are widely used in practice: underground mining and surface mining. In some specific geological conditions (especially due to deposit depth) only the underground method is possible. As an example, hard coal or copper mining in Poland can be given. In the underground method, the raw material is transported from the underground through a complex structure of mining excavations to the surface, firstly by horizontal roadways and then by mostly vertical or sloping excavations.

In this paper we focus on the most complex example of mining which is carried out underground, accompanied with the following natural hazards: fires, roof and rock collapse, methane, coal dust, rock burst, water, and seismic shocks. Accidents related to these risks very often result in serious injuries, or even loss of lives (see Table 1).

An underground mine, nowadays with depth of even 1200 m or more, is a complex, live structure of tunnels and excavations changing the position in time. For this reason, apart the hazards, people localization is of crucial importance, especially in case of occurrence of unforeseen events. Management of safety and occupational health of miners could be supported by various computer systems [14,24,25]. An intelligent response and rescue system for mining industry [26] consist of four major parts: (1) a database, (2) a monitoring center, (3) a fixed underground sensor network, and (4) mobile devices. In such systems, various mobile devices could be used i.e. smartphones or wearable devices [15]: smartwatches, smart eyewear, smart clothing, wearable cameras and others.

Smartphones and smartwatches can be used for workers tracking, communication and navigation purposes, as well as for measurements of physical conditions and underground environment. Smart eyewear and cameras enable additional visualization. Smart clothing, besides measurements of physical conditions

Table 1. Dangerous events and accidents in polish mining in the years 2015-7 ([3])

Event type	Event number			Fatal accident			Total accidents		
	2015	2016	2017	2015	2016	2017	2015	2016	2017
Fires	12	10	13	0	0	0	0	0	0
Roof falls	3	8	3	0	1	0	1	3	1
Rock collapse	3	4	6	2	10	0	4	50	12
Methane ignition	3	5	3	0	1	0	4	1	0
Rock burst	1	0	0	0	0	0	0	0	0
Total	22	27	25	2	12	0	9	54	13

and body tracking, can provide additional monitoring of environmental conditions. These functionalities are enabled by implementing various sensor types in wearables i.e. [15]: environmental sensors, biosensors, location-tracking, communication modules, motion or speed sensors.

Several applications of wearable technology can be found in the mining industry. One of the examples is a smart helmet solution, equipped with methane and carbon monoxide sensors for gas concentration monitoring and alarming before critical atmospheric level [11]. Another proposal provides widening air quality measure (CO, SO_2, NO_2) features of the helmet, with helmet removal sensor and collision (struck by an object) indication [6]. More extensive and flexible solutions are proposed by Deloitte, Vandrico Solutions and Cortex Design [2]. Smart Helmet Clip measures air quality (NO_2, CO, CO_2 and CH_4), as well as temperature and humidity. Modular design of solution allows to customize environmental sensors. It is also equipped with GPS, accelerometer and gyro enabling tracking of the movement of miners and their location, as well as front facing digital camera that can be used to stream video for site observation and remote support. Helmet camera can be also used for miners dust exposure estimation [10]. Camera image, together with instantaneous dust monitoring, are used for collection of dust concentration data. Apart from more obvious features provided by aforementioned solutions, more sophisticated measurements can be done for mining purposes, i.e. monitoring of brain activity and detection of worker fatigue levels [1].

Data collected by sensors is analyzed with commercial or dedicated statistical or/and graphical software, enabling i.e. visualization of measured parameters in relation to time (histograms, plots) or space (3D models). The latter are used mainly for visualization of the tracking and location. The analyses are executed mainly on the platforms provided by computer systems located physically on the surface or in the cloud, but clearly out of the underground environment.

Two main challenges in functioning of the monitoring systems related to underground environment should be emphasized: data transfer [4] and underground positioning (indoor localization) [23]. Problems in data transfer have strong impact on reliability of data analytics carried out on the surface. Data transfer delays or missing data could have serious consequences for workers safety and health. Real-time applications for monitoring of gases or dust concentration, miners tracking and location can use data processing and analysis available for smaller and lighter computing units [20].

Most of wearables are very often paired with smartphones, which have major computing power. The potential of mobile devices is still underestimated and not fully used in industrial application, especially in specific working conditions. Some of the current research work on use of smartphones, also for mining purposes, is related to enhanced navigation (or tracking) of people and assets with ambient wireless received signal strength indication (RSSI) and fingerprinting based localization [16].

3   Challenges for Personnel Monitoring in the Underground Mine

Summarizing some of the most important opportunities for the use of mobile and wearable devices, we identify the following two groups of important use cases.

The first is the *underground localization*. This case very much differs from the surface conditions, as regular GPS is mostly unusable. While Wifi [27] localization could be used in the case of the custom underground WiFi infrastructure, it is known as being often prone to errors. Main problems in this area are related to wave propagation underground [21]. Other approaches could be based on the use of RFID, or BLE.

The second group regards the *health monitoring* of the underground personnel. As peoples's life is the most precious asset, instant reactions might be needed. In the event of the detection of an emergency health condition, the most important issue is to notify the surface monitoring center. However, what might be also important is the possible prediction of risky health condition and notification of the person in danger.

As we stated in the introduction, the operation in the mining environment largely differs from the default operating conditions of regular mobile devices. On the technical level, there are important challenges that should be considered for such systems. We focus on three which – in our opinion – are the most important ones.

The first one regards *imprecise or incomplete sensor data*. All sensor readings have limited precision. However, in the demanding environmental conditions in an underground mine, their operation might be affected, and, furthermore, their precision reduced. Moreover, sensor failures are also common. As monitoring systems in mines often save lives, these issues need to be considered in the design stage.

The second one is related to the consequences of the *poor network connectivity*. Network operation has several uses that have impact on different modules of a monitoring system. The first one is in fact related to data processing. In a typical scenario, mobile devices are used to collect and buffer the sensor data, and then send it to the surface monitoring center. Furthermore, wearable devices transmit their readings to the mobiles. This setting is mostly due to the historically low processing power of mobile devices. The second use of the network signal is related to the localization. This includes positioning based on the RSSI measurements from the underground base stations. In the case of rapid changes of environmental conditions, or an accident, there might be no network connectivity – cutting off communication, or hampering localization.

The third one involves *power consumption* restrictions. In the underground operations, the recharging of mobile devices and wearables might be limited. This is especially relevant in the case of accidents. This is why proper methods should be used not only to minimize, but also to adapt power consumption to the working conditions.

In the following section we discuss architecture of an integrated monitoring system based on our works on the use of mobile devices.

4 Solutions for Activity Monitoring Using Wearables and Mobiles

To address the aforementioned challenges, we propose an approach based on following assumptions and features: (1) context-aware data acquisition framework, (2) physiological data acquisition from wearables, (3) methods for incomplete and imprecise data handling, (4) intelligent data processing and reasoning, (5) computation on the mobile device, (6) hybrid localization using semantic maps, (7) adaptive power management.

Regarding the first module, in [7] we proposed an architecture for a context-aware system using mobile devices, such as smartphones. It allows for designing intelligible, user-centric systems that make use of heterogeneous contextual data and is able to reason upon that information. One of the most important part of the architecture is *context-based controller* that is used to implement communication between the reasoning mechanism, the sensory input and the user. Contextual information could be any information that can characterize the situation of an entity (the user or the system). This can be easily extended to sensory data gathered by one of the devices for monitoring life parameters of miners and environmental conditions underground. In fact, recently in [18] we proposed the extension of the platform with physiological signal monitoring. The main goal of the extension was to exploit information about user's heart-rate, galvanic skin response and others, to detect their emotional state. Such an extension could be of a key importance in industries such as mining, where early detection of high level of stress, or tiredness of a worker could be worth other humans' lives.

Moreover, one of the primary goal of that approach is the handling of uncertainty of data, thus addressing the 2nd requirement. We provide two complementary mechanisms for handling uncertain or vague information. First is based on certainty factors approach. It allows to assign certainty levels both to the input values as well as for the knowledge that was encoded (or automatically discovered) in the system. The second is based on probabilistic interpretation of rule-based models that we use as the primary knowledge representation method [8].

It allows for reasoning to be executed even in cases where the required input is missing.

Furthermore, we proposed the use of a rule-based engine called HEART-DROID [9] for performing high-level reasoning. It is an integral part of the architecture for context-aware systems mentioned before. What is important, taking into the account the possibility of disconnected operation, is the fact that our approach does not depend on the access to the cloud infrastructure. The HEARTDROID reasoning engine is a self-contained mechanism that is able to access data, process knowledge and present output to the user using solely mobile device's resources. Furthermore, we allowed that the rule language that is used by HEARTDROID could be semantically annotated. This allows the end-user to access system's core knowledge component and adjust it accordingly to their needs. This may be a crucial feature in disconnected environments where, in case of system malfunctions, the user should be able to make first fixes.

Semantically annotated knowledge was also used by us to improve automatic localization technique. In our previous works, we demonstrated how our approach can be used for hybrid localization [13]. In this approach we used dead-reckoning mechanism to track user position supported by a mediation algorithm that utilized information about environmental features to disambiguate the localization estimates. We believe that such systems could be implemented in underground mines, where the WiFi infrastructure is not available for some reason, forcing a use of dead-reckoning methods.

Finally, in [17] we proposed a learning middleware that is able to adapt the power consumption based on the contextual information. It used logistic regression to learn situations where the system could release some of the mobile device's resources to save energy. This was achieved by discovering usage patterns of the mobile device sensors and enabling high rate measurements, which are costly in terms of energy consumption, only in situations where they can produce some meaningful reasoning outcomes.

It is worth noting that in our approach we do not use industrial machines' logs directly. Instead, we rely on possibly distributed network of *context providers*, such as beacons [12], or mobile devices. They can be linked with hardware equipment in mines, but also allow to measure many additional parameters in different location of the area of interest. On one hand, this introduces additional burden to build and maintain such a network. On the other hand, it not only provides larger variety of data to analyze, but also allows to partially cope with limited connectivity issue by decentralization of communication nodes.

We investigated example scenarios with proposed solutions for underground mine. First scenario includes normal operation mode in known environmental conditions. Miners have personalized mobile devices which provide reasoning about working conditions and wellness of the worker, with usage of general health measurements, environmental monitoring and localization from various sensors located in i.e. helmet, clothes, gloves and eyewears. Reasoning with rule-based engine is performed locally on the mobile devices and in normal operation mode outputs selected by administrator (i.e. related to anomalies, or critical information such as miner's position) are transferred to the surface for processing and decision-making (including information certainty).

Second scenario includes unforeseen incidents resulting in problems with connectivity and power supply. In this scenario normal operation mode is extended with adaptive power consumption of mobile devices, in order to ensure prolonging the mobile devices operation and transfer the selected data on the surface. Moreover, especially in case of collapsing or gas explosion and connectivity problem, dead-reckoning is used to estimate location of the miners. It is crucial information in rescue actions.

In presented concept, main information processing is performed on mobile devices, saving the computational power on the surface and enabling usage of knowledge by user *in situ* even though connection with surface is failed.

Our proposals do not exhaust the possibilities and benefits of the development of mobile technologies and their application in underground mining, but nevertheless constitute an original attempt to extend the existing response and rescue system with new functionalities.

5   Summary and Future Work

In this paper we discussed the challenges regarding the use of mobile and wearable devices for the monitoring of health of underground mining personnel. Based on the review of the current use of these devices, we presented an idea of an integrated system overcoming some of them. We demonstrated how our previous work in the field of mobile context-aware systems could be applied to solve the problem of creating a self sustainable system for highly dynamic environment such as underground mine. We briefly discussed methods and tools that can address most of the challenges concerned with building such systems, including knowledge representation and reasoning, indoor navigation, health monitoring and energy consumption management. This proposal is a work in progress. The provided use cases are the motivation for the future work. However, it would require close cooperation with the mining industry.

References

1. The smart cup. EdanSafe Pty Ltd. (2017). http://smartcaptech.com/pdf/SmartCapFAQB-2.pdf
2. The smart helmet. Mining World (2017). http://miningworld.com/index.php/2017/09/20/the-smart-helmet/
3. Statistics of dangerous events occurrence and accidents in mines in years 2015–2017. State Mining Authority, Poland, Katowice (2018). http://www.wug.gov.pl/bhp
4. Akyildiz, I.F., Stuntebeck, E.P.: Wireless underground sensor networks: research challenges. Ad Hoc Netw. **4**(6), 669–686 (2006). https://doi.org/10.1016/j.adhoc.2006.04.003, http://www.sciencedirect.com/science/article/pii/S1570870506000230

5. Awolusi, I., Marks, E., Hallowell, M.: Wearable technology for personalized construction safety monitoring and trending: review of applicable devices. Autom. Constr. **85**, 96–106 (2018). https://doi.org/10.1016/j.autcon.2017.10.010, http://www.sciencedirect.com/science/article/pii/S0926580517309184
6. Behr, C.J., Kumar, A., Hancke, G.P.: A smart helmet for air quality and hazardous event detection for the mining industry. In: 2016 IEEE International Conference on Industrial Technology (ICIT), pp. 2026–2031, March 2016. https://doi.org/10.1109/ICIT.2016.7475079
7. Bobek, S., Nalepa, G.J.: Uncertain context data management in dynamic mobile environments. Future Gener. Comput. Syst. **66**(January), 110–124 (2017). https://doi.org/10.1016/j.future.2016.06.007
8. Bobek, S., Nalepa, G.J.: Uncertainty handling in rule-based mobile context-aware systems. Pervasive Mob. Comput. **39**(August), 159–179 (2017). https://doi.org/10.1016/j.pmcj.2016.09.004
9. Bobek, S., Nalepa, G.J., Ślażyński, M.: Heartdroid - rule engine for mobile and context-aware expert systems. Expert Syst. https://doi.org/10.1111/exsy.12328. (in press)
10. Hass, E., Cecala, A., Hoebbel, C.L.: Using dust assessment technology to leverage mine site manager-worker communication and health behavior: a longitudinal case study. J. Progress. Res. Soc. Sci. **3**, 154–167 (2016)
11. Hazarika, P.: Implementation of smart safety helmet for coal mine workers. In: 2016 IEEE 1st International Conference on Power Electronics, Intelligent Control and Energy Systems (ICPEICES), pp. 1–3, July 2016. https://doi.org/10.1109/ICPEICES.2016.7853311
12. Kajioka, S., Mori, T., Uchiya, T., Takumi, I., Matsuo, H.: Experiment of indoor position presumption based on RSSI of Bluetooth LE beacon. In: 2014 IEEE 3rd Global Conference on Consumer Electronics (GCCE), pp. 337–339, October 2014. https://doi.org/10.1109/GCCE.2014.7031308
13. Köping, L., Grzegorzek, M., Deinzer, F., Bobek, S., Ślażyński, M., Nalepa, G.J.: Improving indoor localization by user feedback. In: 2015 18th International Conference on Information Fusion (Fusion), pp. 1053–1060, July 2015
14. Lande, S., Matte, P.: Coal mine monitoring system for rescue and protection using zigbee. Int. J. Adv. Res. Comput. Eng. Technol. (IJARCET) **4**(9), 3704–3710 (2015). https://doi.org/10.1016/j.proeps.2009.09.161, http://www.sciencedirect.com/science/article/pii/S1878522009001623
15. Mardonova, M., Choi, Y.: Review of wearable device technology and its applications to the mining industry. Energies **11**(3) (2018). https://doi.org/10.3390/en11030547, http://www.mdpi.com/1996-1073/11/3/547
16. Mittal, A., Tiku, S., Pasricha, S.: Adapting convolutional neural networks for indoor localization with smart mobile devices. In: Proceedings of the 2018 on Great Lakes Symposium on VLSI, pp. 117–122. GLSVLSI 2018, ACM, New York (2018). https://doi.org/10.1145/3194554.3194594
17. Nalepa, G.J., Bobek, S.: Rule-based solution for context-aware reasoning on mobile devices. Comput. Sci. Inf. Syst. **11**(1), 171–193 (2014)
18. Nalepa, G.J., Kutt, K., Bobek, S.: Mobile platform for affective context-aware systems. Future Gener. Comput. Syst. (2018). https://doi.org/10.1016/j.future.2018.02.033
19. Osswald, S., Weiss, A., Tscheligi, M.: Designing wearable devices for the factory: rapid contextual experience prototyping. In: 2013 International Conference on Collaboration Technologies and Systems (CTS), pp. 517–521. May 2013. https://doi.org/10.1109/CTS.2013.6567280

20. Pasricha, S.: Deep underground, smartphones can save miners' lives. Conversation UK (2016). https://theconversation.com/deep-underground-smartphones-can-save-miners-lives-64653

21. Ranjan, A., Misra, P., Dwivedi, B., Sahu, H.B.: Studies on propagation characteristics of radio waves for wireless networks in underground coal mines. Wirel. Pers. Commun. **97**(2), 2819–2832 (2017). https://doi.org/10.1007/s11277-017-4636-y

22. Scheuermann, C., Heinz, F., Bruegge, B., Verclas, S.: Real-time support during a logistic process using smart gloves. In: Smart SysTech 2017, European Conference on Smart Objects, Systems and Technologies, pp. 1–8, June 2017

23. Thrybom, L., Neander, J., Hansen, E., Landernas, K.: Future challenges of positioning in underground mines. IFAC-PapersOnLine **48**(10), 222–226 (2015). https://doi.org/10.1016/j.ifacol.2015.08.135, http://www.sciencedirect.com/science/article/pii/S2405896315010022. 2nd IFAC Conference on Embedded Systems, Computer Intelligence and Telematics CESCIT 2015

24. Xu, J., Gao, H., Wu, J., Zhang, Y.: Improved safety management system of coal mine based on iris identification and RFID technique. In: 2015 IEEE International Conference on Computer and Communications (ICCC), pp. 260–264 (2015). https://doi.org/10.1109/CompComm.2015.7387578

25. Yi-Bing, Z.: Wireless sensor network's application in coal mine safety monitoring. In: Zhang, Y. (ed.) Future Wireless Networks and Information Systems. LNEE, vol. 144, pp. 241–248. Springer, Heidelberg (2012). https://doi.org/10.1007/978-3-642-27326-1_31

26. Zhang, K., Zhu, M., Wang, Y., Fu, E., Cartwright, W.: Underground mining intelligent response and rescue systems. Proced. Earth Planet. Sci. **1**(1), 1044–1053 (2009). https://doi.org/10.1016/j.proeps.2009.09.161, http://www.sciencedirect.com/science/article/pii/S1878522009001623. Special issue title: Proceedings of the International Conference on Mining Science and Technology (ICMST 2009)

27. Zhang, Y., Li, L., Zhang, Y.: Research and design of location tracking system used in underground mine based on WiFi technology. In: 2009 International Forum on Computer Science-Technology and Applications, vol. 3, pp. 417–419, December 2009. https://doi.org/10.1109/IFCSTA.2009.341

A Taxonomy for Combining Activity Recognition and Process Discovery in Industrial Environments

Felix Mannhardt[1]([☒]) [iD], Riccardo Bovo[2], Manuel Fradinho Oliveira[1], and Simon Julier[2] [iD]

[1] SINTEF Digital, Trondheim, Norway
{felix.mannhardt,manuel.oliveira}@sintef.no
[2] Department of Computer Science, UCL, London, UK
{riccardo.bovo,simon.julier}@ucl.ac.uk

Abstract. Despite the increasing automation levels in an Industry 4.0 scenario, the tacit knowledge of highly skilled manufacturing workers remains of strategic importance. Retaining this knowledge by formally capturing it is a challenge for industrial organisations. This paper explores research on automatically capturing this knowledge by using methods from activity recognition and process mining on data obtained from sensorised workers and environments. Activity recognition lifts the abstraction level of sensor data to recognizable activities and process mining methods discover models of process executions. We classify the existing work, which largely neglects the possibility of applying process mining, and derive a taxonomy that identifies challenges and research gaps.

Keywords: Activity recognition · Process mining · Manufacturing Industrial environment · Tacit knowledge · Literature overview

1 Introduction

The rise of the knowledge worker has contributed to the emphasis on the strategic value of creating, harnessing and applying knowledge within manufacturing environments. With the advent of automation, as part of the Industry 4.0 evolution, the strategic importance of knowledge and high skilled workers has only become more important. However, so did the crippling impact caused by knowledge gaps resulting from the difficulty of managing effectively tacit knowledge garnered through the experience of highly skilled workers once removed from their work environment. In fact, with the continuous advances in technology and increased complexity associated to both the product and the manufacturing processes, tacit knowledge represents by far the bulkiest part of an organization's knowledge. Many of the theories and methodologies associated with the externalization of tacit knowledge require organizational processes and a culture pervading the workplace that facilitate the creation of formal and external knowledge.

© Springer Nature Switzerland AG 2018
H. Yin et al. (Eds.): IDEAL 2018, LNCS 11315, pp. 84–93, 2018.
https://doi.org/10.1007/978-3-030-03496-2_10

The digitization of the workplace through the pervasiveness of sensors, combined with ever more elaborate digital information systems, generates huge amounts of data that may be further enriched when considering the direct placement of sensors on workers in the shopfloor, thus capturing more effectively what is taking place as much of the work entails manual activity, not registered in the supporting information system. With the wealth of data captured, including the human dimension, we envision the approach illustrated in Fig. 1 as a way to externalise tacit knowledge of the operator on the shopfloor. The approach uses sensors and combines activity recognition [4] with process discovery, which automatically derives process models from activity execution sequences [1].

Fig. 1. Overview of the envisioned approach combining activity recognition and process mining.

The purpose of this paper is to conduct a structured literature review on activity recognition applied in industrial contexts with the purpose of externalisation of tacit knowledge. In most, if not all cases, there is no automatic process discovery as the methods and approaches documented in literature are largely dependent on context with supervised learning. Those few unsupervised learning approaches rely on clustering techniques, largely ignoring the benefits of process mining in the discovery of process knowledge. The result of the synthesis of the literature review yielded a preliminary *taxonomy to support the identification of challenges to be addressed*, outlining potential areas of research to develop solutions that leverage activity recognition with process mining towards facilitating externalisation of tacit knowledge.

We structure the remainder as follows. In Sect. 2, activity recognition and process mining are briefly introduced. Section 3 presents our literature search. Based on the results, we present a preliminary taxonomy together with challenges in Sect. 4. We conclude the paper with an outlook for future work in Sect. 5.

2 Background

We give a brief overview of activity recognition and process mining.

2.1 Activity Recognition

Activity recognition (ARC) seeks to accurately identify human activities on various levels of granularity by using sensor readings. In recent years ARC has become an emerging field due to the availability of large amount of data generated by pervasive and ubiquitous computing [2,4,18]. Methods have demonstrated an increased efficiency in extracting and learning to recognise activities in the supervised learning setting using a range of machine learning techniques. Traditional methods often adopt shallow learning techniques such as Decision Trees, Naïve Bayes, Support Vector Machines (SVM), and Hidden Markov Models (HMM) [4] while the more recent methods often use Neural Network architectures, which require less manual feature engineering and exhibit better performance [29,39].

Applications of activity recognition span from smart home (behaviour analysis for assistance) to sports (automatic performances tracking and skill assessment) and even healthcare (medication tracking). The recognition of activities is not an end in itself, but often supports assistance, assessment, prediction and intervention related to the recognised activity. An emerging application field for ARC relates to smart factories and Industry 4.0 where an increasingly sensors-rich environment is generating large amounts of sensor data.

ARC captures activities through the use of sensors such as cameras, motion-sensors, and microphones. Despite the large amount of work, ARC remains a challenging problem due to the complexity and variability of activities as well as due to the context in which activities are meant to be recognised. Data labelling, for instance, is a common challenge related to ARC. Assigning the correct ground truth label is a very time-consuming task. There has been less work on unsupervised [21] or semi-supervised techniques [19] which require fewer annotations [2,39]. Another challenge lies in the emergent topic of transfer learning [29], which helps with the redeployment of an ARC model from one factory floor to another with a different layout, environmental factors, population and activities.

2.2   Process Mining

Process mining is a data analytics method that uses event logs to provide a data-driven view on the actual process execution for analysis and optimisation purposes [1]. Consider, e.g., the order-to-cash process of a manufacturing company. One execution of this process results in a sequence of events (or process trace) being recorded across several information systems. A process trace should contain at least the following: the activities names executed (e.g., *order created*) as well as their execution time. An event log is a set of process traces in which each *process trace* groups together the activities performed in one instance of a recurring process. Process mining can help to uncover the tacit process knowledge of workers by discovering process models from event logs. The discovered models reveal how work is actually performed, including deviations from standard procedures such as workarounds and re-work. Moreover, the actual process execution can be contrasted with existing de-jure models, e.g., to pinpoint deviations to work instructions and analyse performance issues. An in-depth introduction ot process discovery is given in [1] and [6] gives a comprehensive survey of process discovery methods.

However, only very few applications of process mining are reported within the manufacturing domain [13,23]. One reason for this gap might be that in many industrial environments, much of the manual work is not precisely captured in databases or logs. For example, the individual steps performed in an assembly task remain hidden when using event logs from standard information systems only. Thus, the recognition of such manually executed activities is a crucial prerequisite for the successful application of process mining in this context [16].

3   Literature Overview

Based on the premise that activity recognition and process mining can be combined to extract tacit knowledge of operators in industrial processes, we conducted a search of the existing literature on activity recognition in industrial environments. Our goal was to derive a taxonomy that helps to identify the central issues and challenges of using activity recognition and process discovery for externalizing tacit knowledge.

We searched both Google Scholar and Scopus for research on activity recognition that was applied in or is applicable to industrial settings. We used the keywords *events* or *sensors, activity recognition, industrial* or *manufacturing* in our search and followed-up references in the identified work. Furthermore, we widened our search by looking for research on *activity recognition* that mentions one of the keywords *tacit knowledge, process discovery, process elicitation, process analysis.* An initial search revealed that ARC can be decomposed into conceptual work and applied work. For example, in [31] a architecture for process mining in cyber-physical systems is proposed but was not evaluated. Although such work provides useful insights, they have not be validated and might not be applicable in real-work environments. Therefore, we excluded purely conceptual work. Furthermore, we excluded work without connection to an industrial setting. We identified 26 relevant papers that are listed in Table 1. We do not believe our literature review is exhaustive, but we do believe it is representative of existing literature.

We classified the work according to the following criteria.

- Its **recognition type** based on the kind of prior knowledge employed into methods for *supervised recognition, unsupervised recognition,* and *semi-supervised recognition.*
- The **time horizon** of the recognition was categorised into *predictive, online,* and *post-mortem* recognition.
- We distinguished the **sensor type** into *vision-based* (V), *motion-based* (M), *sound-based* (S), and *radiowave-based* (R) sensors. Note that if a RGB camera (vision-based) is used to determine worker movement, we consider it both as vision-based and motion-based sensor.
- Regarding the **sensor location**, we categorise sensors into those attached to *objects* (O), those *ambient* in the environment (A), and those *wearable* (W).
- There is a large variety of activities in industrial settings some of which are more difficult to detect. We categorized the work based on the supported

Table 1. Results of the literature search classified according to the described criteria.

		Recognition	Time	Granularity	Context	Setting	Sensor type				S. location		
							V	M	S	R	W	O	A
2006–2010	[20]	Supervised	Online	Fine	✗	Real	✓	✓			✓		
	[9]	Supervised	Online	Fine	✗	Lab	✓		✓	✓	✓		✓
	[40]	Supervised	Online	Fine	✗	Lab	✓	✓			✓		✓
	[35,36]	Supervised	Online	Fine	State machine	Real	✓		✓	✓	✓		
	[24]	Supervised	Online	Fine	Sequence	Real			✓				✓
	[28]	Supervised	Online	Coarse	✗	Real	✓		✓	✓	✓		✓
	[10]	Semi	Online	Both	Hierarchy	Lab		✓			✓		
	[25]	Semi	Post-mortem	Both	Sequence	Real	✓						✓
2011–2015	[27]	Supervised	Online	Fine	✗	Lab		✓		✓	✓		✓
	[38]	Supervised	Online	Coarse	Workflow	Real	✓						✓
	[37]	Supervised	Online	Both	✗	Real	✓			✓			✓
	[14]	Supervised	Predictive	Fine	State machine	Lab	✓						✓
	[7]	Supervised	Online	Coarse	Workflow	✗				✓		✓	
	[32,33]	Supervised	Online	Both	Hierarchy	Lab	✓	✓					✓
	[8]	Supervised	Post-mortem	Fine	Probabilistic	✗							
	[34]	Supervised	Predictive	Fine	Rules	Lab	✓						✓
2016–2018	[26]	Supervised	Online	Fine	Sequence	Lab	✓	✓					✓
	[21]	Unsupervised	Online	Coarse	Workflow	Real	✓	✓				✓	✓
	[12]	Supervised	Online	Fine	✗	Real	✓				✓		
	[15]	Supervised	Online	Coarse	✗	Real	✓				✓		
	[11]	Unsupervised	Post-mortem	Coarse	Sequence	Lab	✓				✓		
	[17]	Supervised	Online	Fine	✗	Lab	✓						✓
	[30]	Supervised	Online	Fine	Workflow	Real	✓				✓		
	[5]	Semi	Online	Both	Rules	Lab	✓	✓			✓		

granularity of the activities into: *coarse* and *fine*. An example of activity recognition on a coarse granularity level would be recognising that a part of the assembly was installed, whereas on a fine level of granularity recognition would recognise the individual steps required to connect that part, e.g., pick-up screw and fasten screw. Some approaches support *both* coarse-grained and fine-grained activities.

- We also distinguished whether the work takes the **context** of the assembly process into account to improve the detection, e.g., by making use of existing assembly instructions in form of higher-level *workflow models*, *state machines*, *sequences*, or other models.
- Lastly, we distinguished the **setting** in which the method was evaluated into artificial laboratory settings or in real factory environments.

4   Taxonomy for Activity Recognition and Process Discovery in Industrial Environments

Based on our literature study, we derived a taxonomy for knowledge extraction through activity recognition in industrial environments. The taxonomy focuses on the applicability in practical settings and the requirements on activity recognition in a process-mining context. Our goal is to identify challenges for the joint application and help designing new systems for knowledge extraction by describing existing systems in a unified manner. The taxonomy is organised around four major dimensions: *time*, *data*, *process context*, *environment*, and *privacy*. We acknowledge that the taxonomy is still under development. Therefore, we only briefly sketch each of the dimensions with examples from the literature.

Time. In Table 1, we distinguished three major categories of activity recognition regarding the time dimension: *predictive*, *online*, and *post-mortem* activity recognition. Most activity recognition methods in the industrial setting target the *online* setting, in which the activity is detect during its execution. This can be useful to provide up-to-date information for the activity at hand, e.g., in [35] a check list is kept updated. We found much fewer examples for the predictive setting, in which the next activity is predicted before or just when it is about to happen also denoted as intention recognition. A notable exception is [34] which uses state recognition to predict the next activity in a manufacturing application. Such predictive recognition can be useful to provide timely assistance to prevent errors. Lastly, post-mortem activity recognition methods can use both information about past activities as well as future activities to determine the most likely classification. Only two methods in Table 1 take the post-mortem view on activity recognition. This shows that the tacit knowledge discovery angle has been largely neglected. The work in [8] is an exception and, indeed, conceptually close to work on conformance checking and the optimal alignment of event sequences to process models in the process mining literature [22]. Thus, there are clear research gaps regarding the *predictive* and *post-mortem* category of activity recognition in industrial contexts (such as manufacturing) out of which the post-mortem angle is more relevant for our envisioned approach.

Data. The availability of data is a crucial prerequisite for externalising tacit knowledge through process mining and activity recognition. There are several categories in the data dimension: *capture*, *storage* and *processing* of data. Several challenges have to be dealt with in our application scenario. We exemplify one challenge regarding the *data processing* category. Here the availability of ground truth labels is a particular challenge. Since the goal of process discovery is, in fact, to discover the unknown tacit knowledge of workers, it is questionable whether all the activity labels for the use of supervised methods can be determined beforehand. However, as clear from Table 1 there have been only very few unsupervised techniques proposed.

Process Context. Several factors are relevant to the *process context* dimension, such as the *type of activities* executed, the *type of control-flow* in which the activities are embedded, and their *complexity*. For example, Bader et a.l [8] mentions the challenge of considering the teamwork setting in which some activities are of a collaborative nature: multiple workers collaborate on one activity. However, they do not yet provide a solution. Also relevant to the process-context dimension is that some work takes into account a-priori knowledge on the control-flow of the process. For example, in [36] a finite state machine is used to encode this prior knowledge whereas in [7] a higher-level process modelling language is used to define the process. An opportunity for future work might be to leverage on the wealth of higher-level modelling notations used in a process mining context [1]. Lastly the *complexity* of the considered processes and of the activities is worth discussion. In most settings only a few activities are considered (less than 10) and only few consider hierachical dependencies between activities on lower and higher levels. More advanced work in this category are the semi-supervised techniques in [5,10] in which higher-level activities are recognized based on sequences of detected low-level activities.

Environment. The *environment* in which the activities take place is highly relevant to the practical applicability of extracting tacit knowledge through activity recognition and process discovery. For example, the sensor type needs to be carefully selected since there are often several restrictions in a real factory setting [3]: wearable sensors should not interfere with the actual work and safety protocols and ambient sensors are often limited to narrow areas or subject to background noise. Some of the work identified evaluated their method in a realistic factory environment. However, the evaluation mostly takes place in designated areas to avoid costly interruption of production lines. For example, in [36] car assembly activities in a Skoda factory are tracked, but only in a "learning island" that is used for training workers. Thus, the applicability of many techniques on a real production line remains unclear.

Privacy. Activity recognition requires the capture of data, which may include sensors on the operator themselves. This raises important concerns with regards to privacy as the use of the data may have a negative impact on the operator (e.g.: due to poor performance, an operator's employment is terminated). The body of research covered, with exception of [3], focuses very much on the opportunities of processing the data collated, whilst disregarding the potential threats to the operator's well-being [23]. To address the challenges, governments have intervened with regulatory frameworks to safeguard the privacy of the user, such as the General Data Protection Regulation (GDPR) that attempts to place the user in control of their digital selves. Therefore, privacy has become a design requirement and not an afterthought, which may affect how activity recognition research may be realised.

5   Conclusion

We presented a structured literature review on activity recognition from the viewpoint of using the recognised activity data as input to process discovery techniques to reveal tacit knowledge of industrial operators. Based on the identified literature, we contribute a preliminary taxonomy for knowledge extraction from manual industrial processes through activity recognition. Whereas we believe to have included the most relevant literature from the field of activity recognition, we acknowledge that, as future work, this study should be further extended to take into account research from the field of learning organisations and look in more depth at the process discovery task after having recognised relevant activities.

Acknowledgments. This research has received funding from the European Union's H2020 research and innovation programme under grant agreement no. 723737 (HUMAN).

References

1. van der Aalst, W.M.P.: Process Mining - Data Science in Action, 2nd edn. Springer, Heidelberg (2016). https://doi.org/10.1007/978-3-662-49851-4
2. Abdallah, Z.S., Gaber, M.M., Srinivasan, B., Krishnaswamy, S.: Activity recognition with evolving data streams. ACM Comput. Surv. **51**(4), 1–36 (2018)
3. Aehnelt, M., Gutzeit, E., Urban, B.: Using activity recognition for the tracking of assembly processes: challenges and requirements. In: WOAR 2014. Fraunhofer Verlag (2014)
4. Aggarwal, J., Ryoo, M.: Human activity analysis. ACM Comput. Surv. **43**(3), 1–43 (2011)
5. Al-Naser, M., et al.: Hierarchical model for zero-shot activity recognition using wearable sensors. In: ICAART (2), pp. 478–485. SciTePress (2018)
6. Augusto, A., et al.: Automated discovery of process models from event logs: review and benchmark. IEEE Trans. Knowl. Data Eng. (2018)
7. Bader, S., Aehnelt, M.: Tracking assembly processes and providing assistance in smart factories. In: ICAART 2014. SCITEPRESS (2014)
8. Bader, S., Krüger, F., Kirste, T.: Computational causal behaviour models for assisted manufacturing. In: iWOAR 2015. ACM Press (2015)
9. Bannach, D., Kunze, K., Lukowicz, P., Amft, O.: Distributed modular toolbox for multi-modal context recognition. In: Grass, W., Sick, B., Waldschmidt, K. (eds.) ARCS 2006. LNCS, vol. 3894, pp. 99–113. Springer, Heidelberg (2006). https://doi.org/10.1007/11682127_8
10. Blanke, U., Schiele, B.: Remember and transfer what you have learned - recognizing composite activities based on activity spotting. In: ISWC 2010. IEEE (2010)
11. Böttcher, S., Scholl, P.M., Laerhoven, K.V.: Detecting process transitions from wearable sensors. In: iWOAR 2017. ACM Press (2017)
12. Feldhorst, S., Masoudenijad, M., ten Hompel, M., Fink, G.A.: Motion classification for analyzing the order picking process using mobile sensors - general concepts, case studies and empirical evaluation. In: ICPRAM 2016, pp. 706–713. SCITEPRESS (2016)
13. Gonella, P., Castellano, M., Riccardi, P., Carbone, R.: Process mining: a database of applications. Technical report, HSPI SpA - Management Consulting (2017)
14. Goto, H., Miura, J., Sugiyama, J.: Human-robot collaborative assembly by online human action recognition based on an FSM task model. In: Human-Robot Interaction 2013 Workshop on Collaborative Manipulation (2013)
15. Grzeszick, R., Lenk, J.M., Rueda, F.M., Fink, G.A., Feldhorst, S., ten Hompel, M.: Deep neural network based human activity recognition for the order picking process. In: iWOAR 2017. ACM Press (2017)
16. Janiesch, C., et al.: The Internet-of-Things meets business process management: mutual benefits and challenges (2017). arXiv:1709.03628

17. Knoch, S., Ponpathirkoottam, S., Fettke, P., Loos, P.: Technology-enhanced process elicitation of worker activities in manufacturing. In: Teniente, E., Weidlich, M. (eds.) BPM 2017. LNBIP, vol. 308, pp. 273–284. Springer, Cham (2018). https://doi.org/10.1007/978-3-319-74030-0_20

18. Lara, O.D., Labrador, M.A.: A survey on human activity recognition using wearable sensors. IEEE Commun. Surv. Tutor. 15(3), 1192–1209 (2013)

19. Longstaff, B., Reddy, S., Estrin, D.: Improving activity classification for health applications on mobile devices using active and semi-supervised learning. In: ICST 2010. IEEE (2010)

20. Lukowicz, P., et al.: Recognizing workshop activity using body worn microphones and accelerometers. In: Ferscha, A., Mattern, F. (eds.) Pervasive 2004. LNCS, vol. 3001, pp. 18–32. Springer, Heidelberg (2004). https://doi.org/10.1007/978-3-540-24646-6_2

21. Maekawa, T., Nakai, D., Ohara, K., Namioka, Y.: Toward practical factory activity recognition. In: UbiComp 2016. ACM Press (2016)

22. Mannhardt, F., de Leoni, M., Reijers, H.A., van der Aalst, W.M.P.: Balanced multiperspective checking of process conformance. Computing 98(4), 407–437 (2016)

23. Mannhardt, F., Petersen, S.A., de Oliveira, M.F.D.: Privacy challenges for process mining in human-centered industrial environments. In: Intelligent Environments (IE). IEEE Xplore (2018, to appear)

24. Marin-Perianu, M., Lombriser, C., Amft, O., Havinga, P., Tröster, G.: Distributed activity recognition with fuzzy-enabled wireless sensor networks. In: Nikoletseas, S.E., Chlebus, B.S., Johnson, D.B., Krishnamachari, B. (eds.) DCOSS 2008. LNCS, vol. 5067, pp. 296–313. Springer, Heidelberg (2008). https://doi.org/10.1007/978-3-540-69170-9_20

25. Mörzinger, R., et al.: Tools for semi-automatic monitoring of industrial workflows. In: ARTEMIS 2010. ACM Press (2010)

26. Mura, M.D., Dini, G., Failli, F.: An integrated environment based on augmented reality and sensing device for manual assembly workstations. Procedia CIRP 41, 340–345 (2016)

27. Ogris, G., Lukowicz, P., Stiefmeier, T., Tröster, G.: Continuous activity recognition in a maintenance scenario: combining motion sensors and ultrasonic hands tracking. Pattern Anal. Appl. 15(1), 87–111 (2011)

28. Ogris, G., Stiefmeier, T., Lukowicz, P., Troster, G.: Using a complex multi-modal on-body sensor system for activity spotting. In: IWSC 2008. IEEE (2008)

29. Ramamurthy, S.R., Roy, N.: Recent trends in machine learning for human activity recognition-a survey. Wiley Interdiscip. Rev. Data Min. Knowl. Discov. 8(4), e1254 (2018)

30. Raso, R., et al.: Activity monitoring using wearable sensors in manual production processes - an application of CPS for automated ergonomic assessments. In: MKWI 2018. Leuphana Universität Lüneburg (2018)

31. Repta, D., Moisescu, M.A., Sacala, I.S., Stanescu, A.M., Constantin, N.: Generic architecture for process mining in the context of cyber physical systems. Appl. Mech. Mater. 656, 569–577 (2014)

32. Roitberg, A., Perzylo, A., Somani, N., Giuliani, M., Rickert, M., Knoll, A.: Human activity recognition in the context of industrial human-robot interaction. In: APSIPA 2014. IEEE (2014)

33. Roitberg, A., Somani, N., Perzylo, A., Rickert, M., Knoll, A.: Multimodal human activity recognition for industrial manufacturing processes in robotic workcells. In: ICMI 2015. ACM Press (2015)

34. Schlenoff, C., Kootbally, Z., Pietromartire, A., Franaszek, M., Foufou, S.: Intention recognition in manufacturing applications. Robot. Comput. Integr. Manuf. **33**, 29–41 (2015)
35. Stiefmeier, T., Lombriser, C., Roggen, D., Junker, H., Ogris, G., Troester, G.: Event-based activity tracking in work environments. In: IFAWC 2006, pp. 1–10 (2006)
36. Stiefmeier, T., Roggen, D., Ogris, G., Lukowicz, P., Tr, G.: Wearable activity tracking in car manufacturing. IEEE Pervasive Comput. **7**(2), 42–50 (2008)
37. Voulodimos, A., et al.: A threefold dataset for activity and workflow recognition in complex industrial environments. IEEE Multimed. **19**(3), 42–52 (2012)
38. Voulodimos, A.S., Kosmopoulos, D.I., Doulamis, N.D., Varvarigou, T.A.: A top-down event-driven approach for concurrent activity recognition. Multimed. Tools Appl. **69**(2), 293–311 (2012)
39. Wang, J., Chen, Y., Hao, S., Peng, X., Hu, L.: Deep learning for sensor-based activity recognition: a survey. Pattern Recognit. Lett. (2018, in press)
40. Ward, J.A., Lukowicz, P., Troster, G., Starner, T.E.: Activity recognition of assembly tasks using body-worn microphones and accelerometers. IEEE Trans. Pattern Anal. Mach. Intell. **28**(10), 1553–1567 (2006)

Mining Attributed Interaction Networks
on Industrial Event Logs

Martin Atzmueller[1(✉)] and Benjamin Kloepper[2]

[1] Department of Cognitive Science and Artificial Intelligence, Tilburg University,
Warandelaan 2, 5037 AB Tilburg, The Netherlands
`m.atzmuller@uvt.nl`
[2] ABB AG, Corporate Research Center, Wallstadter Str. 59,
68526 Ladenburg, Germany
`benjamin.kloepper@de.abb.com`

Abstract. In future Industry 4.0 manufacturing systems reconfigurability and flexible material flows are key mechanisms. However, such dynamics require advanced methods for the reconstruction, interpretation and understanding of the general material flows and structure of the production system. This paper proposes a network-based computational sensemaking approach on attributed network structures modeling the interactions in the event log. We apply descriptive community mining methods for detecting patterns on the structure of the production system. The proposed approach is evaluated using two real-world datasets.

1 Introduction

In the context of Industry 4.0, future manufacturing systems will be more flexible in order to answer more readily to changing market demands [24] as well as disturbances in the production systems. In particular, this is one of the key aspects in the concept of Industry 4.0 [26] or Cloud Manufacturing [27]. Important capabilities of such flexible and robust manufacturing systems are reconfigurability and flexible material flows [27]. As a consequence, the relationships between elements in the production systems like industrial robots, machining centers and material handling systems become more dynamic as well and the interaction between the (resource) elements becomes thus also more difficult to comprehend. On the other hand, understanding the general material flow and the structure of the production systems is required for continous improvements processes, for instance process mapping is a key activity in the six-sigma process [11].

This paper proposes a network-based approach to recreate the material flow and resource interactions from the log files of the individual components of a production systems. We model log-files as attributed network structures, connecting devices by links labeled with log statements. This allows to detect densely connected groups of devices with an according description of (log) statements. In our experiments, we apply two real-world datasets from serial production systems with a clear hierarchical structure providing a ground truth for evaluating the

© Springer Nature Switzerland AG 2018
H. Yin et al. (Eds.): IDEAL 2018, LNCS 11315, pp. 94–102, 2018.
https://doi.org/10.1007/978-3-030-03496-2_11

performance of the proposed algorithmic approach. Our results show the impact and efficacy of our novel network-based analysis and mining approach.

2   Related Work

Below, we discuss related work concerning the analysis of industrial (alarm) event logs, i. e., in alarm management and in the context of process mining.

2.1   Analysis of Alarm Event Logs

Analysis of event logs has been performed in the context of alarm management systems, where sequential analysis is performed on the alarm notifications. In [13], an algorithm for discovering temporal alarm dependencies is proposed which utilizes conditional probabilities in an adjustable time window. In order to reduce the number of alarms in alarm floods, [2] also performed root cause analysis with a Bayesian network approach and compared different methods for learning the network probabilities. A pattern-based algorithm for identifying causal dependencies in the alarm logs is proposed in [25], which can be used to aggregate alarm information and therefore reduce the load of information for the operator. Furthermore, [6,10] target the analysis of sequential event logs in order to detect anomalies using a graph-based approach. Finally, [21] investigate the prediction of the risk increase factor in nuclear power plants using complex network analysis using topological structure.

In contrast to those approaches, the proposed approach is not about sequential analysis of event logs, nor on the given static network structures. Instead, we provide a network-based approach transforming event logs into (attributed) networks capturing the static interactions and dependencies captured in the event log. The goal is to identify structural dependencies and relations of the production process. Thus, similar to evidence networks in the context of social networks, e. g., [18], we aim to infer the (explicit) structural relations given observed (implicit) interactions between the industrial equipment and devices.

2.2   Analysis of Event Logs Using Process Mining

Process Mining [1] aims at the discovery of business process related events in a sequential event log. The assumption is that event logs contain fingerprints of business processes, which can be identified by sequence analysis. One task of process mining is conformance checking [19,22] which has been introduced to check the matching of an existing business process model with a segmentation of the log entries. Furthermore, for process mining and anomaly analysis there have been approaches based on subgroup discovery, e. g., [23], and subgraph mining, e. g., [14] based on log data; while these neglect the temporal (sequential) dimension, they only focus on the respective patterns not including a priori knowledge, while not including relational, i. e., network modeling.

Compared to these approaches, we do not use any apriori (process) knowledge for our analysis. In contrast, we use a purely data-driven approach, where we perform a feature-rich network-based approach on the event log data. For that, we transform the (event log) interaction data into an attributed interaction network which is then exploited for mining cluster/community structures together with an explicit description – enhancing interpretation and understandability.

3   Method

In Industry 4.0 environments like complex industrial production plants, intelligent data analysis is a key technique for providing advanced data science capabilities. In that context, computational sensemaking [5] aims to develop methods and systems to "make sense" of complex data and information – to make the implicit explicit; important goals are then to comprehensively model, describe and explain the underlying structure in the data [4]. This paper presents a computational sensemaking approach using descriptive pattern mining. The proposed approach consists of three steps: (1) We model the event log as a bimodal network represented as a bipartite graph. (2) We create an attributed graph structure using a projection operator with labels taken from the bimodal structure. (3) Finally, we apply pattern mining (i. e., descriptive community mining) on the attributed graph, in order to detect structural patterns and relations.

3.1   Modeling Attributed Interaction Networks from Event Logs

In the following, we use the data shown in Table 1 as an example for demonstrating the individual steps of the proposed approach. As can be seen in the table, it considers log entries corresponding to a certain *device* and *event_type* in addition to a timestamp. We focus on the *device* and *event_type* information creating a bimodal network. However, first we aggregate the *event_type* information for a device, such that equal *event_type*s for a specific *device* are merged into a single link between *device* and the corresponding *event_type*, respectively. In our example, line #1 and line #13 would thus be merged into a single link. The resulting bipartite graph is shown in Fig. 1. This can already be considered as an attributed graph, where we interpret links between the devices labeled by their common *event_type*s. In our example, every device is connected to every other device with a link labeled with the common $0:0$ ("Safety Stop Activate") and $1:1$ ("System is in Safety Stop") event types.

3.2   Descriptive Community Mining

Community detection [20] aims at identifying densely connected groups of nodes in a graph; using attributed networks, we can additionally make use of information assigned to nodes and/or edges. For mining attributed network structures, we apply the COMODO algorithm presented in [7]: It focuses on *description-oriented community detection* using subgroup discovery [3], and aims at discovering the top-n communities (described by community patterns). COMODO

Table 1. Exemplary (anonymized) log event data, visualized by the bipartite graph to the right.

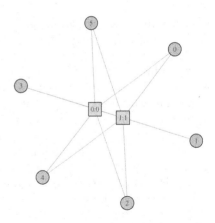

#	Device	Event_Type	Timestamp
1	0	0:0	12.08.12 07:23
2	1	1:1	12.08.12 07:23
3	2	1:1	12.08.12 07:23
4	2	0:0	12.08.12 07:23
5	0	1:1	12.08.12 07:23
6	1	0:0	12.08.12 07:23
7	3	1:1	12.08.12 07:24
8	4	0:0	12.08.12 07:24
9	4	1:1	12.08.12 07:24
10	5	1:1	12.08.12 07:24
11	5	0:0	12.08.12 07:24
12	3	0:0	12.08.12 07:24
13	0	0:0	12.08.12 10:59

Fig. 1. Bipartite graph (example data, left): *devices* (orange circles) and linked *event types* (gray squares). (Color figure online)

utilizes efficient pruning approaches for scalability, for a wide range of standard community evaluation functions. Its results are a set of patterns (given by conjunctions of literals, i. e., attribute–value pairs) that describe a specific subgraph – indicating a specific community consisting of a set of nodes. An example in the context of the analysis of event logs is given by the pattern: $event_1 \, AND \; event_2 \; AND \; event_5$ indicating the event (types) $event_1$, $event_2$, and $event_5$ being jointly connected to the same set of devices. This pattern then directly corresponds to the (covered) subgraph.

Algorithmic Overview. COMODO utilizes both the graph structure, as well as descriptive information of the attributed graph. As outlined above, we transform the graph data into a new dataset focusing on the edges of the graph G: Each data record in the new dataset represents an edge between two nodes. The attribute values of each such data record are the common attributes of the edge's two nodes. For efficiency, COMODO utilizes an extended FP-tree (frequent pattern tree) structure inspired by the FP-growth algorithm [15], which compiles the data into a prefix pattern tree structure, cf. [9,17]. Our adapted tree structure is called the *community pattern tree* (CP-tree) that allows to efficiently traverse the solution space. The tree is built in two scans of the graph dataset and is then mined in a recursive divide-and-conquer manner. Efficient pruning is implemented using optimistic estimates [7]. For community evaluation a set of standard evaluation functions exists, including the Modularity function [20]. As a result, COMODO provides the *top-n* patterns according to a given community evaluation function. For a more detailed description, we refer to [7].

Community Postprocessing. As a final result, we aim at a disjoint partition of the set of nodes in our input graph – which should correspond to the different levels (and category groups). However, the set of communities (or clusters) provided by COMODO can overlap. For the industry 4.0 use case this property is very useful, because overlapping resource communities are expected due to reconfigurability and flexible material flows. In the given dataset, however the devices in the production system are organized in a two-level hierarchy with non-overlapping groups. Thus, we apply a postprocessing step, in order to obtain a disjoint partition of the graph from the given set of top-n patterns. Essentially, given the communities, we construct a similarity graph for the set of nodes: For each pair of nodes, we check the number of times they are contained in a community (pattern), and create a weighted edge accordingly, normalized by the total number of patterns. Then, we uncover (disjoint) communities on the (pruned) similarity graph by a further community detection step.

4   Results

In this section, we first describe the characteristics and context of the applied real-world datasets. After that, we present results and discuss them in detail.

4.1   Datasets

Two real-world datasets from the industrial domain are used in this work. Both datasets are from serial production facilities with several production lines and cell. The first dataset (*Log-Data-A*) contains data from 59 industrial machines and devices from 8 different production lines and 7 production cells. The second dataset (*Log-Data-B*) contains data from 48 machines and devices from 2 production lines with 16 production cells. Basically, each device is assigned to a production line and production cell, where the production lines can be considered as *level 1 categories*, and the production cells as *level 2 categories*, representing the production hierarchy. In the dataset, this information can be used as ground-truth in order to evaluate the mined patterns and communities, respectively. Since the community structures should represent the material flows, this directly corresponds to the respective level 1 and level 2 categories. It is important to note that these categories are a disjoint partitioning of the set of devices, respectively. Therefore, as explained above, we also aim at a disjoint partitioning of the graph given the set of communities.

The event logs contain both normal events, warnings and error events and partially capture the standard activity of the devices (e.g. motor starts and stops, program starts), operator interactions (e.g. safety stops, switching operation modes) and information of interactions with supplementary process like cooling water supply. Due to serial production fashion, products pass through the production lines in a sequential fashion. Consequently, activities of machines and devices are triggered according to the production line and cell structure. Furthermore, the product flow closely interlinks the industrial machines and devices and failure and problems propagate usually forward through the production systems. These features make the two datasets ideal to develop a proof of concept of recovering the flow of material in a production systems from the event log data generated by the individual machines and devices. Table 2 summarizes the characteristics of both datasets.

Table 2. Characteristics of the real-world datasets

Dataset	#Devices	#EventTypes	#Prod. Lines	# Prod. Cells	# Events
Log-data-A	59	356	8	7	50000
Log-data-B	48	102	2	16	50000

4.2 Results and Discussion

First, we take a look at the connectivity structure of our attributed graphs. Figures 2 and 3 depict according (extended) KNC-Plots [8,16] that visualize the number k of common neighbors of the nodes in the original bipartitate graph, as well as the sizes of the largest and 2nd largest components. In our case, k indicates the number of common *event_types* connecting the respective *device* nodes. Overall, the graphs exhibit a strong connectivity structure: As we can see in the figures, there is strong connectivity up to 8 (16) common event_types.

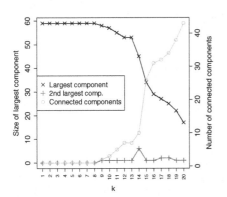

Fig. 2. KNC-plot: log-data-A dataset.   **Fig. 3.** KNC-plot: log-data-B dataset.

For community detection aiming at reconstructing the production system structure in our application scenario, we applied the COMODO algorithm using the modularity evaluation function, with no minimal support threshold. Regarding the only parameter, i. e., determining n for the top-n patterns, we experimented with different selections, where we used $n = 20$ for interpretability. However, with other selections the results as outlined below were quite stable. Finally, for the postprocessing step constructing the similarity graph we pruned edges with a weight below 0.1 such that edges needed to be "supported" by at least 2 community patterns in order to be included in the final similarity graph. For determining the final set of disjoint communities, we utilized the edge betweenness [20] method.

Table 3 shows our results using the Normalized Mutual Information (NMI) measure for comparing community structures using the (production line/cell) category information as ground truth for the different communities/clusters. We compared different baseline methods to our proposed approach using the COMODO algorithm using standard algorithms as included in the igraph [12] software package, i. e., edge betweenness, fast greedy, Infomap, label propagation, leading eigenvector, and louvain. In particular, Infomap and label propagation yielded NMI values of 0, detecting no structure. As we can observe in the table, COMODO outperforms all the other algorithms, while the baselines yield relatively low NMI values discovering no relations. Thus, in comparison to the baselines the proposed approach using COMODO does not only outperform standard community approaches, but also provides descriptive patterns that can be used for inspection, interpretation and explanation.

Table 3. Results: NMI for different community detection approaches

Algorithm/NMI	Log-data-A		Log-data-B	
	Level1	Level2	Level1	Level2
Edge betweenness	0.32	0.20	0.02	0.11
Fast greedy	0.48	0.24	0.01	0.15
Leading eigenvector	0	0	0.01	0.15
Louvain	0.48	0.24	0.01	0.15
COMODO	0.67	0.53	0.19	0.78

5   Conclusions

This paper presented a network-based approach to recreate production system structures and resource interactions from industrial event log data. We modeled those as attributed networks and detected densely connected groups of devices with an according description of (log) statements. For evaluation, we applied two real-world datasets. Our results indicated the impact and efficacy of the proposed network-based approach, outperforming standard community detection baselines while also providing descriptive patterns for interpretation and explanation.

Beyond confirming the applicability of event log analysis for reconstructing resource interactions and material flows, the analysis can also help to detect hotspots in the production process, e. g., segments of the production process in which high amounts of events are generated and thus potentially require special attention in continuous improvement processes like Six Sigma. Thus, advanced hotspot analysis and anomaly detection are interesting directions for future work.

Also, analyzing the evolution of the network – capturing dynamics and temporal dependencies in the event logs – is another interesting direction to consider.

Acknowledgements. This work has been partially funded by the German Research Foundation (DFG) project "MODUS" (under grant AT 88/4-1) and by the EU ECSEL project Productive 4.0.

References

1. Aalst, W.: Process Mining: Discovery Conformance and Enhancement of Business Processes. Springer, Heidelberg (2011). https://doi.org/10.1007/978-3-642-19345-3
2. Abele, L., Anic, M., Gutmann, T., Folmer, J., Kleinsteuber, M., Vogel-Heuser, B.: Combining knowledge modeling and machine learning for alarm root cause analysis. In: MIM, pp. 1843–1848. IFAC (2013)
3. Atzmueller, M.: Subgroup discovery. WIREs DMKD **5**(1), 35–49 (2015)
4. Atzmueller, M.: Onto explicative data mining: exploratory, interpretable and explainable analysis. In: Proceedings of Dutch-Belgian Database Day. TU Eindhoven (2017)
5. Atzmueller, M.: Declarative aspects in explicative data mining for computational sensemaking. In: Seipel, D., Hanus, M., Abreu, S. (eds.) Declarative Programming and Knowledge Management. LNCS, vol. 10997. Springer, Heidelberg (2018). https://doi.org/10.1007/978-3-030-00801-7_7
6. Atzmueller, M., Arnu, D., Schmidt, A.: Anomaly detection and structural analysis in industrial production environments. In: Haber, P., Lampoltshammer, T., Mayr, M. (eds.) Data Science – Analytics and Applications, pp. 91–95. Springer, Wiesbaden (2017). https://doi.org/10.1007/978-3-658-19287-7_13
7. Atzmueller, M., Doerfel, S., Mitzlaff, F.: Description-oriented community detection using exhaustive subgroup discovery. Inf. Sci. **329**, 965–984 (2016)
8. Atzmueller, M., Hanika, T., Stumme, G., Schaller, R., Ludwig, B.: Social event network analysis: structure, preferences, and reality. In: Proceedings of IEEE/ACM ASONAM. IEEE Press, Boston (2016)
9. Atzmueller, M., Puppe, F.: SD-Map – a fast algorithm for exhaustive subgroup discovery. In: Fürnkranz, J., Scheffer, T., Spiliopoulou, M. (eds.) PKDD 2006. LNCS (LNAI), vol. 4213, pp. 6–17. Springer, Heidelberg (2006). https://doi.org/10.1007/11871637_6
10. Atzmueller, M., Schmidt, A., Kloepper, B., Arnu, D.: HypGraphs: an approach for analysis and assessment of graph-based and sequential hypotheses. In: Appice, A., Ceci, M., Loglisci, C., Masciari, E., Raś, Z.W. (eds.) NFMCP 2016. LNCS (LNAI), vol. 10312, pp. 231–247. Springer, Cham (2017). https://doi.org/10.1007/978-3-319-61461-8_15
11. Chen, J.C., Li, Y., Shady, B.D.: From value stream mapping toward a lean/sigma continuous improvement process: an industrial case study. Int. J. Prod. Res. **48**(4), 1069–1086 (2010)
12. Csardi, G., Nepusz, T.: Package igraph: Network Analysis and Visualization (2014)
13. Folmer, J., Schuricht, F., Vogel-Heuser, B.: Detection of temporal dependencies in alarm time series of industrial plants. In: Proceedings of IFAC, pp. 24–29 (2014)

14. Genga, L., Potena, D., Martino, O., Alizadeh, M., Diamantini, C., Zannone, N.: Subgraph mining for anomalous pattern discovery in event logs. In: Appice, A., Ceci, M., Loglisci, C., Masciari, E., Raś, Z.W. (eds.) NFMCP 2016. LNCS (LNAI), vol. 10312, pp. 181–197. Springer, Cham (2017). https://doi.org/10.1007/978-3-319-61461-8_12
15. Han, J., Pei, J., Yin, Y.: Mining frequent patterns without candidate generation. In: Proceedings of SIGMOD, pp. 1–12. ACM Press (2000)
16. Kumar, R., Tomkins, A., Vee, E.: Connectivity structure of bipartite graphs via the KNC-plot. In: Proceedings of WSDM, pp. 129–138. ACM Press (2008)
17. Lemmerich, F., Becker, M., Atzmueller, M.: Generic pattern trees for exhaustive exceptional model mining. In: Flach, P.A., De Bie, T., Cristianini, N. (eds.) ECML PKDD 2012. LNCS (LNAI), vol. 7524, pp. 277–292. Springer, Heidelberg (2012). https://doi.org/10.1007/978-3-642-33486-3_18
18. Mitzlaff, F., Atzmueller, M., Benz, D., Hotho, A., Stumme, G.: Community assessment using evidence networks. In: Atzmueller, M., Hotho, A., Strohmaier, M., Chin, A. (eds.) MSM/MUSE -2010. LNCS (LNAI), vol. 6904, pp. 79–98. Springer, Heidelberg (2011). https://doi.org/10.1007/978-3-642-23599-3_5
19. Munoz-Gama, J., Carmona, J., van der Aalst, W.M.P.: Single-entry single-exit decomposed conformance checking. Inf. Syst. **46**, 102–122 (2014)
20. Newman, M.E., Girvan, M.: Finding and evaluating community structure in networks. Phys. Rev. E Stat. Nonlin. Soft Matter Phys. **69**(2), 1–15 (2004)
21. Rifi, M., Hibti, M., Kanawati, R.: A complex network analysis approach for risk increase factor prediction in nuclear power plants. In: Proceedings of International Conference on Complexity, Future Information Systems and Risk, pp. 23–30 (2018)
22. Rozinat, A., Aalst, W.: Conformance checking of processes based on monitoring real behavior. Inf. Syst. **33**(1), 64–95 (2008)
23. Fani Sani, M., van der Aalst, W., Bolt, A., García-Algarra, J.: Subgroup discovery in process mining. In: Abramowicz, W. (ed.) BIS 2017. LNBIP, vol. 288, pp. 237–252. Springer, Cham (2017). https://doi.org/10.1007/978-3-319-59336-4_17
24. Theorin, A., et al.: An Event-driven manufacturing information system architecture for industry 4.0. Int. J. Prod. Res. **55**(5), 1297–1311 (2017)
25. Vogel-Heuser, B., Schütz, D., Folmer, J.: Criteria-based alarm flood pattern recognition using historical data from automated production systems (aPS). Mechatronics **31**, 89–100 (2015)
26. Weyer, S., Schmitt, M., Ohmer, M., Gorecky, D.: Towards industry 4.0-standardization as the crucial challenge for highly modular, multi-vendor production systems. Proc. IFAC **48**(3), 579–584 (2015)
27. Wu, D., Greer, M.J., Rosen, D.W., Schaefer, D.: Cloud manufacturing: strategic vision and state-of-the-art. JMSY **32**(4), 564–579 (2013)

Special Session on Intelligent Techniques for the Analysis of Scientific Articles and Patents

Evidence-Based Systematic Literature Reviews in the Cloud

Iván Ruiz-Rube[(✉)] ⓘ, Tatiana Person ⓘ, José Miguel Mota ⓘ,
Juan Manuel Dodero ⓘ, and Ángel Rafael González-Toro

Department of Computer Engineering, University of Cádiz, ESI,
Puerto Real (Cádiz), Spain
{ivan.ruiz,tatiana.person,josemiguel.mota,juanma.dodero}@uca.es,
angel.gonzatoro@alum.uca.es

Abstract. Systematic literature reviews and mapping studies are useful research methods used to lay the foundations of further research. These methods are widely used in the Health Sciences and, more recently, also in Computer Science. Despite existing tool support for systematic reviews, more automation is required to conduct the complete process. This paper describes CloudSERA, a web-based app to support the evidence-based systematic review of scientific literature. The tool supports researchers to carry out studies by additional facilities as collaboration, usability, parallel searches and search integration with other systems, The flexible data scheme of the tool enables the integration of bibliographic databases of common use in Computer Science and can be easily extended to support additional sources. It can be used as a service in a cloud environment or as on-premises software.

Keywords: Systematic literature review · Mapping study
Information retrieval · Bibliographic data

1 Introduction

Research and development activity usually requires a preliminary study of related literature to know the up-to-date state of the art about issues, techniques and methods in a given research field. Digital libraries, bibliographical repositories and patent databases are extensively used by researchers to systematically collect the inputs required to lay the foundations of the intended research outputs. Also, bibliometric [16], science mapping [1,2] and science of science [7] analysis need a dataset that, although it is usually retrieved with a query launched in a bibliographic database (e.g. Web of Science or Scopus), sometimes the records obtained must be manually reviewed in order to select only those ones related with the field.

In order to assure the accuracy and reproducibility of reviews and obtain unbiased results, it is helpful to systematically follow a set of steps and shared guidelines when performing the review process. Following these steps often

© Springer Nature Switzerland AG 2018
H. Yin et al. (Eds.): IDEAL 2018, LNCS 11315, pp. 105–112, 2018.
https://doi.org/10.1007/978-3-030-03496-2_12

requires executing certain tasks that, without automation support, may be a daunting and dull task.

Systematic Reviews (SR) allow researchers to identify, evaluate and interpret the existing research that is relevant for a particular Research Question (RQ) or phenomenon of interest [3]. Some reasons for performing SRs are to summarize the existing evidence concerning a given topic, to identify gaps in current research and suggest areas for further investigation, and to provide a background to position new research activities [11]. Updated reviews of empirical evidence conducted in a systematic and rigorous way are valuable for researchers and stakeholders of different fields.

As other knowledge areas, Computer Science has also benefited from the evidence-based SR approach. Kitchenham and Charters published a set of guidelines for performing Systematic Literature Reviews (SLR) in this field [10]. These authors defined a process consisting of three stages, namely *Planning*, *Conducting* and *Reporting*. The SLR method has become a popular research methodology for conducting literature review and evidence aggregation in Software Engineering. However, there are concerns about the required time and resources to complete an SLR and keep it updated, so it is important to seek a balance between methodological rigour, required effort [20] and ways to automate the process. Similarly, systematic mapping studies or scope studies allow researchers to obtain a wide overview of a research area, providing them with an indication of the quantity of the evidence found [18].

This paper introduces CloudSERA, a web-based app to support evidence-based systematic reviews of scientific literature. The tool aims at making the review process easier. The rest of the paper is structured as follows: Sect. 2 describes tool support for literature reviews. CloudSERA is described in Sect. 3. Finally, conclusions and future works are drawn in the last section.

2   Literature Review Tool Support

Several authors have studied which are the needs and the features that tools to support systematic reviews should fulfil. Marshall and Brereton [13] identified various SLR tools in the literature, which were mostly evaluated only thorough small experiments, reflecting hence the immaturity of this research area. Most papers present tools based on text mining and discuss the use of visualisation techniques. However, few tools which support the whole SLR process were found. These tools were analysed and compared for enumerate a set of desirable features that would improve the development of SLRs [14]. In a third study [15], these authors explored the scope and practice of tool support for systematic reviewers in other disciplines. Reference management tools were the most commonly used ones. Afterwards, Hassler et al. [9] conducted a community workshop with software engineering researchers in order to identify and prioritise the necessary SLR tool features. According to their results, most of the high-priority features are not well-supported in current tools. From the previous studies, the major features required for an integrated tool for supporting SLR are presented in Table 1.

Table 1. Main desirable features in SLR tools

Feature	Description
Non functional features	
Status	Availability. Up-to-date maintenance
Platform	Installation (desktop/cloud)
Cost	Open source/privative
Usability	Ease of installation and use: user and installation guides, tutorials, etc.
Overall protocol	
Data sharing & Collaboration	Data sharing between processes and tasks, and collaboration among the SLR team, including role management, security, dispute resolution and coordination. and among the SLR team
Automate tasks	Automation of the processes and tasks
Data maintenance	Data maintenance and preservation functions to access past research questions, protocols, studies, data, metadata, bibliographic data and reports
Traceability	Forward and backward traceability to link goals, actions, and results for accountability, standardization, verification and validation
Search & selection & quality assessment	
Integrated search	Ability to search multiple databases without having to perform separate searches
Study selection	Selection of primary studies using inclusion/exclusion criteria
Quality assessment	Evaluation of primary studies using quality assessment criteria
Analysis & presentation	
Automated analysis	Ability to automatically analyse the extracted data
Visualisation	The visualisation mechanisms can support selection, analysis, synthesis, and conveyance of results

Currently, there are several tools for supporting the whole SLR process. These are applicable to different fields of research [12]. CloudSERA, in its current version, is centred on SLR process applied on Computer Science. The existing tools that have been identified on this discipline in the literature are: Parsifal [8], REviewER [4], SLuRP [19], SLR-Tool [6] and StArt [5]. These tools were analysed taking into account the previous features. Below, the major weaknesses of these tools are described:

- Some of the tools reviewed enable to export the bibliography in BibTeX format, but none of them provides integration with reference managers such as Mendeley or Zotero.
- Only PARSIFAL provides a cloud-version that can be easily accessed by users.
- Only the SLuRP tool allows reviewers to follow the traceability of decisions made, but none of the tools enables users to follow the steps to be performed in an SLR.
- All the tools enable to visualise graphs of the obtained results and some allow to export these results in Excel format, as SLR-Tool. However, only PARSIFAL allows reporting the results in a suitable format to be included in a paper.
- Most of the tools allow users to extract bibliographic data, but only PARSI-FAL enables to automatically issue queries in external search engines. However, it does not retrieve neither import the metadata of the papers. In addition, selecting the sections of the papers (title, abstract, full-text, etc.) where find the keywords is not also allowed.

3   The CloudSERA Tool

None of the current SLR tools provide complete support for the major needs described in the previous section. For this reason, a web application, called CloudSERA, has been developed. Below, a description of the main features of this tool are presented. Unlike other SLR tools, CloudSERA is a web application, requiring hence no installation, and is available for free use at http://slr.uca.es. Also, the application code has been released[1] as open source to foster its further evolution and, if required, to deploy it on an on-premises instance. The application architecture follows the common MVC pattern and has been developed using the Grails framework for Java servlet containers.

The user interface is based on the Bootstrap toolkit for providing a responsive and enriched user experience. The tool is provided with documentation and some tutorials. In short, the tool satisfies the considered non-functional aspects, namely status, cost and usability.

With regard to the overall protocol, users can create systematic reviews (mapping studies or literature reviews) and define the research questions to be answered. CloudSERA enables to automate several tasks during the review process and includes a step-by-step wizard (see Fig. 1) to guide researchers. The tool provides a role management module for the whole application and per systematic review. Thus, CloudSERA enables researches to collaborate during the review process, distinguishing between performers and supervisors. Due to the own web nature of the tool and its authentication and authorization system, data sharing is assured between the SLR's team members. Furthermore, under consent of the users, the protocols and results of the SLRs developed may be easily accessed from the application for preservation and replicability purposes. Also, CloudSERA enables users to follow each other's activities, creating hence

---

[1]   https://github.com/spi-fm/CloudSERA.

communities of users. In addition, CloudSERA is provided with a logging system to trace all the actions performed by the users in the context of a SLR process. To sum up, the tool proposed in this research give users the required tools for managing the overall protocol, ranging from data sharing and collaboration to task automation, including data maintenance and traceability.

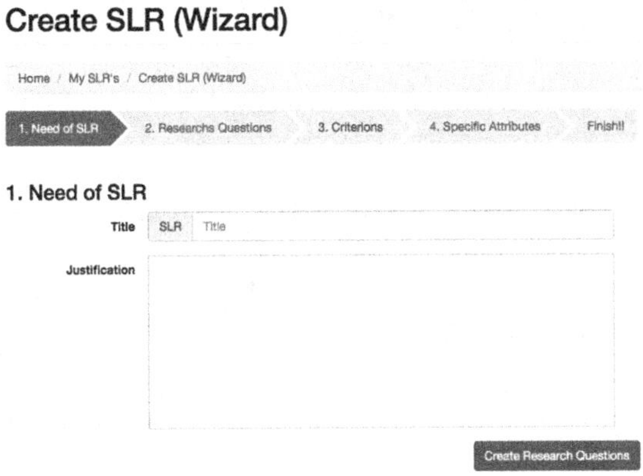

Create SLR (Wizard)

Fig. 1. CloudSERA wizard to create a new systematic literature review

The tool is capable to automate the searching tasks, by triggering the proper queries to the pre-configured digital databases. Currently, the databases supported are ACM Digital Library, IEEE Computer Society, Springer Link and Science Direct. However, CloudSERA has been designed in such way that supports the inclusion of new digital databases with relative ease. In this way, the user has not to either perform separate searches or deal with the particularities of each library. These searching tasks run in background to avoid halt the user and once the set of references have been received, the user is notified.

The software relies on Mendeley, the popular reference management tool, to authenticate users and the consolidate found references. These references are automatically annotated with the common metadata, such as publication year, authors, journal, etc. Moreover, the tool enables to add specific attributes to design data extraction forms and quality assessment instruments. In this way, users are able to collect all the information needed from the primary studies to address the review questions by using textual or nominal—in a range of admitted values—attributes. In addition, the users may evaluate the quality of each study by using a scale based on numerical attributes.

Moreover, CloudSERA enables to define inclusion and exclusion criteria which can be applied to the found references. To do this, users can easily visualise and refine the references by using a set of facets according to the automatically retrieved metadata and the manually entered values for the above attributes. In this way, user are able to easily tag the references with the corresponding inclusion or exclusion criteria (see Fig. 2). As can be noted, all the main features related to the search, selection and quality assessment can be implemented with the tool.

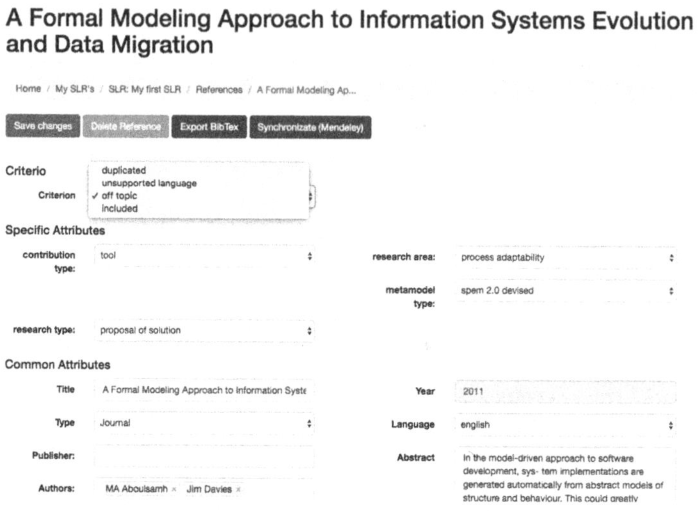

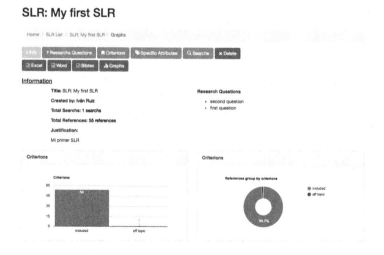

Fig. 2. Screen to view and annotate data of a given paper in CloudSERA

CloudSERA includes some charts to visualise the extracted data according some aspects, such as inclusion or exclusion criteria, document type and language, among others. An example can be observed in Fig. 3. On the other hand, in addition to the two-way communication between CloudSERA and Mendeley,

Fig. 3. Main screen of CloudSERA for accessing/editing a systematic review

the tool handles different formats of data exportation: (i) BibTeX, in order cite the primary papers from a Latex file; (ii) Word, to generate a template of dissemination report; and (iii) Excel, for further calculation. The two latter formats also provide pages or sheets including the research questions, used specific attributes, search history, primary studies and some charts, such as the annual trends. In brief, the tool provides support for automated analysis and data visualisation.

4 Conclusion

Conducting reviews (SLR or mapping studies) in a systematic way are vital for ensuring the new researches are well-founded in the current state of the art. However, at the present time, there are no tools that cover and provide automation for the whole review process. CloudSERA allows users to plan, to conduct and to report systematic reviews in a integrated tool which satisfies most of the needs of the researchers. Among other features, this tool enables collaboration via a web environment, a single search interface for several databases and capabilities for selecting and assessing studies thorough a flexible data scheme. Currently, the tool is mostly aimed at CS researchers, but it is easily extensible to support other digital databases or search engines.

As a future work, we plan to enrich the platform with several advances features, such as, a workflow engine to orchestrate the execution of the tasks in a SLR process, some machine learning algorithms for clustering data in mapping studies, and an On-Line Analytical Processing (OLAP) viewer based on a drag and drop interface for doing multi-dimensional analysis, which may be particularly useful for mapping studies. A general heuristic evaluation will be also performed. Thus, a usability test is being designed by following the heuristics proposed by Nielsen [17]. The study will be conducted on a set of experts previously selected between the authors who have published a SLR in CS.

Acknowledgements. This work was funded by the Spanish Government under the VISAIGLE Project (grant TIN2017-85797-R).

References

1. Börner, K., Chen, C., Boyack, K.W.: Visualizing knowledge domains. Ann. Rev. Inf. Sci. Technol. **37**(1), 179–255 (2003)
2. Cobo, M.J., López-Herrera, A.G., Herrera-Viedma, E., Herrera, F.: Science mapping software tools: review, analysis, and cooperative study among tools. J. Am. Soc. Inf. Sci. Technol. **62**(7), 1382–1402 (2011)
3. Collaboration, C., et al.: The Cochrane Reviewers' Handbook Glossary. Cochrane Collaboration, London (2001)
4. (ESEG), E.S.E.G.: REviewER (2013). https://sites.google.com/site/eseportal/tools/reviewer. Accessed 10 July 2018
5. Fabbri, S., Silva, C., Hernandes, E., Octaviano, F., Di Thommazo, A., Belgamo, A.: Improvements in the start tool to better support the systematic review process. In: Proceedings of the 20th International Conference on Evaluation and Assessment in Software Engineering, p. 21. ACM (2016)

6. Fernández-Sáez, A.M., Bocco, M.G., Romero, F.P.: SLR-tool: a tool for performing systematic literature reviews. In: ICSOFT, vol. 2, pp. 157–166 (2010)
7. Fortunato, S., et al.: Science of science. Science **359**(6379) (2018)
8. Freitas, V.: Persifal (2014). https://parsif.al/. Accessed 10 July 2018
9. Hassler, E., Carver, J.C., Hale, D., Al-Zubidy, A.: Identification of SLR tool needs - results of a community workshop. Inf. Softw. Technol. **70**(Supplement C), 122–129 (2016)
10. Kitchenham, B., Charters, S.: Guidelines for performing systematic literature reviews in software engineering. Technical Report EBSE 2007–001, Keele University and Durham University Joint Report (2007)
11. Kitchenham, B.A., Budgen, D., Brereton, P.: Evidence-Based Software Engineering and Systematic Reviews, vol. 4. CRC Press, Boca Raton (2015)
12. Kohl, C., et al.: Online tools supporting the conduct and reporting of systematic reviews and systematic maps: a case study on cadima and review of existing tools. Environ. Evid. **7**(1), 8 (2018)
13. Marshall, C., Brereton, P.: Tools to support systematic literature reviews in software engineering: a mapping study. In: 2013 ACM/IEEE International Symposium on Empirical Software Engineering and Measurement, pp. 296–299, October 2013
14. Marshall, C., Brereton, P., Kitchenham, B.: Tools to support systematic reviews in software engineering: a feature analysis. In: Proceedings of the 18th International Conference on Evaluation and Assessment in Software Engineering, EASE 2014, pp. 13:1–13:10. ACM, New York (2014)
15. Marshall, C., Brereton, P., Kitchenham, B.: Tools to support systematic reviews in software engineering: a cross-domain survey using semi-structured interviews. In: Proceedings of the 19th International Conference on Evaluation and Assessment in Software Engineering, EASE 2015, pp. 26:1–26:6. ACM, New York (2015)
16. Moed, H.F., Glänzel, W., Schmoch, U.: Handbook of Quantitative Science and Technology Research. Springer, Dordrecht (2015)
17. Nielsen, J.: Ten usability heuristics (2005)
18. Petersen, K., Vakkalanka, S., Kuzniarz, L.: Guidelines for conducting systematic mapping studies in software engineering: an update. Inf. Softw. Technol. **64**, 1–18 (2015)
19. Barn, B.S., Raimondi, F., Athappian, L., Clark, T.: SLRtool: a tool to support collaborative systematic literature reviews. In: Proceedings of the 16th International Conference on Enterprise Information Systems - Volume 2, ICEIS 2014, pp. 440–447. SCITEPRESS - Science and Technology Publications, LDA, Portugal (2014)
20. Zhang, H., Babar, M.A.: Systematic reviews in software engineering: an empirical investigation. Inf. Softw. Technol. **55**(7), 1341–1354 (2013)

Bibliometric Network Analysis to Identify the Intellectual Structure and Evolution of the Big Data Research Field

J. R. López-Robles[1(✉)] , J. R. Otegi-Olaso[1] , I. Porto Gomez[2] ,
N. K. Gamboa-Rosales[3] , H. Gamboa-Rosales[3] ,
and H. Robles-Berumen[3]

[1] Department of Graphic Design and Engineering Projects,
University of the Basque Country (UPV/EHU), Alameda Urquijo,
S/N, 48013 Bilbao, Spain
jrlopez005@ikasle.ehu.eus, joserra.otegi@ehu.eus
[2] Deusto Business School, Deusto University, Bilbao, Spain
[3] Academic Unit of Electric Engineering, Autonomous University of Zacatecas,
Zacatecas, Mexico

Abstract. Big Data has evolved from being an emerging topic to a growing research area in business, science and education fields. The Big Data concept has a multidimensional approach, and it can be defined as a term describing the storage and analysis of large and complex data sets using a series of advanced techniques. In this respect, the researches and professionals involved in this area of knowledge are seeking to develop a culture based on data science, analytics and intelligence. To this end, it is clear that there is a need to identify and examine the intellectual structure, current research lines and main trends. In this way, this paper reviews the literature on Big Data evaluating 23,378 articles from 2012 to 2017 and offers a holistic approach of the research area by using SciMAT as a bibliometric and network analysis software. Furthermore, it evaluates the top contributing authors, countries and research themes that are directly related to Big Data. Finally, a science map is developed to understand the evolution of the intellectual structure and the main research themes related to Big Data.

Keywords: Big data · Bibliometric network analysis
Information management · Strategic intelligence · SciMAT

1 Introduction

In today's data age, organizations are under pressure to face challenges such as capturing, curation, storage, analysis, search, sharing, visualization, querying, privacy and sources of data and information. In this way, organizations are integrating a Strategic Intelligence approach into their core processes, with special attention to the concept of Big Data [1, 2].

There are many definitions of Big Data, but most of them coincide that the Big Data concept has three main characteristics: volume, velocity and variety. Nevertheless, for

© Springer Nature Switzerland AG 2018
H. Yin et al. (Eds.): IDEAL 2018, LNCS 11315, pp. 113–120, 2018.
https://doi.org/10.1007/978-3-030-03496-2_13

many authors, the technologies associated with these characteristics are another aspect of the Big Data concept to be taken into account. In view of both approaches, Big Data can be defined as a term describing the storage and analysis of large and complex data sets using a series of advanced techniques. In any case, it is clear that organizations are facing the challenge of doing something with the new data that appears every day [3, 4].

To really understand Big Data, it is helpful to have a holistic background. To do that, a bibliometric network analysis is a suitable framework to conduct an objective, integrative, and comparative analysis of the main themes related to Big Data and evaluate its evolution. In addition, it makes it possible to include prospective support for future decisions and identification of research gaps [5–8].

Considering the above, the main aim of this paper is to identify the intellectual structure and evolution of the Big Data research field using SciMAT [9]. To do that, the main indicators associated to bibliometric performance such as published publications, received citations, data on geographic distribution of publications have been measured. Finally, a conceptual thematic analysis is carried out.

2   Methodology and Dataset

The bibliometric methodology implemented in this publication combines both the performance analysis and science mapping approaches of Big Data. The performance analysis is focused on the citation-based impact of the scientific output, while the science mapping represents the evolution of the themes that built the filed using SciMAT [10–12].

In order to successfully develop the bibliometric performance and science mapping analysis, the publications have been collected from Web of Science Core Collection (WoS) using the following advance query: TS = ("big data" OR "big-data") on July 4, 2018.

This advance query retrieved a total of 28,494 publications from 1993 to 2018, of which 25,658 are for the period 2012–2017. To evaluate the evolution of the Big Data research field the entire time period was divided into three comparable subperiods: 2012–2013, 2014–2015 and 2016–2017 (Table 1).

Table 1. Distribution of publications and its citations by subperiod (2012 to 2017).

Subperiod	Publications	Sum of times cited (without self citations)	Citing articles (without self citations)
2012–2013	1,643	15,289 (15,130)	11,817 (11,709)
2014–2015	8,552	48,130 (45,957)	34,742 (33,488)
2016–2017	15,463	26,745 (24,497)	21,954 (20,812)

3 Performance Bibliometric Analysis of the Big Data

The bibliometric performance analysis is structured in two sections. The first section evaluates the publications and their citations with the aim of testing and evaluating the scientific growth and the second one analysis the authors, publications and research areas to assess the impact of the these [13].

The distribution of publications and citations related to Big Data per year are shown in Fig. 1.

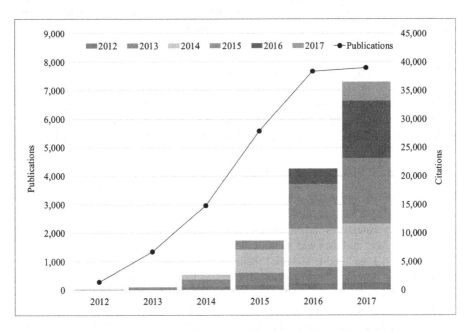

Fig. 1. The solid line corresponds to the number of publications (Left) and the bars represents the number of citations per each year (Right) published in individual years from 2012 to 2017.

Based on the results shown above, the publications have been increasing year by year and this evolution reveals the growing interest in the Big Data concept and its main components.

In line with the foregoing, the most productive authors and the most cited authors are shown in Table 2.

Table 2. Most productive and cited authors from 2012 to 2017.

Publications	Author(s)	Cites	Author(s)
119	Li, Y.	863	Chen, H. C.
117	Zhang, Y.	842	Chaing, R.H.L.; Storey, V.C.
101	Liu, Y.	559	Wu, X. D.
98	Wang, Y.	558	Wu, G. Q.
97	Wang, J.	320	Murdoch, T. B.

It is important to highlight that all most productive authors are not among the most cited authors during the evaluated period.

On the other hand, the most productive countries related to the Big Data were United States of America (7,504 publications), China (6,646 publications), England (1,631 publications), India (1,558 publications) and Germany (1,157 publications).

Finally, Computer Sciences (13,986 publications), Information Sciences (7,617 publications), Engineering (2,154 publications), Telecommunications (2,154 publications), Business Economics (1,372 publications) and Automation Control Systems (806 publications) were the main subject areas identified.

Up to that point, it is clear that the Big Data research field is experiencing rapid growth and it is interesting for the academic, scientific and business community.

4   Science Mapping Analysis of Big Data

As a further step in the analysis of the Big Data field, an overview of the science mapping and the relations between core themes is carried out. To do that, the analysis of the content of the publications and the conceptual evolution map are developed.

4.1   Conceptual Structure of Big Data Research Field

In order to identify and analyze the most highlighted themes of the Big Data field for each subperiod, we set out these in several strategic diagrams (see Fig. 2) that are divided in four categories according to their relevance: Motor themes (upper-right), Highly developed and isolated themes (upper-left), Emerging or declining themes (lower-left) and Basic and transversal themes (lower-right) [10]. The research themes within strategic diagrams are represented as spheres and its size is proportional to the number of published documents associated with each research themes. In addition, it includes the number of citations achieved by each theme in parenthesis.

Consistent with the earlier points, the first subperiod recorded ten research themes (Fig. 2(a)), the second subperiod twelve (Fig. 2(b)) and the third subperiod fourteen research themes (Fig. 2(c)) related to the Big Data.

In this way, the *Motor themes* and *Basic and transversal themes* are considered key to structure the field of research. These are presented in Table 3.

Table 3. Motor themes and basic and transversal themes (key themes) per subperiod

Subperiod	Themes
2012–2013	DATA-MINING, MAPREDUCE, NOSQL, PRIVACY, RECOMMENDER-SYSTEMS
2014–2015	BIG-DATA-ANALYTICS, DATABASE, FEATURE-SELECTION, MAPREDUCE, NEURAL-NETWORKS, PRIVACY, SOCIAL-NETWORKS
2016–2017	COLLABORATIVE-FILTERING, DATA MINING, DATA-WAREHOUSE, DECISION-MAKING, MAPREDUCE, NEURAL-NETWORKS, PRIVACY, SOCIAL-MEDIA

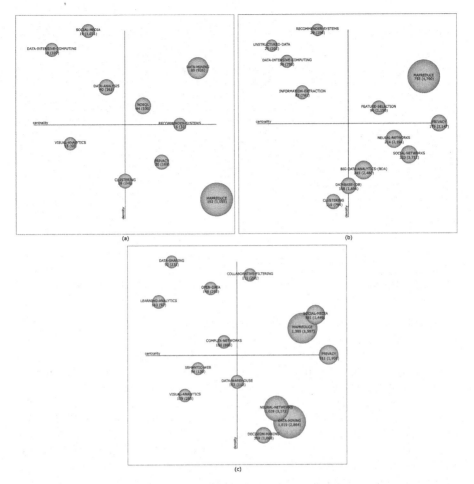

Fig. 2. Strategic diagrams. (a) Subperiod 2012–2013. (b) Subperiod 2014–2015. (c) Subperiod 2016–2017.

On the basis of the above, the next step is to visualize and analyze the evolution of the research themes and their relationship from 2013 to 2017.

4.2   Conceptual Evolution Map

In view of the foregoing, the Fig. 3 exposes the pattern of development within the Big Data field throughout the period analyzed and the relationship between each research theme. Moreover, the characteristics of the line define the quality of the relation (i.e. the solid line indicates the type of link that exists between two themes and the thickness of it is proportional to the inclusion index).

Furthermore, in the Big Data evolution map, four thematic areas can be identified: *Data Management, Decision Support, Privacy* and *WEB* and *Social Networks*.

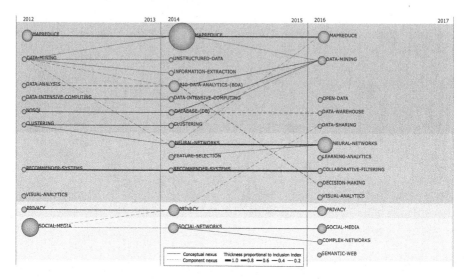

Fig. 3. Conceptual and thematic evolution of the Big Data research field from 2012 to 2017.

In view of the above, the structure of these thematic areas and the key performance indicators are shown below.

Data Management is the most representative thematic area within the thematic evolution map. It accounts for 2,932 documents and 8,028 citations. In terms of structure and thematic composition, it mainly integrates a *Motor themes* and *Basic and transversal* themes in all periods. Although, it also has a presence in the rest of the quadrants. These themes cover research lines such as DATA-ARCHITECTURES, DATA-INTEGRATION, INFORMATION-EXTRACTION, DISTRIBUTED-COMP UTING, DATA-ANALYSIS, HADOOP, SPARK, SQL, NOSQL, BUSINESS-INTELLIGENCE, among others.

Decision Support is the second representative thematic area with 2,161 documents and 2,013 citations. In terms of structure and thematic composition, it is composed of all themes, but mainly by *Basic and transversal themes*. These themes cover lines such as HUMAN-INFORMATION-INTERACTION, INFORMATION-SYSTEMS, PREDICTIVE-ANALYTICS, INFORMATION-AND-DATA VISUALIZATION, PERSONALIZATION, EXPLORATORY-DATA-ANALYSIS, mainly.

WEB and Social Networks is the third thematic area within the map. It accounts for 1,084 documents and 1,653 citations. In terms of structure and thematic composition, it remains all the quadrants. These themes cover research lines such as SENTIMENT-ANALYSIS, NATURAL-LANGUAGE-PROCESSING, COMPLEX-NETWORKS, SOCIAL MEDIA, LINKED DATA, among others.

Privacy is the last representative thematic area within the thematic evolution map in terms of number of documents. It has 852 documents and 1,030 citations. This thematic area remains as the *Basic and transversal themes* in all periods evaluated. These themes cover research lines such as DATA-PROTECTION, DATA-PRIVACY, DATA-ANONYMIZATION, ETHICS, CYBERSECURITY, predominantly.

5   Conclusions

The literature related to Big Data shows a noticeable increase from 1993 to date. Bearing in mind the large volume of publications and citations received in this field, it is expected that the Big Data concept and its main elements will grow up in the coming years.

The presented research has provided a complete view of the intellectual structure of the Big Data research field, giving the researches and professionals the ability to uncover the different themes researched by Big Data community from 2012 to 2017.

Bearing in mind the results of the strategic diagrams and the evolution map, four thematic areas were identified: *Data Management, Decision Support, Privacy* and *WEB and Social Networks*. Included in these thematic areas are the following topics that can be highlighted as relevant for the development of the knowledge area: MAPREDUCE, DATA-MINING, PRIVACY, NEURAL-NETWORKS and SOCIAL-MEDIA.

Finally, it is important to highlight that the Big Data research field is developing at a fast pace. Some of the recent scientific documents are relevant to the state of the art in terms of content and number of citations. In addition, the bibliometric network analysis of these documents is interesting to identify of common themes that can be used to reach other knowledge areas.

Acknowledgements. The authors J. R. López-Robles, N. K. Gamboa-Rosales, H. Gamboa-Rosales and H. Robles-Berumen acknowledge the support by the CONACYT-Consejo Nacional de Ciencia y Tecnología (Mexico) and DGRI-Dirección General de Relaciones Exteriores (México) to carry out this study.

References

1. Liebowitz, J.: Strategic Intelligence: Business Intelligence, Competitive Intelligence, and Knowledge Management. Auerbach Publications, Boca Raton (2006)
2. Manyika, J., et al.: Big Data: The Next Frontier for Innovation, Competition, and Productivity. McKinsey & Company, New York (2011)
3. Gandomi, A., Haider, M.: Beyond the hype: big data concepts, methods, and analytics. Int. J. Inf. Manag. **35**, 137–144 (2015)
4. Sagiroglu, S., Sinanc, D.: Big data: a review. In: 2013 International Conference on Collaboration Technologies and Systems (CTS), pp. 42–47. IEEE (2013)
5. Glänzel, W.: Bibliometric methods for detecting and analysing emerging research topics. Prof. Inf. **21**, 194–201 (2012)
6. Glänzel, W.: The role of core documents in bibliometric network analysis and their relation with h-type indices. Scientometrics **93**, 113–123 (2012)
7. Sivarajah, U., Kamal, M.M., Irani, Z., Weerakkody, V.: Critical analysis of big data challenges and analytical methods. J. Bus. Res. **70**, 263–286 (2017)
8. Van Raan, A.F.: The use of bibliometric analysis in research performance assessment and monitoring of interdisciplinary scientific developments. Technol. Assess.-Theory Practice **1**, 20–29 (2003)
9. Cobo, M.J., Lopez-Herrera, A.G., Herrera-Viedma, E., Herrera, F.: SciMAT: a new science mapping analysis software tool. J. Am. Soc. Inf. Sci. Technol. **63**, 1609–1630 (2012)

10. Cobo, M.J., Lopez-Herrera, A.G., Herrera-Viedma, E., Herrera, F.: An approach for detecting, quantifying, and visualizing the evolution of a research field: a practical application to the fuzzy sets theory field. J. Informetr. **5**, 146–166 (2011)
11. Zupic, I., Čater, T.: Bibliometric methods in management and organization. Organ. Res. Methods **18**, 429–472 (2015)
12. Cobo, M.J., Lopez-Herrera, A.G., Herrera-Viedma, E., Herrera, F.: Science mapping software tools: review, analysis, and cooperative study among tools. J. Am. Soc. Inf. Sci. Technol. **62**, 1382–1402 (2011)
13. Gutiérrez-Salcedo, M., Martínez, M.Á., Moral-Munoz, J., Herrera-Viedma, E., Cobo, M.J.: Some bibliometric procedures for analyzing and evaluating research fields. Appl. Intell. **48**, 1275–1287 (2018)

A New Approach for Implicit Citation Extraction

Chaker Jebari[1,2(✉)], Manuel Jesús Cobo[3], and Enrique Herrera-Viedma[4]

[1] Computer Science Department, Tunis El Manar University, Tunis, Tunisia
[2] Information Technology Department, Colleges of Applied Sciences, Ibri, Oman
jebarichaker@yahoo.fr
[3] Department of Computer Science and Engineering, University of Cádiz,
Cádiz, Spain
manueljesus.cobo@uca.es
[4] Department of Computer Science and Artificial Intelligence, University of Granada,
Granada, Spain
viedma@decsai.ugr.es

Abstract. The extraction of implicit citations becomes more important since it is a fundamental step in many other applications such as paper summarization, citation sentiment analysis, citation classification, etc. This paper describes the limitations of previous works in citation extraction and then proposes a new approach which is based on topic modeling and word embedding. As a first step, our approach uses LDA technique to identify the topics discussed in the cited paper. Following the same idea of Doc2Vec technique, our approach proposes two models. The first one called Sentence2Vec and it is used to represent all sentences following an explicit citation. This sentences are candidates to be implicit citation sentences. The second model called Topic2Vec, used to represent the topics covered in the cited paper. Based on the similarity between Sentence2Vec and Topic2Vec representations we can label a candidate sentence as implicit or not.

Keywords: Implicit citation extraction · Topic modeling · Doc2Vec
Sentence2Vec · Topic2Vec

1 Introduction

In the last few years, citation context analysis has gained more attention by many researchers in bibliometrics field [8,9]. It has been considered as a fundamental step in many other applications such as citation summarization, citation sentiment analysis, survey generation, citation recommendation and author co-citation analysis [6,20]. Despite the existence of many works to extract citation context, it is still a research question to be solved as it has limitations. All previous works on citation context extraction either explored explicit citation sentences only [1] or used a fixed-size text window to recognize citation context [13] which leads to a lot of noises [4]. It is noted that the sense of a word depends

© Springer Nature Switzerland AG 2018
H. Yin et al. (Eds.): IDEAL 2018, LNCS 11315, pp. 121–129, 2018.
https://doi.org/10.1007/978-3-030-03496-2_14

on its domain. For example the word *chips* refer to *potato chips* if the paper is about nutrition and refers to *electronic chips* if the paper is about electronics. In order to map each word to its correct sense it is necessary to identify the topics covered by a given paper. For this reason, it is very important to model the topics involved in the cited papers. In our approach we propose to use Latent Dirichlet Allocation (LDA) technique [7] to generate the latent topics from papers before starting the extraction of implicit citations.

In all previous studies, citation context extraction have been formalized as a supervised classification problem that produces the citation context using a collection of annotated papers. Until now, all used corpora are manually annotated and no one can label citation context automatically [9]. To deal with this issue, we propose to use an unsupervised learning technique that do not needs an annotated corpus. Our approach uses all sentences appearing next to an explicit citation sentence. This sentences are candidates to be implicit citation sentences. Based on the similarity between each candidate sentence and the cited paper, we can decide whether to consider it as an implicit citation or not a citation. Following the same idea of Doc2Vec model [15], our approach proposes two new word embedding models to represent the candidate sentences and the cited paper as well. The first one, called Sentence2Vec is used to represent all sentences following an explicit citation, while the second model, called Topic2Vec is used to represent the topics covered in the cited paper and generated using LDA technique. One major drawback in citation context analysis is the lack of standard benchmark corpus, therefore, researchers cannot compare their results. So far, ACL Antology Reference Corpus (ACL ARC)[1] [16] is the most common source of data used in citation extraction. ACL ARC includes 22878 articles belonging to only computational linguistics field. To generalize the performance of citation context extraction, a large multidisciplinary corpus is needed.

This paper is organized as follows. Section 1 explains citation extraction process. Section 2 describes the proposed approach. Section 3 presents a real example about implicit citation extraction. Section 4 concludes the paper and presents the future work.

2   Citation Context Extraction

2.1   Definition

A citation context can be defined as a block of text within a citing paper that mentions a cited paper. A citation context is defined by as a block of text composed of one or more consecutive sentences surrounding references in the body of a publication [17]. A citation sentence can be classified as explicit or implicit. Explicit citation is a citation sentence that contains one or more citation references [5]. An implicit citation sentence appear next to the explicit citation sentence and do not attach any citation reference but supply additional information of the content of the cited paper [14]. The following examples illustrates

---

[1] https://acl-arc.comp.nus.edu.sg/.

the two types of citations, where the sentence in bold is an explicit citation and the italic sentence is an implicit citation.

...In order to improve sentence-level evaluation performance, several metrics have been proposed, including $ROUGE - W$ **and** $METEOR$ **[4].** *METEOR is essentially a unigram based metric, which prefers the monotonic word alignment between MT output and the references by penalizing crossing word alignments....*

2.2 Related Works

In the last few years, many studies have been proposed to extract the citation context. These studies differ with respect to the following three main factors: (i) the features used (ii) the machine learning technique and (iii) the corpus used in the experimentation. This section presents in chronological order the different works developed in citation extraction field.

Kaplan et al. [11] used coreference chains to identify citation context. They first trained an SVM coreference resolver, then applied it to the sentences surrounding an explicit citation. The sentences that contain an antecedent to be implicit citation sentences are deemed. For experimentation, they manually created a corpus of 38 research papers taken from the field of computational linguistics. Their experiments show a big difference between Micro and Macro averaged F1-score and therefore their approach is not stable.

Radev et al. [16] employed a window of four sentences to summarize citations. They used a belief propagation mechanism to identify context sentences before classifying them using SVM technique. By combining explicit and implicit citations, they achieved encouraging results using ACL ARC corpus.

Sugiyama et al. [19] classified citations into two types: citation and not-citation using Maximum Entropy (ME) and SVM classifiers. In their study, they employed n-grams, next and previous sentence, proper nouns, orthographic characteristics and word position as classification features. Using ACL ARC corpus, they reported an accuracy of 0.882.

Athar and Teufel [5] noted that considering implicit and explicit citation sentences can effectively detect the author sentiment rather than using only one explicit citation sentence. They stated that implicit citations can cover many sentiment words. To do this, they used many features such as n-grams, number of negation phrases, number of adjectives, grammatical relationships between words, negation words, etc. Using a manually annotated corpus composed of 20 papers selected from ACL ARC corpus and SVM classifier, they reported better results compared to using only explicit citation sentences.

Jochim and Schutze [10] stated that detecting the optimal number of sentences to be considered in citation context extraction affects citation sentiment analysis. To do this, they used many features including word-level features, n-grams, sentence location, comparatives, lexicon, etc. To evaluate their approach, they used Stanford Maximum Entropy classifier (SME) and they build their own corpus that comprises 84 scientific papers.

To identify the author sentiment polarity of a citation, Abu-Jbara et al. [2] used only four citation sentences surrounding an explicit citation. These sentences are annotated as positive or negative citation. Afterwards, they applied a regular expression to clean the citation context. They used many features such as citation count, POS tags, self-citation, negation, dependency relations, etc. Using SVM classifier, they achieved an accuracy of 0.814 and macro-F of 0.713.

Sondhi and Zhai [18] used a constrained Hidden Markov Model (HMM) approach that independently trains a separate HMM for each citation and then performs a constrained joint inference to label non-explicit citing sentences. Using a subset of 10 articles selected from ACL ARC corpus, they achieved better results in comparison with other existing works.

Kim et al. [12] extracted citation sentences by using word statistics, author names, publication years, and citation tags in a sentence. For experimentations, they collected 5848 biomedical papers. By applying SVM with a rule-based approach as a post-processing step, they achieved an F1-score of 0.970.

3   Proposed Approach

Our approach aims to extract implicit citation sentences. It consists of two main steps as shown in Fig. 1. The first step generates a list of latent topics in the whole corpus, where each topic is represented by a list of words. This list of topics will help to deal with multi-sense words. In the second step, our approach deals with the extraction implicit citations as a classification problem. In this step, our approach proposes two word embedding techniques named Sentence2Vec and Topic2Vec to represent the citation sentence and the topics covered in the cited paper.

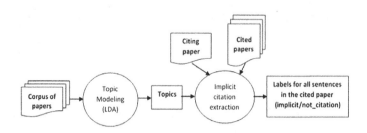

Fig. 1. The overall architecture of the proposed approach.

3.1   Step1: Topic Modeling

Topic Modeling provides an unsupervised way to analyze big unclassified corpus of documents. A topic contains a cluster of words that frequently occurs together. A topic modeling can connect words with similar meanings and distinguish between uses of words with multiple meanings. Many topic modeling methods have been proposed in the literature such as Latent Semantic Analysis (LSA), Probabilistic Latent Semantic Analysis (PLSA), Latent Dirichlet Allocation (LDA), and Correlated Topic Model (CTM) [3]. All topic modeling methods are built on the distributional hypothesis, suggesting that similar words occur in similar contexts. To this end, they assume a generative process (a sequence of steps), which is a set of assumptions that describe how the documents are generated. Given the assumptions of the generative process, inference is done, which results in learning the latent variables of the model. In our approach, we use LDA technique [7]. It takes a corpus of papers as an input and it models each paper as a mixture over K latent topics, each of which describes a multinomial distribution over a word vocabulary W. For example, the sports topic has word *football, hockey* with high probability and the computer topic has word *data, network* with high probability. Then, a collection of papers has probability distribution over topics, where each word is regarded as drawn from one of those topics. With this probability distribution over each topic, we will know how much each topic is involved in a paper, meaning which topics a paper is mainly talking about. Figure 2 shows a graphical representation of LDA technique.

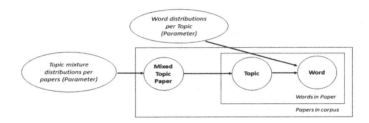

Fig. 2. Graphical representation of LDA.

For each paper p in a corpus D, the generative process for LDA is depicted in the following algorithm. Here, Gibbs sampling estimation[2] is used to obtain paper-topic and topic-word probability matrices noted Θ and Φ respectively.

Algorithm 1. LDA generative process

1: **for each** the N word w_n in paper p **do**
2:     Sample a topic $t_n \sim Multinomial(\theta_p)$
3:     Sample a word $w_n \sim Multinomial(\phi_{z_n})$
4: **end for**

---
[2] http://gibbslda.sourceforge.net/.

3.2 Step2: Implicit Citation Extraction

The extraction of implicit citation sentences can be formalized as a classification problem where each sentence of a citing paper can be classified, in relation to a given cited paper, as one of the following categories: (1) explicit citation (2) implicit citation and (3) not a citation. Explicit citation sentences are tagged with citation references, and therefore they are very easy to extract. In this paper we only consider the other two categories as an output for the problem. Our approach considers all sentences appearing after an explicit citation sentence as candidate sentences. By calculating the similarity between a candidate sentence and a cited paper we can consider the more similar sentences to be implicit citation sentences. To represent each candidate sentence and the cited paper, we propose two new word embedding models named Sentence2Vec and Topic2Vec respectively. Both of them are based one the same idea of Doc2Vec model. Algorithm 2 describes the different steps to extract implicit citations.

Algorithm 2. Implicit citation extraction algorithm

Input:

- p_i: A paper
- $Y_i = \{p_1, ..., p_m\}$: list of their cited papers
- $CitingTopics_i$: topics of the citing paper
- $CitedTopics_i$: topics of the cited papers

Output: labels for all sentences in the citing paper p_i (implicit or not a citation)
Processing:
1: Extract the list of references $refp_i$ from the citing paper p_i
2: Split the citing paper into sentences
3: **for each** reference $ref \in refp_i$ **do**
4:    Extract the topics covered by ref
5:    Topic2Vec: representation of the covered topics using Doc2Vec model
6:    Extract candidate sentences $Candidate_i$ appearing after ref
7:    **for each** reference $s \in Candidate_i$ **do**
8:       Sen2Vec: representation of each candidate sentence using Doc2Vec model
9:       Compute the similarity between Topic2Vec and Sen2Vec representations
10:    **end for**
11:    Label the most similar candidate sentences as implicit and the rest as not a citation
12: **end for**

---

4 Real Example About Implicit Citation Extraction

In this section we present a real example to show how our approach works. The cited and citing papers are shown in Fig. 3. In the citing paper, the explicit citation is marked in bold face and the sentences coming after are candidates to

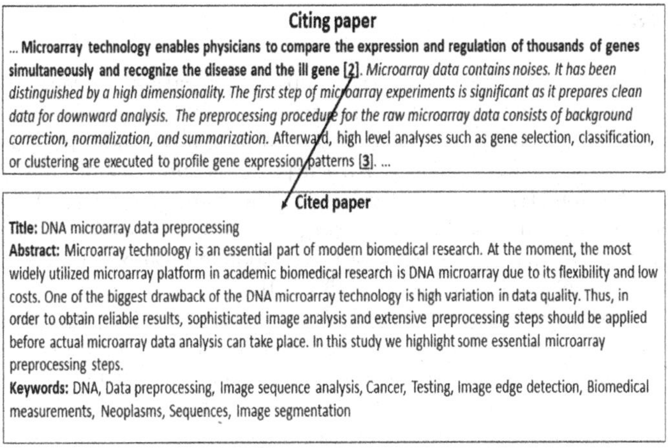

Citing paper
... **Microarray technology enables physicians to compare the expression and regulation of thousands of genes simultaneously and recognize the disease and the ill gene [2]**. *Microarray data contains noises. It has been distinguished by a high dimensionality. The first step of microarray experiments is significant as it prepares clean data for downward analysis. The preprocessing procedure for the raw microarray data consists of background correction, normalization, and summarization. Afterward, high level analyses such as gene selection, classification, or clustering are executed to profile gene expression patterns [3].* ...

Cited paper
Title: DNA microarray data preprocessing
Abstract: Microarray technology is an essential part of modern biomedical research. At the moment, the most widely utilized microarray platform in academic biomedical research is DNA microarray due to its flexibility and low costs. One of the biggest drawback of the DNA microarray technology is high variation in data quality. Thus, in order to obtain reliable results, sophisticated image analysis and extensive preprocessing steps should be applied before actual microarray data analysis can take place. In this study we highlight some essential microarray preprocessing steps.
Keywords: DNA, Data preprocessing, Image sequence analysis, Cancer, Testing, Image edge detection, Biomedical measurements, Neoplasms, Sequences, Image segmentation

Fig. 3. Example of implicit citation extraction.

be implicit citations. For the cited paper, the topics covered can be extracted from the title, the abstract and the keywords. It is clear that the most covered topic in this paper is *microarray*. From this example, we can observe that the sentences marked in italic are more similar to the topic *microarray*, and therefore they are labeled as implicit citations. The last sentence can be marked as an explicit citation to be treated in the same way.

5   Conclusion and Future Work

This paper proposed a new approach to extract implicit citations between citing and cited papers. As a first step, our approach generates latent topics covered in the cited paper using LDA technique. In our approach, the sentences appearing after an explicit citation are considered candidates to be implicit citations. To represent the candidate sentences and the cited paper, we propose two word embedding models named Topic2Vec and Sentence2Vec. Based on the similarity between the sentence and topics vectors, our approach labels the most similar sentences as implicit citations, while the rest of sentences will be labelled as not citations. In contrast to previous studies, our approach suggests an unsupervised technique that does not require annotated training corpus.

As a future work, we will implement and evaluate the performance of our approach using a large and multidisciplinary corpus. More precisely, we will show the importance of word embedding in representing citation sentences. Moreover, we will show how topic modeling can handle multi-sense words in citation sentences and hence it can improve the performance of citation extraction.

References

1. Abu-Jbara, A., Radev, D.R.: Reference scope identification in citing sentences. In: Conference of the North American Chapter of the Association for Computational Linguistics: Human Language Technologies, Montréal, Canada, pp. 80–90 (2012)
2. Abu-Jbara, A., Ezra, J., Radev, D.R.: Purpose and polarity of citation: towards NLP-based bibliometrics. In: Proceedings of the North American Association for Computational Linguistics, Atlanta, Georga, USA, pp. 596–606 (2013)
3. Alghamdi, R., Alfalqi, K.: A survey of topic modeling in text mining. Int. J. Adv. Comput. Sci. Appl. (IJACSA) **6**, 147–153 (2015)
4. Athar, A.: Sentiment analysis of citations using sentence structure-based features. In: Proceedings of the ACL 2011 Student Session, pp. 81–87 (2011)
5. Athar, A., Teufel, S.: Context-enhanced citation sentiment detection. In: Proceedings of the Conference of the North American Chapter of the Association for Computational Linguistics: Human Language Technologies, Montreal, Canada, pp. 587–601 (2012)
6. Bu, Y., Wang, B., Huang, W.B., Che, S., Huang, Y.: Using the appearance of citations in full text on author co-citation analysis. Scientometrics **116**(1), 275–289 (2018)
7. David, M.B., Andrew, Y.N., Michael, I.J.: Latent dirichlet allocation. J. Mach. Learn. Res. **3**, 993–1022 (2003)
8. Fortunato, S. et al.: Science of science. Science, **359**(1007) (2018)
9. Hernandez-Alvarez, M., Gomez, J.M.: Survey about citation context analysis: tasks, techniques, and resources. Nat. Lang. Eng. **22**(3), 327–349 (2015)
10. Jochim, C., Schutze, H.: Improving citation polarity classification with product reviews. In: Proceedings of 52nd Annual Meeting of the Association for Computational Linguistics, pp. 42–48. ACL, Baltimore (2014)
11. Kaplan, D., Iida, R., Tokunaga, T.: Automatic extraction of citation contexts for research paper summarization: a coreference-chain based approach. In: Proceedings of the 2009 Workshop on Text and Citation Analysis for Scholarly Digital Libraries, Singapore, pp. 88–95 (2009)
12. Kim, I.C., Le, D.X., Thoma, G.R.: Automated method for extracting citation sentences from online biomedical articles using SVM-based text summarization technique. In: Paper Proceedings of the IEEE International Conference on Systems, Man, and Cybernetics (SMC), San Diego, CA, USA, pp. 1991–1996 (2014)
13. O'Connor, J.: Citing statements: computer recognition and use to improve retrieval. Inf. Process. Manag. **18**(3), 125–131 (1982)
14. Qazvinian, V., Radev, D.R.: Identifying non-explicit citing sentences for citation-based summarization. In: Proceedings of 48th Annual Meeting of the Association for Computational Linguistics, Uppsala, Sweden, pp. 555–564 (2010)
15. Quoc., L.E., Tomas. M.: Distributed representations of sentences and documents. In: Proceedings of the 31st International Conference on International Conference on Machine Learning, Beijing, China (2014)
16. Radev, D.R., Muthukrishnan, P., Qazvinian, V.: The ACL anthology network corpus. Lang. Resour. Eval. **47**(4), 919–944 (2013)
17. Small, H.: Interpreting maps of science using citation context sentiments: a preliminary investigation. Scientometrics **87**, 373–388 (2011)
18. Sondhi, P., Zhai, C.X.: A constrained hidden Markov model approach for non-explicit citation context extraction. In: Proceedings of the 2014 SIAM International Conference on Data Mining, pp. 361–369 (2014)

19. Sugiyama, K., Kumar, T., Kan, M.Y., Tripathi. R.C.: Identifying citing sentences in research papers using supervised learning. In: Proceedings of the 2010 International Conference on Information Retrieval and Knowledge Management, Malaysia, pp. 67–72 (2010)
20. Yousif, A.: A survey on sentiment analysis of scientific citations. Artif. Intell. Rev. 1–34 (2017). https://doi.org/10.1007/s10462-017-9597-8

Constructing Bibliometric Networks
from Spanish Doctoral Theses

V. Duarte-Martínez[1]([envelope])[iD], A. G. López-Herrera[2,3][iD], and M. J. Cobo[3][iD]

[1] Facultad de Ingeniería en Electricidad y Computación, Escuela Superior Politécnica del Litoral, ESPOL, Guayaquil, Ecuador
vealduar@espol.edu.ec
[2] Department of Computer Science and Artificial Intelligence, University of Granada, Granada, Spain
lopez-herrera@decsai.ugr.es
[3] Department of Computer Science and Engineering, University of Cádiz, Cádiz, Spain
manueljesus.cobo@uca.es

Abstract. The bibliometric networks as representations of complex systems provide great information that allow discovering different aspects of the behavior and interaction between the participants of the network. In this contribution we have built a fairly large bibliometric network based on data from Spanish doctoral theses. Specifically, we have used the data of each theses defense committee to build the network with its members and we have conducted a study to discover how the nodes of this network interact, to know which are the most representative and how they are grouped within communities according to their participation in theses defense committee.

Keywords: Science mapping analysis · Bibliographic network
Co-committee members · Spanish theses · Computer science

1 Introduction

In Spain, for an applicant to obtain a PhD title, it is necessary to support their research work before a group of experts gathered in a committee to evaluate the work done. Subsequently, the information of a doctoral thesis is stored in a public access web repository that has been set up by the Secretariat of the Council of Universities (TESEO[1]). This web page contains all the Spanish theses that have been approved since 1970. Therefore, this is a great source of information that can be exploited for different purposes of bibliometric analysis.

Among the different types of bibliometric analysis, we can mention bibliometric networks, which are those that allow us to represent interactions between different entities, not only in the bibliographic field, but also its usefulness can

---

[1] https://www.educacion.gob.es/teseo/irGestionarConsulta.do.

© Springer Nature Switzerland AG 2018
H. Yin et al. (Eds.): IDEAL 2018, LNCS 11315, pp. 130–137, 2018.
https://doi.org/10.1007/978-3-030-03496-2_15

be extended to other sources of information, for example: patents [9], financing information [7], or even information related to doctoral theses, modeling this information as a complex evolutionary system [19]. These networks include: co-word networks [6,8], coauthor networks [14], co-citation networks [15] and bibliographic link networks [13], among others. The choice of network type will depend on the context of the study [2].

There is evidence of some bibliometric analysis works done using doctoral thesis information as input [4,5,10,12,16–18,20]; However, there are no studies carried out using other information units related to a doctoral thesis. This is why, in this study, the extraction of Theses defense committee members is proposed to build bibliometric networks that allow modeling the behavior of these entities in a network similar to a co-authoring network. This means that this network will be formed by experts who have participated in a committee, and two entities will be related as long as they have met in the same defense committee. Then, the most relevant members of the committee will be identified, repetitive collaboration patterns will be discovered, and hidden groups will be detected through community detection.

The rest of the paper is organized as follows: Sect. 2 shows the methodology followed to carry out the analysis. Section 3 shows the output of applying the steps established in the previous stage. Finally, in Sect. 4 some conclusions are thrown.

2   Methodology

The science mapping analysis is usually carried out according to the following work-flow [3]: data retrieval, data pre-processing, network extraction, network normalization, mapping, analysis and visualization. These stages are described in detail, as follows:

- In *Data Retrieval* stage, we have used data from TESEO database. These collected data have been stored into a relational database to ensure the preservation of raw data.

  After that, with cleaned data, we have counted how many times a person appears as a committee member. This value has been called *Frequency* and it has been established to assign a level of relevance to defense committee members.

  Next, there have been built the co-occurrences list among members. In this case, it has been defined the *Weight* property, which represents the number of times that an interaction between two members is found.
- Afterwards, the co-occurrences list gotten in the previous stage has been used to build an undirected graph. With the network built, we have used the Louvain algorithm [1] to detect communities with the aim of discovering the structure of the network. This is an efficient algorithm, especially for large networks, because of its good processing capacity and its speed of execution.

– Subsequently, the Analysis and Visualization stage was carried out. In this step, we have used visualization techniques to show the structure of the built network. As part of the analysis, the nodes have been analyzed using centrality measures [11] in order to discover the most important and influential members of the network. Hence, *Degree*, *Closeness*, *Betweenness* and *Eigenvector* centrality have been calculated with network data. Referring to the concepts of centrality, it has been established some attributes:

- *Best connected member* (Degree). This means that a member has a privileged position within the network because he/she has been on a committee with more variety of people than the others.
- *Member with the most closeness* (Closeness). This means the importance of the member is defined by the ties of his/her neighbors.
- *The most intermediary member* (Betweenness). This means that the member is more accessible by others.
- *The most influential member* (Eigenvector). It says a member is important if he/she is linked to more important members.

  Moreover, we have modified the network to show each group of detected communities with different colors. This makes it easier to understand the distribution of the nodes in the network.

– When the science mapping analysis has finished, the analyst has to interpret the results and maps using their experience and knowledge. In the interpretation step, the analyst looks to discover and extract useful knowledge that could be used to make decisions on which policies to implement.

3 Results

Following all the steps described in the methodology section we have obtained a dataset with 237.187 theses records that have been downloaded from TESEO web page. The distribution of number of theses per year is shown in Fig. 1. The amount of thesis for the year 2017 is very low due to the fact that Teseo data were downloaded until March 2017.

Using the above dataset a series of analyzes has been carried out. First, we have calculated the frequency of appearance of each defense committee member. In Table 1, the members with highest frequency are shown.

Then, using the co-occurrences list, the network has been built. The result obtained has been an undirected graph with 210.936 nodes and 1'612.012 edges as can be seen in Fig. 2. From here on, the whole process of analysis has been carried out.

At granular level, we have applied centrality measures in the network and we have discovered who are the best connected members using the measure of degree centrality. Furthermore, we have calculated closeness centrality to find members with the most closeness to other nodes in the network. After, we have used betweenness centrality and we have determined the most intermediary members. And finally, we have applied eigenvector centrality in the network in order to find the most influential members. The results are shown in Table 2.

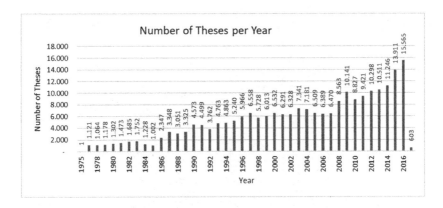

Fig. 1. Number of Theses per Year.

Table 1. Frequency of occurrence of individuals.

Theses defense committee member	Frequency
PRIETO PRIETO JOSE	391
ALVAREZ DE MON SOTO MELCHOR	324
DE LA FUENTE DEL REY MARIA MONICA	323
BALIBREA CANTERO JOSE LUIS	318
RODRÍGUEZ MONTES JOSÉ ANTONIO	306
VILARDELL TARRES MIQUEL	304
GUILLÉN MARTÍNEZ GABRIEL	302
SÁNCHEZ GUIJO PEDRO	302
GIL DE MIGUEL ANGEL	282
FERNANDEZ RODRIGUEZ TEODOSIO	252

Table 2. Centrality measures results for the network.

Measure	Member
Best connected member	DE LA FUENTE DEL REY MARIA MONICA
Member with the most closeness	ALVAREZ DE MON SOTO MELCHOR
The most intermediary member	DE LA FUENTE DEL REY MARIA MONICA
The most influential member	ALVAREZ DE MON SOTO MELCHOR

With these results, we can observe that most of the times the measures point to the same nodes, placing these as the best nodes located in the network. Its importance lies in its capacity to spread information, influence communities, serve as a communication bridge between diverse groups and other activities related to the interaction of network members.

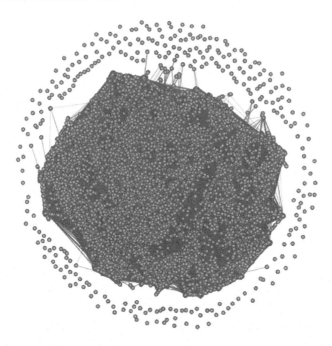

Fig. 2. Graph of theses defense committee members.

Additionally, we have applied Louvain clustering algorithm [1]. Louvain method is based on recursive tasks composed of two steps, the first one assigns each node to its own community and the second one is measuring the modularity to go grouping nodes in new communities, in this way, the process ends when the value of modularity can not be improved.

We have used the network to make visualization and understanding of results easier and we have discovered 328 communities. In Fig. 3 is shown how the communities are formed where each one is represented by an unique color.

Uncovered communities are formed in large majority by 4 or 5 members, this means that they have been courts formed by members who have not met again together in a defense committee. Only a few large communities have been detected, less than a dozen, with many more members, where we can really see the purpose of this study, the interrelation between different theses defense committees where one expert often co-exists with another to evaluate a student in his doctoral research work. Of all these communities, the largest component is composed of 35.142 nodes, which is about sixteenth part of this large global network that was initially built. In Fig. 4, can be seen the distribution of communities size detected on the main network.

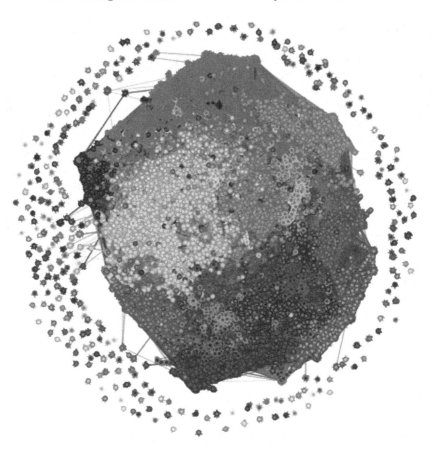

Fig. 3. Communities detected with Louvain algorithm. (Color figure online)

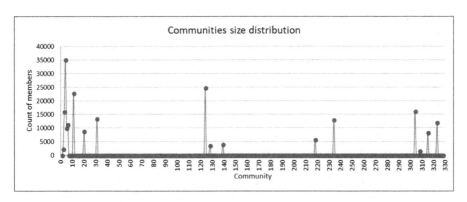

Fig. 4. Communities size distribution.

4   Conclusions

In this contribution, we have made a bibliometric study about data from Spanish doctoral theses. We have extracted data from TESEO database website with the aim of uncover some novel information. We have followed a set of steps to find communities inside the network of theses defense committee members, and finally we have uncover some interesting results and statistics.

As has been demonstrated, the use of other data sources can serve different bibliometric purposes. In this case, TESEO is a great data source that offers the possibility of making a wide variety of science mapping analysis and given that this is a free access source, it would be a waste not take advantage of these data and the amount of diverse analyzes that can be done with them. The idea is to broaden the field of possible data sources and arouse curiosity to look for more sources of information than traditionally used one (more oriented to analyze scientific papers or patents).

As future work, we propose to include in the analysis another kinds of social relations such as the relationships between theses directors and members of the defense committee and relationships between theses directors and keywords where we will surely find interesting patterns of collaboration.

Acknowledgements. The authors would like to acknowledge FEDER fund under grant TIN2016-75850-R.

References

1. Scalable Community Detection with the Louvain Algorithm, pp. 28–37. IEEE, May 2015. https://doi.org/10.1109/IPDPS.2015.59
2. Batagelj, V., Cerinšek, M.: On bibliographic networks. Scientometrics **96**(3), 845–864 (2013). https://doi.org/10.1007/s11192-012-0940-1
3. Börner, K., Chen, C., Boyack, K.W.: Visualizing knowledge domains. Annu. Rev. Inf. Sci. Technol. **37**(1), 179–255 (2003). https://doi.org/10.1002/aris.1440370106
4. Botterill, D., Haven, C., Gale, T.: A survey of doctoral theses accepted by universities in the UK and Ireland for studies related to tourism, 1990–1999. Tour. Stud. **2**(3), 283–311 (2002). https://doi.org/10.1177/14687976020023004
5. Breimer, L.H.: Age, sex and standards of current doctoral theses by Swedish medical graduates. Scientometrics **37**(1), 171–176 (1996). https://doi.org/10.1007/BF02093493
6. Callon, M., Courtial, J.P., Turner, W.A., Bauin, S.: From translations to problematic networks: an introduction to co-word analysis. Soc. Sci. Inf. **22**(2), 191–235 (1983). https://doi.org/10.1177/053901883022002003
7. Chen, X., Chen, J., Wu, D., Xie, Y., Li, J.: Mapping the research trends by co-word analysis based on keywords from funded project. Procedia Comput. Sci. **91**, 547–555 (2016). https://doi.org/10.1016/j.procs.2016.07.140
8. Coulter, N., Monarch, I., Konda, S.: Software engineering as seen through its research literature: a study in coword analysis. J. Am. Soc. Inf. Sci. **49**(13), 1206–1223 (1998). https://doi.org/10.1002/(SICI)1097-4571(1998)49:13⟨1206::AID-ASI7⟩3.3.CO;2-6

9. Courtial, J.P., Callon, M., Sigogneau, A.: The use of patent titles for identifying the topics of invention and forecasting trends. Scientometrics **26**(2), 231–242 (1993). https://doi.org/10.1007/BF02016216

10. Finlay, C.S., Sugimoto, C.R., Li, D., Russell, T.G.: LIS dissertation titles and abstracts (19302009): where have all the librar* gone? Libr. Q. **82**(1), 29–46 (2012). https://doi.org/10.1086/662945

11. Freeman, L.C.: Centrality in social networks conceptual clarification. Soc. Netw. **1**(3), 215–239 (1978). https://doi.org/10.1016/0378-8733(78)90021-7

12. Huang, S.: Tourism as the subject of China's doctoral dissertations. Ann. Tour. Res. **38**(1), 316–319 (2011). https://doi.org/10.1016/j.annals.2010.08.005

13. Kessler, M.M.: Bibliographic coupling between scientific papers. Am. Doc. **14**(1), 10–25 (1963). https://doi.org/10.1002/asi.5090140103

14. Peters, H.P.F., Raan, A.F.J.: Structuring scientific activities by co-author analysis. Scientometrics **20**(1), 235–255 (1991). https://doi.org/10.1007/BF02018157

15. Small, H.: Co-citation in the scientific literature: a new measure of the relationship between two documents. J. Am. Soc. Inf. Sci. **24**(4), 265–269 (1973). https://doi.org/10.1002/asi.4630240406

16. Sugimoto, C.R., Li, D., Russell, T.G., Finlay, S.C., Ding, Y.: The shifting sands of disciplinary development: analyzing North American library and information science dissertations using latent Dirichlet allocation. J. Am. Soc. Inf. Sci. Technol. **62**(1), 185–204 (2011). https://doi.org/10.1002/asi.21435

17. Villarroya, A., Barrios, M., Borrego, A., Frías, A.: PhD theses in Spain: a gender study covering the years 1990–2004. Scientometrics **77**(3), 469–483 (2008). https://doi.org/10.1007/s11192-007-1965-8

18. Yaman, H., Atay, E.: PhD theses in Turkish sports sciences: a study covering the years 1988–2002. Scientometrics **71**(3), 415–421 (2007). https://doi.org/10.1007/s11192-007-1679-y

19. Zeng, A., et al.: The science of science: from the perspective of complex systems. Phys. Rep. **714–715**, 1–73 (2017). https://doi.org/10.1016/j.physrep.2017.10.001

20. Zong, Q.J., Shen, H.Z., Yuan, Q.J., Hu, X.W., Hou, Z.P., Deng, S.G.: Doctoral dissertations of library and information science in China: a co-word analysis. Scientometrics **94**(2), 781–799 (2013). https://doi.org/10.1007/s11192-012-0799-1

Measuring the Impact of the International Relationships of the Andalusian Universities Using Dimensions Database

P. García-Sánchez$^{(\boxtimes)}$ (iD) and M. J. Cobo (iD)

Department of Computer Science and Engineering, University of Cádiz, Cádiz, Spain
{pablo.garciasanchez,manueljesus.cobo}@uca.es

Abstract. Researchers usually have been inclined to publish papers with close collaborators: same University, region or even country. However, thanks to the advancements in communication technologies, members of international research networks can cooperate almost seamlessly. These networks usually tend to publish works with more impact than the local counterparts. In this paper, we try to demonstrate if this assumption is also valid in the region of Andalusia (Spain). The Dimensions.ai database is used to obtain the articles where at least one author is from an Andalusian University. The publication list is divided into 4 geographical areas: local (only one affiliation), regional (only Andalusian affiliations), national (only Spanish affiliations) and International (any affiliation). Results show that the average number of citations per paper increases as the author collaboration networks increases geographically.

Keywords: Bibliometric analysis · International collaboration
Andalusian universities · Dimensions.ai

1 Introduction

Science is usually developed in teams that could be considered domestic, if the researchers are all from the same country, or international, if researchers belong to different countries. According to [1], the fourth age of research is driven by international collaborations, which could be motivated by two kinds of factors [2] related to the diffusion of scientific capacity, or related to the interconnectedness of researchers. That is, in the current global knowledge society [3], researchers tend to collaborate with colleagues from other countries in order to advance in their own fields.

Furthermore, papers developed within international teams, used to be more cited than paper developed within domestic teams [2,4]. Indeed, recently has been demonstrated that researchers to develop their career in different countries tend to be more cited that those who still in the same country in all their academic life [5].

So, the main aim of this contribution is to determine if there an increase in the number of citations when researchers from different geographical areas

© Springer Nature Switzerland AG 2018
H. Yin et al. (Eds.): IDEAL 2018, LNCS 11315, pp. 138–144, 2018.
https://doi.org/10.1007/978-3-030-03496-2_16

collaborate. To answer this question, we develop a bibliometric analysis [6–8] we have focused on the 9 Universities from the region of Andalusia (Spain). We compare the number of citations in the next scenarios: when the publications are only signed by members of the same Andalusian university, when they are only signed by members of Andalusian Universities, when they are signed by members of Andalusian Universities and other affiliations from Spain, and finally when they are signed by at least one member from an Andalusian University. The number of publications, citations and citations per publication is compared.

The Dimensions database has been used to obtain the publications and number of citations. We have chosen this database because it is freely available for academic purposes, it includes a large corpus (i.e., 89m publications and 4 billion references), and it has a powerful API that allows advanced analytic by means of their own DSL (Domain Specific Language) query language, and using a programming language such as Python.

The rest of the paper is structured as follows: in Sect. 2 the methodology to obtain the dataset of publications is explained. Then, a discussion of the results is presented in Sect. 3. Finally, conclusions and future work are presented.

2   Methodology

This section describes the steps followed to obtain the dataset of articles published by researchers from Andalusian Universities. The data is sourced from Dimensions, an inter-linked research information system provided by Digital Science (https://www.dimensions.ai). This database has been chosen not only because its large amount of data available, including the number of citations by publication, but also because it offers the possibility to use an API to perform queries using a DSL. This SQL-like petitions can be performed from any programming language and used to obtain a large batch of specific results in JSON format, thus facilitating their processing and analysis. Python language has been used to call the Dimensions API and to plot and analyse the results.

The data we are interested in are the publications signed by at least one member from one Andalusian University. As we want to compare the citation number when different regional and international collaborations appear, we divided the dataset as follows, from local to international co-authoring:

- P_{One}: Papers from Andalusia (only one affiliation). This dataset includes the papers where all the authors belong to the same Andalusian University.
- P_{And}: Papers from Andalusia (only with Andalusian Universities collaborators). This dataset includes the papers where all the authors belong to Andalusian Universities.
- P_{Spa}: Papers from Andalusia (only with Spanish Entities collaborators). This dataset includes the papers where all authors belong to a Spanish institution, and at least one is from Andalusia.
- P_{All}: All papers from Andalusia universities. All papers where at least one author is from a University from Andalusia.

Every dataset is included in the one above, therefore $P_{One} \subseteq P_{And} \subseteq P_{Spa} \subseteq P_{All}$.

The rest of parameters for the queries are:

- The selected universities to perform the analysis are the 9 public universities of Andalusia. To perform the queries, their associate Global Research Identifier Database (GRID), available from https://grid.ac/, has been used[1].
- Date range: from 2010 to 2015.
- Only publications of type "article" are used.
- The queries were performed on 25th July 2018.

To obtain all the papers from Andalusia Universities (P_{All}), the query used is:

```
search publications
 where (year in [2010:2015]
 and research_orgs.id in
  ["grid.4489.1","grid.7759.c","grid.9224.d","grid.15449.3d","grid.18803.32",
   "grid.411901.c","grid.10215.37","grid.21507.31","grid.28020.38"]
 and type="article"
return publications[id+year+times_cited+research_orgs+FOR_first+funders]
sort by id limit 1000 skip 0
```

The query to obtain the papers of P_{One} directly from dimensions, the modifier and count(research_orgs) = 1 can be used:

```
search publications
 where (year in [2010:2015]
 and research_orgs.id in
  ["grid.4489.1","grid.7759.c","grid.9224.d","grid.15449.3d","grid.18803.32",
   "grid.411901.c","grid.10215.37","grid.21507.31","grid.28020.38"]
 and count(research_orgs) = 1)
 and type="article"
return publications[id+year+times_cited+research_orgs+FOR_first+funders]
sort by id limit 1000 skip 0
```

To select all the papers from Andalucia Universities with only collaborators from Spain, the modifier used is count(research_org_countries) = 1.

```
search publications
 where (year in [2010:2015]
 and research_orgs.id in
  ["grid.4489.1","grid.7759.c","grid.9224.d","grid.15449.3d","grid.18803.32",
   "grid.411901.c","grid.10215.37","grid.21507.31","grid.28020.38"]
 and count(research_org_countries) = 1
 and type="article"
return publications[id+year+times_cited+research_orgs+FOR_first+funders]
sort by id limit 1000 skip 0
```

---

[1] The list of Universities and their and associated id are: University of Cádiz (grid.7759.c), University of Sevilla (grid.9224.d), Pablo de Olavide University (grid.15449.3d), University of Huelva (grid.18803.32), University of Córdoba (grid.411901.c), University of Málaga (grid.10215.37), University of Jaén (grid.21507.31) and University of Almería (grid.28020.38).

Dimensions does not offer the possibility to filter directly using "only from" a specific attribute (for example, only from the list of universities). That is the reason we have filtered the P_{All} dataset to obtain P_{And} iterating for every publication and removing those that have authors whose affiliation is not in the Andalusian universities list.

3   Results

Obtained results and corpus sizes are summarized in Table 1. As it can be seen, a great number of articles (39.81%) are only signed by researchers of the same affiliation (P_{One}). On the other side, Andalusian researchers tend to collaborate with researchers from affiliations outside Andalusia: 1426 publications with other Andalusian Universities, and 5536 with other Spanish affiliations (without counting Andalusian Universities, that is $P_{Spa} - P_{And}$). But it is remarkable that 13944 publications ($P_{All} - P_{Spa}$) are signed by Andalusian Universities and foreign affiliations (40.14% of the total), more than with regional and national collaborators. Results in Table also show that the average number of citations increases when including publications from other affiliations, clearly being the P_{All} dataset the one with the larger value (15.675). This can be explained because large research projects involving different countries are more ambitious than regional ones.

Table 1. Summary of citations and publications per corpus. All publications are included in the dataset below: $P_{One} \subseteq P_{And} \subseteq P_{Spa} \subseteq P_{All}$

Corpus	#Publications	#Citations	Average citations per publication
P_{One}	13,832	162,211	11.727
P_{And}	15,258	185,525	12.159
P_{Spa}	20,794	265,997	12.792
P_{All}	34,738	544,559	15.676

Plotting the citation histogram of all publications also show clear differences between the datasets. Figure 1 shows that the publication citations follow a long-tail scheme, where the majority of the publications are not cited, or below of 200 citations, while a few highest cited papers are cited up to 594 times. When increasing the geographical ambit, higher cited papers appear, for example for P_{And} and P_{Spa} (Figs. 2 and 3), the highest cited paper has 1089 citations, clearly being a paper from P_{And}. However, the average citation per paper is still greater if we take into account the Spain geographical ambit. Although another highly cited paper appears in Fig. 4 with 839 citations, the differences between both datasets are not so clear. It is when plotting all the papers with Andalusian authors and the rest of the world P_{All} (Fig. 4), where the highest cited papers appears (4231, 2919, 2309 and 1515 respectively), but also, the group at the beginning of the x-axis moves to higher amount of citations.

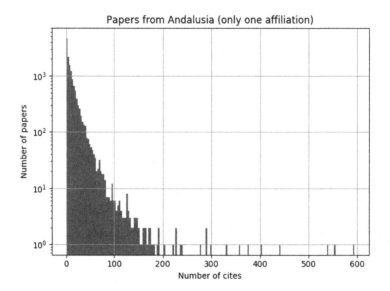

Fig. 1. Citation histogram for P_{One}. Y-axis uses a logarithmic scale.

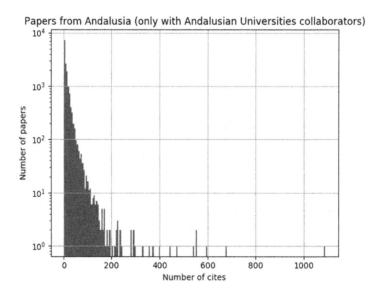

Fig. 2. Citation histogram for P_{And}. Y-axis uses a logarithmic scale.

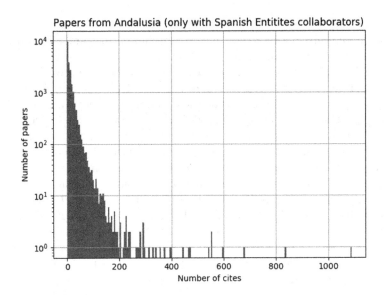

Fig. 3. Citation histogram for P_{Spa}. Y-axis uses a logarithmic scale.

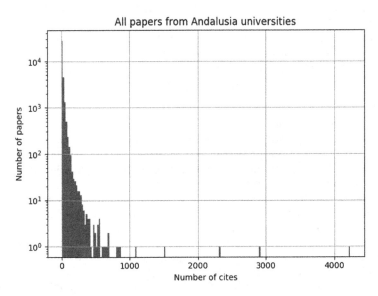

Fig. 4. Citation histogram for P_{All}. Y-axis uses a logarithmic scale.

4   Conclusions

In this paper, the number of citations of articles from Andalusian Universities is analysed taking into account the geographical collaboration network of the authors. The Dimensions.ai database has been used to obtain the articles from

Andalusian researchers, divided into 4 different geographical areas: only one affiliation, only Andalusian affiliations, only Spanish affiliations and all publications.

Results show that Andalusian publications are clearly divided into two groups: articles signed by researchers within the same affiliation (39.8%) and signed with researchers from foreign countries (40.14%). The average number of citations per paper also increases when the collaboration network geographically increases, meaning that publications with international collaborations obtain more citations than the ones with only one affiliation.

Future studies will include more complete information by separating the presented datasets into disjoint sets, or limiting by specific University. Another kind of analysis, such as collaboration graphs between countries or universities, may give more insight to determine the members of each network or the quality of the research. Furthermore, Dimensions API also allows to obtain the funding agencies of each publication, so a study to compare the impact of projects funded by different countries can be performed. Patents and clinical trials are also available in Dimensions, so a comparison of the different types of publications may also be relevant to the issue.

Acknowledgements. This contribution has been made possible thanks to Dimensions.ai database. Also, the authors would like to acknowledge FEDER funds under grants TIN2016-75850-R and TIN2017-85727-C4-2-P and Program of Promotion and Development of Research Activity of the University of Cádiz (Programa de Fomento e Impulso de la actividad Investigadora de la Universidad de Cádiz).

References

1. Adams, J.: The fourth age of research. Nature **497**(7451), 557–560 (2013)
2. Wagner, C.S., Leydesdorff, L.: Network structure, self-organization, and the growth of international collaboration in science. Res. Policy **34**(10), 1608–1618 (2005)
3. Moed, H.F., Aisati, M., Plume, A.: Studying scientific migration in Scopus. Scientometrics **94**(3), 929–942 (2013)
4. Persson, O., Glänzel, W., Danell, R.: Inflationary bibliometric values: the role of scientific collaboration and the need for relative indicators in evaluative studies. Scientometrics **60**(3), 421–432 (2004)
5. Sugimoto, C.R., Robinson-Garcia, N., Murray, D.S., Yegros-Yegros, A., Costas, R., Larivière, V.: Scientists have most impact when they're free to move. Nature **550**(7674), 29–31 (2017)
6. Cobo, M.J., López-Herrera, A.G., Herrera-Viedma, E., Herrera, F.: Science mapping software tools: review, analysis, and cooperative study among tools. J. Am. Soc. Inf. Sci. Technol. **62**(7), 1382–1402 (2011)
7. Fortunato, S., et al.: Science of science. Science **359**(6379) (2018)
8. Gutiérrez-Salcedo, M., Martínez, M.Á., Moral-Munoz, J.A., Herrera-Viedma, E., Cobo, M.J.: Some bibliometric procedures for analyzing and evaluating research fields. Appl. Intell. **48**(5), 1275–1287 (2018)

Special Session on Machine Learning for Renewable Energy Applications

Gaussian Process Kernels for Support Vector Regression in Wind Energy Prediction

Víctor de la Pompa, Alejandro Catalina$^{(\boxtimes)}$, and José R. Dorronsoro

Departmento de Ingeniería Informática and Instituto de Ingeniería del Conocimiento,
Universidad Autónoma de Madrid, Madrid, Spain
`alejandro.catalina@uam.es`

Abstract. We consider wind energy prediction by Support Vector
Regression (SVR) with generalized Gaussian Process kernels, proposing
a validation–based kernel choice which will be then used in two prediction problems instead of the standard Gaussian ones. The resulting model
beats a Gaussian SVR in one problem and ties in the other. Furthermore,
besides the flexibility this approach offers, SVR hyper–parameterization
can be also simplified.

1 Introduction

It is well known that effective Support Vector Machine (SVM) models have to be
built on positive definite kernels. As such, many different kernel choices could be
possibly made, as methods to build kernels on top of other previous kernels are
well known [2]. However, in practice this is usually not done and the Gaussian
(or RBF) kernel is used by default almost universally. There are good reasons
for this, such as the embedding it implies into a possibly infinite dimensional
feature space or the natural interpretation of the resulting models as a linear
combination of Gaussians centered at the support vectors. Another reason may
be the lack of a principled approach to the selection of a concrete kernel for the
problem at hand.

On the other hand, a totally different situation appears for Gaussian Process
Regression (GPR), often presented as a natural alternative to Support Vector Regression (SVR). Many different kernel proposals appear in GPR applications, a reason certainly being the need to work with effective enough kernels to get good models but another being the relatively simple learning of the
hyper-parameters Θ of a GPR kernel k_Θ by maximizing the marginal likelihood
$p(y|X, \Theta, \sigma) = \mathcal{N}(0, K_\Theta + \sigma^2 I)$, where X denotes the data matrix, $y = f(x) + n$
is the target vector, n denotes 0–mean Gaussian noise with variance σ^2 and K_Θ
is the sample covariance matrix associated to an underlying kernel k_Θ.

Usually different roles are assigned to the kernel in GPR and SVR. In the
latter the kernel contains the dot products of the non-linear extension of the
patterns while in GPR the kernel contains the covariance of the basis expansion

© Springer Nature Switzerland AG 2018
H. Yin et al. (Eds.): IDEAL 2018, LNCS 11315, pp. 147–154, 2018.
https://doi.org/10.1007/978-3-030-03496-2_17

implicit in the GPR model. Since the basis expansion is guaranteed by the positive definiteness of the kernel function, it is clear that SVR kernels could also be seen under this light and this suggests the following natural question: does the Gaussian kernel give the most appropriate covariance for a SVR problems or there might be better covariances derived from other kernels?

In this work we will address this question in the following way. Given a sample data matrix X, a target vector y and a kernel family $\mathcal{K} = \{k^p = k^p_{\Theta^p}\}$ containing the RBF one, we first identify the best fitting kernel in $\mathcal{K}$ under a GPR setting, adjusting each kernel's hyper–parameters Θ^p by maximizing the log–likelihood $\log p(y|X, \Theta^p)$ but selecting as the best kernel not the one with the largest likelihood but instead the one that results in a smaller Mean Absolute Error (MAE) on a validation subset. The MAE choice is motivated by its popularity as error metric in renewable energy prediction problems. Once this optimal kernel k^* is chosen, we can revert to a SVR setting where we now consider the standard C and ϵ SVR hyper–parameters as well as those of k^*_Θ, which we re–hyper–parameterize jointly with C and ϵ.

We shall apply the previous ideas to two wind energy prediction problems, a local one for the Sotavento wind farm and a global one where the entire Peninsular Spain is considered. The growing presence of renewable sources in the energy mix of many countries, Spain among them, combined with the current non–dispatchable nature of wind and solar energy imply an acute need to have accurate forecasts, either for individual farms or over the large geographical areas where transmission system operators (TSO), Red Eléctrica de España (REE) in Spain's case, must work. Clearly, a high penetration and very difficult storage can only be compensated by adequate planning which, in turn, requires accurate enough forecasting methods. Because of this, Machine Learning (ML) methods are gaining a strong application in renewable energy prediction and SVR are often the models of choice for such predictions [3–6], generally using Gaussian kernels. However, and as we shall see, the proposed GPR approach to kernel selection results in kernels whose SVR models either tie or outperform the ones built using the RBF kernel. Of course, it is likely that the best kernel SVR choice will greatly depend on the concrete farm or area to be studied but our proposed use of general GPR kernels offers a simple and principled approach for the best SVR kernel selection. The paper is organized as follows. In Sect. 2 we will briefly review GP for regression as well as kernel SVR models; we also briefly discuss the kernels we will work with. Our wind energy experimental results are in Sect. 3 where we discuss our procedure for optimal kernel selection and the SVR models on the optimal kernel. The paper ends with a brief discussion section where we also give pointers to further work.

2   GPR and SVR

Our GPR description largely follows [8], Sects. 2.2 and 5.4. Zero centered Gaussian Processes (GP) are defined by functions $f : \mathbf{R}^d \to \mathbf{R}$ such that for any finite subset $S = \{x_1, \ldots, x_N\}$ the vector $f(x_1), \ldots, f(x_N)$ follows a Gaussian distribution $\mathcal{N}(0, K^S)$, where the covariance matrix K^S is defined as

$K_{i,j}^S = k(x_i, x_j)$, with k being a positive definite kernel. When applied in regression problems, a sample (x_i, y_i) is assumed where $y_i = f(x_i) + \epsilon_i$, with f following a GP prior and $\epsilon_i \sim \mathcal{N}(0, \sigma^2)$. It thus follows that the marginal likelihood is $p(y|X) \sim \mathcal{N}(0, K^S + \sigma^2\ I)$. If we want to predict the output $\widehat{y}$ over a new pattern x we can marginalize $p(\widehat{y}|x, X, y)$ on f

$$p(\widehat{y}|x, X, y) = \int p(\widehat{y}|x, f, X)p(f|X)df$$

and take advantage of all the densities being Gaussians to arrive at $p(\widehat{y}|x, X, y) \sim \mathcal{N}(\widehat{\mu}, \widehat{\sigma}^2)$, where

$$\widehat{\mu} = (K_x^S)^t \left(K^S + \sigma^2\ I \right)^{-1} y,$$
$$\widehat{\sigma}^2 = k(x, x) - (K_x^S)^t \left(K^S + \sigma^2\ I \right)^{-1} K_x^S + \sigma^2,$$

with K_x^S denoting the vector $(k(x, x_1), \dots, k(x, x_N))^t$.

Often the kernel $k = k_\Theta$ is parameterized by a vector $\Theta = (\theta_1, \dots, \theta_M)^t$ and training the GP is equivalent to finding an optimal Θ^*, i.e, to "learn" the kernel. Given a sample data matrix X and its target vector y, the conditional likelihood $p(y|X, \Theta)$ is $\sim \mathcal{N}(0, K_\Theta^S + \sigma^2 I)$, and its log–likelihood becomes

$$\log p(y|X, \Theta) = -\frac{1}{2} \log \det(K_\Theta^S + \sigma^2 I) - \frac{1}{2} y^t (K_\Theta^S + \sigma^2 I)^{-1} y + \text{const.} \quad (1)$$

We can thus find the optimal Θ^* (and estimate σ^2 along the way) by maximizing the log–likelihood of the pair X, y considering various restarts to control for local maxima. Observe that there is a trade–off between the target term $y^t(K_\Theta^S + \sigma^2 I)^{-1}y$ (smaller for complex models and possible overfit) and the model complexity term $\log \det(K_\Theta^S + \sigma^2 I)$.

The GPR model estimate at a new x is given by

$$\widehat{\mu} = \widehat{f}(x) = y^t \left(K^S + \sigma^2\ I \right)^{-1} K_x^S = \sum_i \beta_i k(x_i, x), \quad (2)$$

which is also the model returned by kernel Ridge Regression (KRR). In contrast with the GPR approach, the KRR model is derived by the direct minimization of the square error plus the Tikhonov regularizer $\frac{\widehat{\sigma}}{2}\|\beta\|^2$, and KRR only aims in principle to the estimate (2), while GPR yields a generative model through the posterior $p(f|X, y)$ as well as associated error intervals. Another important difference is that in KRR the optimal Tikhonov parameter $\widehat{\sigma}$ is estimated through cross validation, while in GPR it can be learned by maximizing the sample's likelihood along other parameters that may appear in the kernel k.

While kernel SVR models can also be written as in (2), they are built minimizing

$$\ell_S(W, b) = \sum_i [y^i - W \cdot \Phi(x_i) - b]_\epsilon + \frac{1}{C} \|\mathbf{w}\|_2^2, \quad (3)$$

where $[z]_\epsilon = \max(0, |z| - \epsilon)$, Φ is the feature map from $\mathbf{R}^d$ into a Reproducing Kernel Hilbert Space (RKHS) $\mathcal{H}$ and W has the form $W = \sum_i \alpha_i \Phi(x_i)$; recall that the feature map verifies $\Phi(x) \cdot \Phi(z) = k(x, z)$. Instead of problem (3), one solves the much simpler dual problem

$$\Theta(\alpha, \beta) = \frac{1}{2} \sum_{p,q} (\alpha_p - \beta_p)(\alpha_q - \beta_q)\Phi(x_p) \cdot \Phi(x_q) +$$
$$\epsilon \sum_p (\alpha_p + \beta_p) - \sum_p y^p(\alpha_p - \beta_p) \tag{4}$$

subject to $0 \leq \alpha_p, \beta_q \leq C$, $\sum \alpha_p = \sum \beta_q$; here α and β are the multipliers of the Lagrangian associated to (3). Since the optimal W^* can be written as $W^* = \sum(\alpha_p^* - \beta_p^*)\Phi(x_p)$ and the optimal b^* can also be derived from the optimal α^* and β^*, the final model becomes

$$f(x) = b^* + \sum(\alpha_p^* - \beta_p^*)\Phi(x_p) \cdot \Phi(x) = b^* + \sum(\alpha_p^* - \beta_p^*)k(x_p, x). \tag{5}$$

The dual problem (4), the SMO algorithm usually applied to solve it and the final model (5) only involve dot products in $\mathcal{H}$ which, in turn, only involve the kernel; thus, we do not have to work explicitly with Φ at any moment. The most usual and effective kernel is the Gaussian one $k(x, z) = e^{-\gamma \|x-z\|^2}$; its optimal parameter γ^* as well as C^* and ϵ^* are obtained by cross–validation. Observe that in the GPR literature, γ is written as $\gamma = \frac{1}{2\ell^2}$ and ℓ is customarily used instead of γ. See [9] for a thorough discussion of SVR.

Table 1. Validation MAE and log likelihood values over 2014 for the kernels considered.

Kernel	Sotavento		REE	
	MAE (%)	Log lik.	MAE(%)	Log lik.
RBF	8.215	10,941.3	3.645	22,303.168
2 RBF	7.296	11,254.6	3.570	22,356.146
RationalQuadratic	7.322	**11,281.8**	3.522	**22,369.9**
Matérn 0.5	**7.139**	11,031.4	**3.305**	19,537.3
Matérn 1.5	7.473	11,232.8	3.451	22,158.3
Matérn 2.5	7.652	11,166.7	3.511	22,321.4

3   Wind Energy Experiments

We will consider models to predict the wind energy of the Sotavento wind farm in northwestern Spain and that of the entire peninsular Spain. Sotavento's production data are available in their web; those of peninsular Spain have been kindly provided by REE. We will refer as those problems as the Sotavento and REE ones. They represent two different and interesting wind energy problems:

the local prediction at a given farm in Sotavento's case, and that of the energy aggregation over a wide area in REE's case.

We will use as features the NWP forecasts of the operational model of the European Centre for Medium Weather Forecasts (ECMWF) of 8 weather variables, namely the U and V wind components at 10 and 100 m as well as their module, 2 meter temperature and surface pressure. They are given every 3 h and we also add two 2 wind–turbine power estimates obtained from wind speeds using a generic wind–to–power conversion curve.

The lower left and upper right coordinates for REE are $(35.5°, -9.5°)$ and $(44.0°, 4.5°)$, respectively. The number of features are thus 1,200 for Sotavento and a much larger 5,220 for REE. Wind energy values are normalized to a 0–100% range. Recall that we consider zero mean GPRs; thus, we center the targets for the GPR models but do not do so for the MLP or SVR ones.

We deal first with the selection of optimal GPR kernels for which we shall consider the following five kernels:

- RBF: $k(x, x') = \theta \exp\left(-\frac{1}{2\ell^2}\|x - x'\|^2\right)$.
- 2 RBF: $k(x, x') = \theta_1 \exp\left(-\frac{1}{2\ell_1^2}\|x - x'\|^2\right) + \theta_2 \exp\left(-\frac{1}{2\ell_2^2}\|x - x'\|^2\right)$.
- RationalQuadratic: $k(x, x') = \theta \left(\frac{1 + \|x - x'\|^2}{2\alpha\ell^2}\right)^{-\alpha}$.
- Matérn 0.5: $k(x, x') = \theta \exp\left(-\frac{\|x - x'\|}{\ell^2}\right)$.
- Matérn 1.5: $k(x, x') = \theta \left(1 + \frac{\sqrt{3}\|x - x'\|}{\ell^2}\right) \exp\left(-\frac{\sqrt{3}\|x - x'\|}{\ell^2}\right)$.
- Matérn 2.5: $k(x, x') = \theta \left(1 + \frac{\sqrt{5}\|x - x'\|}{\ell^2} + \frac{5\|x - x'\|^2}{3\ell^4}\right) \exp\left(-\frac{\sqrt{5}\|x - x'\|}{\ell^2}\right)$.

To all of them we add a white noise kernel $\sigma^2\delta(x, x')$. This can also be seen as a kind of Tikhonov regularization which also helps with numerical issues while training. We use the `GaussianProcessRegressor` class in scikit–learn [7] and the kernel hyper–parameters are optimized by maximizing the sample's log-marginal-likelihood that we can compute by (1). Observe that the log likelihood maximization is done by gradient ascent starting at a random initial hyper–parameter set; because of this we will restart the ascent five times and retain the hyper–parameters yielding the largest likelihood. We use for this the year 2013 and in principle we would select the kernel for which $p(y|X, \Theta)$ is largest. According to Table 1, these would be the 2 RBF and RationalQuadratic kernels for REE and Sotavento respectively. However, we choose the Matérn 0.5 kernel whose GPR has the smallest Mean Absolute Error (MAE) for the 2014 validation year in both REE and Sotavento.

Once the Matérn 0.5 kernel has been so chosen, we turn our attention to use it as a kernel for SVR. To do so we must first obtain its best hyperparameters, namely the C and ϵ standard in SVR plus the length scale ℓ. We drop the θ parameter as it can be compensated by the SVR multipliers. We do so now by a grid search of these three parameters using again 2013 for training and the 2014 MAEs for validation. In contrast with the relatively fast, gradient ascent based maximization of the log likelihood used previously, the current grid search

Table 2. Matérn 0.5 GPR and SVR hyper–parameters.

	GPR			SVR		
Problem	θ	ℓ	σ^2	C	ℓ	ϵ
Sotavento	0.537	880	1e−05	0.25	138.56	2.116e−04
REE	1.877	9.16e+04	1e−05	16.0	3.699e+04	1.445e−04

Table 3. Sotavento and REE 2015 test results of MLPs, Gaussian SVR, Matérn 0.5 SVR and Matírn 0.5 GPR.

Problem	MLP	SVR	SVR Matérn 0.5	GPR Matérn 0.5
Sotavento	5.86	5.80	5.70	6.01
REE	2.76	2.54	2.55	2.65

takes considerably longer. Although we will not do so here, notice that an option would be to leave ℓ at the optimal value obtained by log likelihood optimization at the previous step; this would leave only C and ϵ to be estimated.

The optimal hyper–parameters are given in Table 2, which contains both the ℓ, θ and σ hyperparameters of the Matérn 0.5 GPR as well as the C, ϵ and ℓ ones of the Matérn 0.5 SVR. Once obtained, we train the corresponding REE and Sotavento SVRs over the 2013 and 2014 years combined and test them over 2015. Recall that the SVR problem is convex and, hence, has a unique solution; thus a single train pass is enough. The resulting MAEs are given in Table 3; its MLP and SVR results are taken from [1]; they correspond to a MLP with 4 hidden layers with 100 units for Sotavento and 1,000 units for REE and a standard Gaussian SVR. As it can be seen, the Matérn 0.5 SVR gives the best test MAE for Sotavento and essentially ties with the Gaussian SVR for REE. The Matérn 0.5 GPR falls clearly behind in Sotavento but less so in REE, where it beats the MLP.

These comparisons can be made more precise using non–parametric hypothesis testing. The usual first choice is Wilcoxon's signed rank test [10] to check whether the population mean ranks of two paired samples differ. Table 4, left, shows its p values when the paired samples are the REE and Sotavento hourly absolute errors; a p value of 0.000 means that the returned p value was below 5×10^{-4}. As it can be seen, the null hypothesis of the error differences following a symmetric distribution around zero can be safely rejected in all cases except for the MLP vs GPR comparison in Sotavento. In particular, when comparing the SVR and Matérn 0.5 SVR models the p value is below the 5% threshold in REE and much lower in Sotavento.

As a further illustration, Table 4 also shows (right) the p values when the Wilcoxon–Mann–Whitney rank sum test [10] is applied. Notice that an assumption of this test is that both samples are independent, something that here would need further study for our sample choice of hourly absolute errors. The null hypothesis H_0 is now the equal likelihood of a random value from the first sample being less than or greater than a random value from the second one. As it can be seen, H_0 can be rejected with a p value below 5% in all cases except the SVR vs Matérn 0.5 SVR for REE and Sotavento, and the MLP vs GPR for Sotavento; H_0 cannot be rejected at the 5% level when comparing the Matérn 0.5 SVR vs GPR in REE but it could be rejected at the 10% level.

In any case, based on our practical experience, the improvement in Sotavento of the Matérn 0.5 SVR performance over the Gaussian one would be relevant for the operation of a wind energy prediction system and even more over the GPR model. For REE we would retain the SVR model although its difference with the Matérn 0.5 SVR model may not be statistically significant.

Table 4. Wilcoxon signed rank (left) and Wilcoxon-Mann-Whitney rank sum (right) tests p values.

Comparison	Signed rank		Rank sum	
	REE	Sotavento	REE	Sotavento
SVR vs MLP	0.000	0.000	0.000	0.000
SVR vs SVR Mat.	0.042	0.000	0.560	0.431
SVR vs GPR	0.000	0.000	0.026	0.000
MLP vs SVR Mat.	0.000	0.000	0.000	0.000
MLP vs GPR	0.000	0.550	0.000	0.660
SVR Mat. vs GPR	0.000	0.000	0.094	0.000

4   Discussion and Conclusions

In contrast to the standard Gaussian kernel used in SVR, much more varied kernels are routinely considered in GPR. This motivates this work, where we have considered kernels other than the Gaussian one for SVR, choosing for this those which show an optimal behavior when the underlying problem is tackled by GPR. However, instead of choosing the GPR kernel which maximizes the marginal log–likelihood, we have used likelihood maximization to learn the kernel hyper–parameters but have selected the kernel whose GPR forecasts have the lowest MAE on the validation set. We have applied this approach to two wind energy prediction problems, for which the GPR–optimally chosen kernel has given the smallest test MAE in one problem and tied with the Gaussian SVR model in the other. We point out that, while not being too practical for big data problems, for moderate size regression problems such as wind (or solar) energy prediction, a properly tuned Gaussian SVR model is usually quite hard to beat.

This suggests that the combination of SVR models with kernels that are optimal when the underlying problem is solved via GPR may yield a simple way of enhancing SVR models that beat the Gaussian ones. This also opens the way to the consideration of quite complex kernels or combinations of them. The usually costly training of kernel SVR makes it prohibitive to use grid–search hyper–parameterization when multiple parameter kernels are involved. On the other hand, hyper–parameter learning of GPR kernels is much faster as it relies on gradient–based ascent on the sample log–likelihood function. Thus, general kernel SVR hyper–parameterization can be split into two parts, a relatively fast kernel–specific parameter search by likelihood maximization and a second one where the SVR C and ϵ hyper–parameters are found by grid search over a much more manageable 2–dimensional grid. This combination will be faster than the standard C, γ and ϵ grid search for Gaussian SVR and, if the GPR kernels are properly chosen, may result in better models not only for wind energy forecasts but also in general SVR problems.

Acknowledgements. With partial support from Spain's grants TIN2016-76406-P and S2013/ICE-2845 CASI-CAM-CM. Work supported also by project FACIL–Ayudas Fundación BBVA a Equipos de Investigación Científica 2016, and the UAM–ADIC Chair for Data Science and Machine Learning. We thank Red Eléctrica de España for making available wind energy data and gratefully acknowledge the use of the facilities of Centro de Computación Científica (CCC) at UAM. We also thank the Agencia Estatal de Meteorología, AEMET, and the ECMWF for access to the MARS repository.

References

1. Catalina, A., Dorronsoro, J.R.: NWP Ensembles for Wind Energy Uncertainty Estimates. In: Woon, W.L., Aung, Z., Kramer, O., Madnick, S. (eds.) DARE 2017. LNCS, vol. 10691, pp. 121–132. Springer, Cham (2017). https://doi.org/10.1007/978-3-319-71643-5_11
2. Cristianini, N., Shawe-Taylor, J.: An Introduction to Support Vector Machines and Other Kernel-Based Learning Methods. Cambridge University Press, Cambridge (2003)
3. Foley, A.M., Leahy, P.G., Marvuglia, A., McKeogh, E.J.: Current methods and advances in forecasting of wind power generation. Renew. Energy **37**(1), 1–8 (2012)
4. Heinermann, J.P.: Wind power prediction with machine learning ensembles. Ph.D. thesis. Universität Oldenburg (2016)
5. Kramer, O., Gieseke, F.: Short-term wind energy forecasting using support vector regression. In: Corchado, E., Snášel, V., Sedano, J., Hassanien, A.E., Calvo, J.L., Ślezak, D. (eds.) SOCO 2011. AINSC, vol. 87, pp. 271–280. Springer, Berlin Heidelberg (2011). https://doi.org/10.1007/978-3-642-19644-7_29
6. Mohandes, M., Halawani, T., Rehman, S., Hussain, A.A.: Support vector machines for wind speed prediction. Renew. Energy **29**(6), 939–947 (2004)
7. Pedregosa, F., et al.: Scikit-learn: machine learning in Python. J. Mach. Learn. Res. **12**, 2825–2830 (2011)
8. Rasmussen, C.E., Williams, C.K.I.: Gaussian Processes for Machine Learning. Adaptive Computation and Machine Learning. MIT Press, Cambridge (2006)
9. Schölkopf, B., Smola, A.: Learning with Kernels: Support Vector Machines, Regularization, Optimization and Beyond. Adaptive Comunication and Machine Learning. MIT Press, Cambridge (2002)
10. Wilcoxon, F.: Individual comparisons by ranking methods. Biometr. Bull. **1**, 80–83 (1945)

Studying the Effect of Measured Solar Power on Evolutionary Multi-objective Prediction Intervals

R. Martín-Vázquez, J. Huertas-Tato, R. Aler, and I. M. Galván[✉]

Computer Science Departament, Carlos III University of Madrid,
Avda. Universidad, 30, 28911 Leganés, Spain
igalvan@inf.uc3m.es

Abstract. While it is common to make point forecasts for solar energy generation, estimating the forecast uncertainty has received less attention. In this article, prediction intervals are computed within a multi-objective approach in order to obtain an optimal coverage/width trade-off. In particular, it is studied whether using measured power as an another input, additionally to the meteorological forecast variables, is able to improve the properties of prediction intervals for short time horizons (up to three hours). Results show that they tend to be narrower (i.e. less uncertain), and the ratio between coverage and width is larger. The method has shown to obtain intervals with better properties than baseline Quantile Regression.

Keywords: Solar energy · Prediction intervals
Multi-objective optimization

1 Introduction

In recent years, solar energy has shown a large increase in its presence in the electricity grid energy mix [1]. Having accurate point forecasts is important for solar energy penetration in the electricity markets and most of the research has focused on this topic. However, due to the high variability of solar radiation, it is also important to estimate the uncertainty around point forecasts. A convenient way of quantifying the variability of forecasts is by means of Prediction Intervals (PI) [2]. A PI is a pair of lower and upper bounds that contains future forecasts with a given probability (or reliability), named Prediction Interval Nominal Coverage (PINC). There are several methods for computing PI [3]: Delta method, Bayesian technique, Mean-Variance, and Bootstrap method. However, a recent evolutionary approach has shown better performance than the other methods in several domains [3], including renewable energy forecasting [4,5]. This approach, known as LUBE (Lower Upper Bound Estimation), uses artificial neural networks with two outputs, for the lower and upper bound of the interval, respectively. The network weights are optimized using evolutionary computation techniques such as Simulated Annealing (SA) [6] or Particle Swarm Optimization (PSO) [7].

© Springer Nature Switzerland AG 2018
H. Yin et al. (Eds.): IDEAL 2018, LNCS 11315, pp. 155–162, 2018.
https://doi.org/10.1007/978-3-030-03496-2_18

The optimization of PI is inherently multi-objective, because of the trade-off between the two main properties of intervals: coverage and width. Coverage can be trivially increased by enlarging intervals, but obtaining high coverage with narrow intervals is a difficult optimization problem. Typical approaches to LUBE aggregate both goals so that optimization can be carried out by single-objective optimization methods (such as SA or PSO). In [8], it was proposed to use a multi-objective approach (using the Multi-objective Particle Swarm Optimization evolutionary algorithm or MOPSO [9]) for this purpose. The main advantage of this approach is that in a single run, it is able to obtain not one, but a set of solutions (the Pareto front) with the best trade-offs between coverage and width. If a solution with a particular PINC value is desired, it can be extracted from the front. In that work, due to the nature of the data, the aim was to estimate solar energy PI on a daily basis, and using a set of meteorological variables forecast for the next day.

In some works, meteorological forecasts are combined with measurements for training the models with the purpose of improving predictions [10,11]. In particular, in [12] it was observed that using measured power (in addition to meteo variables) was helpful to improve point forecasts for short term horizons. In the present work, we apply the MOPSO approach for estimating solar power PI, studying the influence of using measured solar power, in addition to meteorological forecasts, on the quality of intervals (coverage and interval width). In a similar way to point forecasts where using measured values improves the accuracy of the forecast if the prediction horizon is close to the measurement, we expect that using measured values will have an effect of reducing the uncertainty of prediction intervals. For that purpose, short prediction horizons of up to three hours will be considered in the experiments, in steps of 1 h. Linear Quantile Regression method [13] is also used as baseline with the purpose to comparison, using the R quantreg package [14].

The rest of the article is organized as follows: Sect. 2 describes the dataset used for experiments, Sect. 3 summarizes the evolutionary multi-objective approach for interval optimization, as well as the baseline method Quantile Regression. Section 4 describes the experimental setup and the results. Conclusions are finally drawn in Sect. 5.

2   Data Description

The data used in this work is obtained from the Global Energy Forecasting Competition 2014 (GEFCom2014), specifically from task 15 of the probabilistic solar power forecasting problem [15]. The data provided included measured solar power and meteorological forecasts. Solar power was provided on an hourly basis, from 2012-04-01 01:00 to 2014-06-01 00:00 UTC (for training) and from 2014-06-01 01:00 to 2014-07-01 00:00 UTC for testing. The meteorological forecasts included 12 weather variables that had been obtained from the European Centre for Medium-range Weather Forecasts (ECMWF) [15]. Those variables were issued everyday at midnight UTC for each of the next 24 h. These 12 variables are: Total column liquid water (kg m-2), Total column ice water (kg m-2),

Surface pressure (Pa), Relative humidity at 1000 mbar (r), Total cloud cover (0'1), 10-metre U wind component (m s-1), 10-metre V wind component (m s-1), 2-metre temperature (K), Surface solar rad down (J m-2), Surface thermal rad down (J m-2), Top net solar rad (J m-2), and Total precipitation (m).

Data was provided for three power plants in Australia, although their exact location was not disclosed. In this article we are using station number 3 and the short term forecasting horizons (+1, +2, and +3 h). For some of the work carried out in this article, it is necessary to separate the available training data into training and validation. The validation set is used for model selection tasks, such as choosing the best neural network architecture or the best number of optimization iterations. In this work, the first 80% of the dataset has been used for training and the remaining 20% for validation.

3   Multi-objective Optimization for Prediction Intervals

The purpose of this section is to summarize the multi-objective evolutionary optimization of PI reported in [8]. This approach is based on LUBE [3], where a 3-layered neural network is used to estimate the lower and upper of bounds of prediction intervals for a particular input, but using a multi-objective evolutionary approach. The network receives as inputs meteorological variables. The outputs are the lower and upper bounds estimated by the network for those particular inputs. Although the observed irradiance for some particular inputs is available in the dataset, the upper and lower bounds are not. Hence, the standard back-propagation algorithm cannot be used to train the network (i.e. it is not a standard supervised regression problem, because the target output is not directly available). Therefore, in this approach, an evolutionary optimization algorithm is used to adjust the weights of the neural network by optimizing the two most relevant goals for PI: reliability and interval width. A prediction intervals is reliable if it achieves some specified reliability level or nominal coverage (i.e. PINC). This happens when irradiance observations lay inside the interval at least as frequently as the specified PINC. It is always possible to have high reliability by using very wide intervals. Therefore, the second goal used to evaluate PI is interval width (with the aim of obtaining narrow intervals). Reliability and interval width are formalized in the following paragraphs.

Let $M = \{(X_i, t_i)_{i=0 \cdots N}\}$ be a set of observations, where X_i is a vector with the input variables and t_i is the observed output variable. Let $PI_i = [Low_i, Upp_i)$ be the prediction interval for observation X_i (Low_i, Upp_i would be the outputs of the neural network). Then, the reliability (called Prediction Interval Coverage Probability or PICP) is computed by Eq. 1 and the Average Interval Width (AIW) by Eq. 2.

$$PICP = \frac{1}{N} \sum_{i=0}^{N} \chi_{PI_i}(X_i) \tag{1}$$

$$AIW = \frac{1}{N} \sum_{i=0}^{N} (Upp_i - Low_i) \qquad (2)$$

where N is the number of samples, $\chi_{PI_i}(X_i)$ is the indicator function for interval PI_i (it is 1 if $t_i \in PI_i = [Low_i, Upp_i)$ and 0 otherwise). Upp_i and Low_i are the upper and lower bounds of the interval, respectively.

Given that there is a trade-off between reliability and width (PICP and AIW, respectively), the multi-objective approach (MOPSO) proposed in [8] is used to tackle the problem studied in this article. The MOPSO particles encode the weights of the networks, and the goals to be minimized are $1 - PICP$ (Eq. 1) and AIW (Eq. 2). In this work, the inputs to the networks are the meteorological variables given in the dataset (see Sect. 2) or any other information that may be useful for the estimation of solar irradiance (as solar power measurements).

The final result of MOPSO optimization is a non-dominated set of solutions (a Pareto front) as shown in Fig. 1. Each point (or solution) in the front represents the $x = AIW$ and $y = 1 - PICP$ of a particular neural network that achieves those values on the training dataset. If a particular target PINC is desired, then the closest solution in the Pareto front to that PINC is selected. That solution corresponds to a neural network that can be used on new data (e.g. test data) in order to compute PI for each of the instances in the test data. Figure 1 shows the solution that would be extracted from a Pareto front for PINC = 0.9.

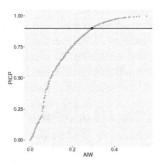

Fig. 1. Pareto front of solutions. Selected solution for PICP = 0.90.

In order to have a baseline to compare MOPSO results, Linear Quantile Regression (QR) has been used [13]. QR is a fast technique for estimating quantiles using linear models. While the standard method of least squares estimates the conditional mean of the response variable, quantile regression is able to estimate the median or other quantiles. This can be used for obtaining PI. Let q_1 and q_2 be the $\frac{1-PINC}{2}$ and $\frac{1+PINC}{2}$ quantiles, respectively. Quantile q_1 leaves a $\frac{1-PINC}{2}$ probability tail to the left of the distribution and quantile q_2 leaves $\frac{1-PINC}{2}$ probability tail to the right of the distribution. Therefore, the interval

$[q_1, q_2]$ has a coverage of PINC. Quantile Regression is used to fit two linear models that, given some particular input X_i, returns q_1 and q_2 with which the interval $[q_1, q_2]$ can be constructed.

4   Experimental Validation

As mentioned in the introduction, one of the goals of this article is to study the influence on the quality of intervals, of using measured solar power at the time of prediction $t_0 = 00 : 00$ UTC, in addition to the meteorological forecasts (that have already been described in Sect. 2). t_0 corresponds to 10:00 AM at the location in Australia, which is the time when meteo forecasts are issued everyday. With that purpose two derived datasets have been constructed, one with only the 12 meteorological variables and another one with those variables and the measured solar power at t_0. The latter (meteo + measured power) will be identified with $+P_{t_0}$. In both cases, the day of the year (from 1 to 365) has also been used as input, because knowing this information might be useful for computing the PI.

Table 1. Best combination of parameters for each prediction horizon

Horizon	MOPSO		MOPSO + P_{t_0}	
	Neurons	Iterations	Neurons	Iterations
1 h	15	8000	50	8000
2 h	8	8000	10	8000
3 h	6	8000	15	8000

We have followed a methodology similar to that of [8]. Different number of hidden neurons for the neural network (2, 4, 6, 8, 10, 15, 20, 30, and 50) and different number of iterations for PSO (4000, 6000, and 8000) has been tested. The process involves running PSO with the training dataset and using the validation dataset to select the best parameters (neurons and iterations). Given that PSO is stochastic, PSO has been run 5 times for each number of neurons and iterations, starting with different random number generator seeds. Similarly to [8], the measure used to select the best parameter combination has been the average hypervolume of the front on the validation set (the validation front is computed by evaluating each neural network from the training Pareto front, on the validation set). It is important to remark that, differently to [8], this has been done for each different prediction horizon. That means that this parameter optimization process has been carried out independently for each of three forecasting horizons considered in this work (+1 h, +2 h, +3 h). Table 1 displays the best combination of parameters for each horizon and whether P_{t_0} is used or not. It can be observed that the number of hidden neurons depends on the horizon and that the number of iterations is typically the maximum value tried (8000). We have not extended the number of iterations for PSO because no further change was observed in the Pareto fronts by doing so.

In order to evaluate the experimental results, three target nominal coverage values (PINC) have been considered: 0.9, 0.95, and 0.99. The Quantile Regression approach must be run for each desired PINC value. The MOPSO approach needs to be run only once, because it provides a set of solutions (the Pareto front), out of which the solutions for particular PINCs can be extracted, as it has been explained in Sect. 3 (see Fig. 1).

Table 2. Evaluation measures on the test set for the four different approaches (QR, QR + P_{t_0}, MOPSO, MOPSO + P_{t_0}). Left: *Delta coverage*. Middle: Average interval width (AIW). Right: PICP/AIW ratio.

	Delta coverage			AIW			*PICP/AIW* ratio		
PINC	0.99	0.95	0.90	0.99	0.95	0.90	0.99	0.95	0.90
QR	0.027	0.072	0.089	0.756	0.611	0.495	1.291	1.438	1.640
QR + P_{t_0}	**0.020**	0.052	0.084	0.732	0.571	0.487	1.373	1.609	1.693
MOPSO	0.036	0.074	0.142	0.715	0.561	0.461	1.344	1.568	1.652
MOPSO + P_{t_0}	**0.020**	**0.051**	**0.061**	**0.646**	**0.495**	**0.427**	**1.530**	**1.861**	**2.018**

The performance of the solutions for each horizon, has been evaluated using three evaluation measures. The first one, named *delta coverage* in Table 2, measures how much the solution PICP fails to achieve the target PINC (on the test set). If the PICP fulfills the PINC ($PICP >= PINC$) then *delta coverage* is zero, otherwise it is computed as $PINC - PICP$ (in other words: *delta coverage* = max(0, $PINC - PICP$)). The latter measure evaluates PINC fulfillment, but it only tells part of the performance because it is trivial to obtain small (or even zero) *delta coverage* by using very wide intervals. Thus, the second evaluation measure uses the ratio between PICP and the average interval width (AIW), which is calculated as *PICP/AIW*. Solutions that achieve high PICPs by means of large intervals will obtain low values for this ratio. Good solutions, with an appropriate tradeoff between PICP and width will obtain high values on this measure. Additionally, the average interval width (AIW) will also be shown.

Table 2 display the values of the three evaluation measures averaged over the 5 runs and the 3 horizons, for the three different values of PINC (0.99, 0.95, and 0.90). In Table 2 it can be seen that *delta coverage* (left) is larger than zero for all methods, which means that there are horizons for which PINC is not achieved. The best *delta coverage* values for all PINC values are obtained by MOPSO + P_{t_0} (this means that PICP is closer to the target PINC). It is also observed that the use of the measured solar power at 00:00 UTC helps MOPSO + P_{t_0} to obtain smaller *delta coverage*. Using P_{t_0} also helps QR in this regard. The same trend can be observed with respect to the AIW (see Table 2 middle) and the *PICP/AIW* ratio (Table 2 right). Therefore, MOPSO + P_{t_0} obtains the best coverage, using the narrowest intervals, and reaching the best tradeoff between PICP and AIW.

Next, we will compare both approaches (MOPSO and QR) breaking down results by horizon. Table 3 shows the $PICP/AIW$ ratio and the AIW for all methods, horizons and PINC values. In the case of MOPSO and MOPSO + P_{t_0}, the average and standard deviation of the 5 runs are displayed. With respect to the ratio, it can be seen that using the P_{t_0} helps MOPSO for all horizons and target PINC values. In the QR case, it helps for the first horizon but not (in general) for the rest. The best performer for all horizons and target PINC is MOPSO + P_{t_0}, except for the second horizon and PINC = 0.99, where it is slightly worse than QR without P_{t_0}. The same trend can be observed for the AIW except for the third horizon and PINC = 0.90, where MOPSO and MOPSO + P_{t_0} are very similar.

Finally, for MOPSO, the improvement in the $PICP/AIW$ ratio by using P_{t_0} is larger for the first horizon than for the rest. For horizon 1, the improvement in ratio is 25%, 39%, and 40% for PINC values 0.99, 0.95, and 0.90, respectively. The reduction in AIW follows a similar behavior: 18%, 20%, and 18%, respectively. For the rest of horizons, there is also improvement, but smaller in size, and the larger the horizon, the smaller the improvement.

Table 3. Average and standard deviation of the $PICP/AIW$ ratio and AIW per prediction horizon (1 h, 2 h, 3 h). PINC values = 0.99, 0.95, 0.90.

Horizon	Method	$PICP/AIW$ ratio			AIW		
		0.99	0.95	0.90	0.99	0.95	0.90
1	QR	1.258	1.416	1.561	0.742	0.589	0.491
1	QR + P_{t_0}	1.648	1.773	1.959	0.607	0.447	0.422
1	MOPSO	1.415 (0.085)	1.569 (0.114)	1.635 (0.098)	0.671 (0.048)	0.538 (0.040)	0.428 (0.011)
1	MOPSO + P_{t_0}	**1.762** (0.127)	**2.174** (0.117)	**2.294** (0.194)	**0.552** (0.055)	**0.430** (0.039)	**0.351** (0.031)
2	QR	**1.452**	1.481	1.761	**0.666**	0.585	0.473
2	QR + P_{t_0}	1.375	1.574	1.565	0.677	0.613	0.485
2	MOPSO	1.283 (0.099)	1.508 (0.114)	1.571 (0.044)	0.747 (0.058)	0.596 (0.051)	0.484 (0.026)
2	MOPSO + P_{t_0}	1.446 (0.108)	**1.734** (0.114)	**1.918** (0.184)	0.691 (0.057)	**0.515** (0.048)	**0.451** (0.062)
3	QR	1.162	1.417	1.599	0.861	0.659	0.521
3	QR + P_{t_0}	1.097	1.481	1.555	0.911	0.652	0.554
3	MOPSO	1.333 (0.057)	1.628 (0.106)	1.749 (0.235)	0.727 (0.031)	0.550 (0.027)	**0.470** (0.040)
3	MOPSO + P_{t_0}	**1.383** (0.092)	**1.674** (0.216)	**1.842** (0.209)	**0.696** (0.047)	**0.540** (0.065)	0.478 (0.026)

5   Conclusions

In this article, we have used a multi-objective approach, based on Particle Swarm Optimization, to obtain prediction intervals with an optimal tradeoff between interval width and reliability. In particular, the influence on short prediction horizons, of using measured solar power as an additional input, has been studied. This has shown to be beneficial, because prediction interval tend to be narrower (hence, less uncertainty on the forecast), and the ratio between coverage and width is larger. This is true for the three short prediction horizons studied, but the improvement is larger for the shortest one (+1 h). Results have been compared to Quantile Regression and shown to be better for all evaluation criteria.

While Quantile Regression also benefits from using measured solar radiation, this happens only for the 1 h horizon, but not for +2 or +3 h.

Acknowledgements. This work has been funded by the Spanish Ministry of Science under contract ENE2014-56126-C2-2-R (AOPRIN-SOL project).

References

1. Raza, M.Q., Nadarajah, M., Ekanayake, C.: On recent advances in pv output power forecast. Sol. Energy **136**, 125–144 (2016)
2. Pinson, P., Nielsen, H.A., Møller, J.K., Madsen, H., Kariniotakis, G.N.: Nonparametric probabilistic forecasts of wind power: required properties and evaluation. Wind. Energy **10**(6), 497–516 (2007)
3. Khosravi, A., Nahavandi, S., Creighton, D., Atiya, A.F.: Lower upper bound estimation method for construction of neural network-based prediction intervals. IEEE Trans. Neural Netw. **22**(3), 337–346 (2011)
4. Wan, C., Xu, Z., Pinson, P.: Direct interval forecasting of wind power. IEEE Trans. Power Syst. **28**(4), 4877–4878 (2013)
5. Khosravi, A., Nahavandi, S.: Combined nonparametric prediction intervals for wind power generation. IEEE Trans. Sustain. Energy **4**(4), 849–856 (2013)
6. Kirkpatrick, S., Gelatt, C.D., Vecchi, M.P.: Optimization by simulated annealing. Science **220**(4598), 671–680 (1983)
7. Eberhart, R.C., Shi, Y., Kennedy, J.: Swarm Intelligence. Elsevier, Amsterdam (2001)
8. Galván, I.M., Valls, J.M., Cervantes, A., Aler, R.: Multi-objective evolutionary optimization of prediction intervals for solar energy forecasting with neural networks. Inf. Sci. **418**, 363–382 (2017)
9. Coello Coello, C.A., Lechuga, M.S.: MOPSO: a proposal for multiple objective particle swarm optimization. In: Proceedings of the 2002 Congress on Proceedings of the Evolutionary Computation on CEC 2002, vol. 2, pp. 1051–1056. IEEE Computer Society, Washington (2002)
10. Aguiar, L.M., Pereira, B., Lauret, P., Díaz, F., David, M.: Combining solar irradiance measurements, satellite-derived data and a numerical weather prediction model to improve intra-day solar forecasting. Renew. Energy **97**, 599–610 (2016)
11. Wolff, B., Kühnert, J., Lorenz, E., Kramer, O., Heinemann, D.: Comparing support vector regression for pv power forecasting to a physical modeling approach using measurement, numerical weather prediction, and cloud motion data. Sol. Energy **135**, 197–208 (2016)
12. Martín-Vázquez, R., Aler, R., Galván, I.M.: Wind energy forecasting at different time horizons with individual and global models. In: Iliadis, L., Maglogiannis, I., Plagianakos, V. (eds.) AIAI 2018. IAICT, vol. 519, pp. 240–248. Springer, Cham (2018). https://doi.org/10.1007/978-3-319-92007-8_21
13. Koenker, R.: Quantile Regression. Econometric Society Monographs, vol. 38. Cambridge University Press, Cambridge (2005)
14. Koenker, R.: quantreg: Quantile Regression. R package version 5.36 (2018)
15. Hong, T., Pinson, P., Fan, S., Zareipour, H., Troccoli, A., Hyndman, R.J.: Probabilistic energy forecasting: global energy forecasting competition 2014 and beyond. Int. J. Forecast. **32**(3), 896–913 (2016)

Merging ELMs with Satellite Data and Clear-Sky Models for Effective Solar Radiation Estimation

L. Cornejo-Bueno[1], C. Casanova-Mateo[2,3], J. Sanz-Justo[3], and S. Salcedo-Sanz[1(✉)]

[1] Department of Signal Processing and Communications, Universidad de Alcalá, Madrid, Spain
sancho.salcedo@uah.es
[2] Department of Civil Engineering: Construction, Infrastructure and Transport, Universidad Politécnica de Madrid, Madrid, Spain
[3] LATUV, Laboratorio de Teledetección, Universidad de Valladolid, Valladolid, Spain

Abstract. This paper proposes a new approach to estimate Global Solar Radiation based on the use of the Extreme Learning Machine (ELM) technique combined with satellite data and a clear-sky model. Our study area is the radiometric station of Toledo, Spain. In order to train the Neural Network proposed, one complete year of hourly global solar radiation data (from the 1st of May 2013 to the 30th of April 2014) is used as the target of the experiments, and different input variables are considered: a cloud index, a clear-sky solar radiation model and several reflectivity values from Meteosat visible images. To assess the results obtained by the ELM we have selected as a reference a physical-based method which considers the relation between a clear-sky index and a cloud cover index. Then a measure of the Root Mean Square Error (RMSE) and the Pearson's Correlation Coefficient (r^2) is obtained to evaluate the performance of the suggested methodology against the reference model. We show the improvement of the results obtained by the ELM with respect to those obtained by the physical-based method considered.

Keywords: Solar radiation estimation · Extreme learning machines Meteosat data

1 Introduction

Solar radiation is currently the second most important renewable resource, behind wind energy [1]. It is expected, however, an exponential expansion of solar energy facilities in the next decades, specially in those areas with more solar potential, such as mid-east and southern Europe and Australia [2]. An accurate estimation of the solar energy resource is key in order to promote the integration of this renewable resource in the electrical system [3,4].

In recent years, different techniques have been applied to solar energy prediction, many of them based on machine learning algorithms [5]. They used

© Springer Nature Switzerland AG 2018
H. Yin et al. (Eds.): IDEAL 2018, LNCS 11315, pp. 163–170, 2018.
https://doi.org/10.1007/978-3-030-03496-2_19

different inputs for the prediction, such as latitude, longitude or sunshine duration, as well as atmospheric parameters such as temperature, wind speed and direction or daily global irradiation among others [6]. There are different works dealing with the application of Extreme Learning Machines to Solar radiation estimation problems. For example, in [7] a case study of solar radiation prediction in Arabia Saudi is discussed comparing the performance of artificial neural networks with classical training and Extreme Learning Machines (ELM). In [8] a hybrid wavelet-ELM approach is tested in a problem of solar irradiation prediction for application in a photovoltaic power station. In [9] a comparison of a support vector regression algorithm and an ELM is carried out in a problem of direct solar radiation prediction, with application in solar thermal energy systems. In [10] a hybrid Coral Reefs Optimization with ELMs was proposed for a problem of solar global radiation prediction.

Satellite data have been previously used together with artificial neural networks in solar radiation prediction. We discuss here two main works, which are closely related to our approach: first, [11] proposes an artificial neural network where meteorological and geographical data (latitude, longitude, altitude, month, mean diffuse radiation and mean beam radiation) are used as inputs for the neural network. This work proposes a comparison of the results obtained with those by a physical model from satellite measurements over Turkey, including clear sky and cloud index values. More recently in [12], the ELM approach is applied to a solar radiation prediction problem over Turkey from satellite data and geographic variables. The ELM results were compared with that by Multi-Layer Perceptron showing improvements in performance, and a high improvement in computational time for the network training.

In this paper, we further explore the capacity prediction of ELM in a problem of solar radiation estimation from Meteosat data. We consider a cloud index, a clear-sky model and several satellite reflectivity values as ELM inputs. No geographical nor meteorological variables are considered in this study, aiming at evaluate the real performance of Meteosat observations in solar radiation estimation problems, without alternative contributions. Meteosat measurements are considered over a radiometric station located in the center of Spain. There, the cloud index is also calculated from reflectivity values obtained from Meteosat images. Specifically, we have extracted the reflectivity information from the nearest pixel to the location of the radiometric station. Additionally, reflectivity information from the 8 pixels surrounding the central one has also been extracted. The solar estimation obtained is then compared to that of a physical model proposed in [17] and also used in [11] as comparison method.

The rest of the paper is structured as follows: next section defines the problem of solar radiation estimation tackled, with details on the satellite variables and methodology followed. Section 3 briefly describes the ELM approach used in this work. Section 4 presents the results obtained in this problem of solar radiation estimation with the ELM and the physical model considered for comparison. Section 5 closes the paper with some final conclusions and remarks.

2   Data Description and Methodology

In this study we use information from the Meteosat satellite. This geostationary satellite orbiting at 36.000 km above the equator is one of the most famous weather satellites around the world. Operated by EUMETSAT (the European Organisation for the Exploitation of Meteorological Satellites), its information has become an essential element in the provision of reliable and up-to-date meteorological information both for maintaining a continuous survey of meteorological conditions over specific areas and for proving invaluable information to support weather prediction models output.

The basic payload of this satellite consists of the following instruments:

- The Spinning Enhanced Visible and Infrared Image (SEVIRI) is the main instrument on board Meteosat Second Generation Satellites. Unlike its predecessor (MVIRI) with only 3 spectral channels, SEVIRI radiometer has 12 spectral channels with a baseline repeat cycle of 15 min: 3 visible and near infrared channels centered at 0.6, 0.8 and 1.6 µm, 8 infrared channels centered at 3.9, 6.2, 7.3, 8.7, 9.7, 10.8, 12.0 and 13.4 µm and one high-resolution visible channel [13]. The nominal spatial resolution at the sub-satellite point is 1 km^2 for the high-resolution channel, and 3 km^2 for the other channels [14,15].
- The Geostationary Earth Radiation Budget Experiment (GERB) is a visible-infrared radiometer for earth radiation budget studies [16]. It makes accurate measurements of the shortwave and longwave components of the radiation budget at the top of the atmosphere [15].

Considering the purpose of this work, we have used reflectivity information obtained from SEVIRI spectral channels VIS 0.6 and VIS 0.8. This magnitude is obtained at LATUV Remote Sensing Laboratory (Universidad de Valladolid) from Level 1.5 image data considering the Sun's irradiance, the Sun's zenith angles and the Earth-Sun distance for each day.

Our study area is the radiometric station of Toledo, Spain (39° 53'N, 4° 02'W, altitude 515 m). One complete year of hourly global solar radiation data (from the 1st of May 2013 to the 30st of April 2014) was available. In order to estimate the solar radiation at this location using Meteosat data, a closeness criterion was applied to determine which satellite information would be used. Specifically, we have extracted the reflectivity information from the nearest pixel to the location of the radiometric station (meaning the minimum Euclidean distance considering latitude and longitude values). Additionally, reflectivity information from the eight pixels surrounding the central one has also been extracted. This way, 18 reflectivity values (9 for each visible channel) every 15 min were available to calculate the cloud index for each pixel. Finally, because global solar radiation data at Toledo were only available at 1-h temporal resolution, we calculated the mean hourly value for the each of the 18 pixels considered.

With the reflectivity information we have calculated the cloud index following the HELIOSAT-2 method [17]. In this model, the cloud index $n(i, j)$ is defined at instant t and for pixel (i, j) as follows:

$$n(i, j) = \frac{\rho(i, j) - \rho_g(i, j)}{\rho_{cloud}(i, j) - \rho_g(i, j)}$$

In this equation, $\rho(i, j)$ is the reflectivity, or apparent albedo, observed by the sensor for the time t and the pixel (i, j), $\rho_{cloud}(i, j)$ is the apparent albedo of the clouds, and $\rho_g(i, j)$ is the apparent albedo of the ground under clear sky.

Following the approach suggested by [11] we have chosen the following physical-based estimation model as a reference to assess the performance of the methodology suggested: the clear-sky index, K_{clear}, is equal to the ratio of the global solar radiation at ground, G and the same quantity but considering a *clear sky* model, G_{clear}:

$$K_{clear} = \frac{G}{G_{clear}}$$

With the K_{clear} parameter we can obtain G, because G_{clear} are known values obtained from the clear-sky model. Hence, following the indications in [17], K_{clear} will be calculated, depending on cloud index, as:

$n < 0.2$, $K_{clear} = 1.2$
$0.2 < n < 0.8$, $K_{clear} = 1 - n$
$0.8 < n < 1.1$, $K_{clear} = 2.0667 - 3.6667n + 1.6667n^2$
$n > 1.1$, $K_{clear} = 0.05$

With this procedure we can obtain the values of G for the pixel near the measuring station (physical-based model) and then compare them with our proposal using the ELM estimation.

The complete list of input and target variables considered in the ELM are summarized in Table 1. Time series of hourly data go from 05:00 a.m. to 08:00 p.m. Missing reflectivity values in each visible channel (0.6 and 0.8 μm) are detected through a preprocessing task carried out before doing the experiments.

Table 1. Predictive variables and target used in the experiments (ELM).

Predictive variables	Units
Reflectivity	[%]
Clear sky radiance	[W/m^2]
Cloud index	[%]
Target	Units
Global solar radiation	[W/m^2]

Note that the ELM estimates the Solar radiation using 37 input values (18 reflectivity values, 18 values for the cloud index and the clear-sky value for Toledo station). We compare this case with the ELM using 19 input values, clear-sky plus cloud index and also, with the physical-based model described above.

3 The Extreme-Learning Machine

An extreme-learning machine [18] is a novel and fast learning method based on the structure of multi-layer perceptrons that trains feed-forward neural networks with a perceptron structure. The most significant characteristic of the training of the extreme-learning machine is the random setting of the network weights from which a pseudo-inverse of the hidden-layer output matrix is obtained. The advantage of this technique is its simplicity, which makes the training algorithm extremely fast, while comparing excellently with cutting-edge learning methods, as well as other established approaches, such as classical multi-layer perceptrons and support-vector-regression algorithms. Both the universal-approximation and classification capabilities of the extreme-learning-machine network have been demonstrated in [19].

The extreme-learning-machine algorithm is summarized by taking a training set,

$$\mathbb{T} = (\mathbf{x}_i, \boldsymbol{\vartheta}_i) | \mathbf{x}_i \in \mathbb{R}^n, \boldsymbol{\vartheta}_i \in \mathbb{R}, i = 1, \cdots, l,$$

where x_i are the inputs and ϑ_i is the target (Global Solar Radiation), an activation function $g(x)$, and the number of hidden nodes (N), and applying the following steps:

1. Randomly assign input weights $\mathbf{w}_i$ and the bias b_i, where $i = 1, \cdots, N$, using a uniform probability distribution in $[-1, 1]$.
2. Calculate the hidden-layer output matrix $\mathbf{H}$, defined as

$$\mathbf{H} = \begin{bmatrix} g(\mathbf{w}_1\mathbf{x}_1 + b_1) & \cdots & g(\mathbf{w}_N\mathbf{x}_1 + b_N) \\ \vdots & \cdots & \vdots \\ g(\mathbf{w}_1\mathbf{x}_l + b_1) & \cdots & g(\mathbf{w}_N\mathbf{x}_N + b_N) \end{bmatrix}_{l \times N}. \tag{1}$$

3. Calculate the output weight vector β as

$$\beta = \mathbf{H}^\dagger \mathbf{T}, \tag{2}$$

where $\mathbf{H}^\dagger$ is the Moore-Penrose inverse of the matrix $\mathbf{H}$ [18], and $\mathbf{T}$ is the training output vector, $\mathbf{T} = [\boldsymbol{\vartheta}_1, \cdots, \boldsymbol{\vartheta}_l]^T$.

Note that the number of hidden nodes (N) is a free parameter to be set before the training of the extreme-learning machine, and must be estimated for obtaining good results. In this problem, The mechanism used to obtain N consists of a search of the best number of neurons among a range of values. Usually the range of values is set from 50 until 150 and depending on the set of samples (in the validation set) we will obtain one value or another.

We use the extreme-learning machine implemented in Matlab by Huang, which is freely available at [20].

4   Experiments and Results

In order to compare the proposed ELM approach with the physical-based model for global solar radiation estimation, we describe the methodology carried out to obtain the final results. Table 2 summarizes the comparative results between the proposed approaches, in terms of RMSE (in the ELM case, we indicate the RMSE for the training (TrS) and test set (TS)) and the Pearson's Correlation Coefficient (r^2). As previously mentioned, the physical-based model has been compared with 2 ELM scenarios: the first one takes into account the clear sky radiation and the cloud index, and the second one is obtained when the clear sky radiation, the cloud index and the reflectivity values are used as input variables in the ELM algorithm. We can observe how the best results are obtained by the ELM, with a RMSE in the test set of 112.46 W/m^2 against 146.06 W/m^2 in the case of the physical-based model for the first scenario. Moreover, the r^2 is around 85% in the ELM approach whereas the physical-based model only gets a 76%. In the second scenario (37 variables in the ELM), the results are improved with the use of reflectivity values from the satellite as part of the predictors in the ELM algorithm, obtaining in this case a best RMSE of 101.45 W/m^2 and a r^2 of 87%.

Table 2. Comparative results of the global solar radiation estimation by the ELM and the physical-based model. Scenario 1 (19 input variables): clear sky radiation and cloud index as predictors. Scenario 2 (37 input variables): clear sky radiation, cloud index and reflectivities as predictors.

Experiments	RMSE [W/m^2]: TrS	RMSE [W/m^2]: TS	r^2
Scenario 1			
Physical model	-	146.06	0.7606
ELM	102.85	**112.46**	**0.8544**
Scenario 2			
Physical model	-	146.06	0.7606
ELM	91.95	**101.45**	**0.8738**

Figure 1 shows the prediction of the global solar radiation by the ELM approach over 100 test samples (randomly selected, without keeping the time series structure). This figure shows how the prediction obtained by the ELM is highly accurate in this problem, both for hours with high values of radiation as well as for hours in which the solar radiation reaching the study area is low.

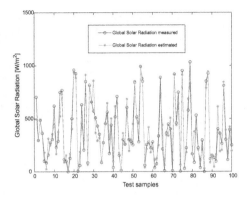

Fig. 1. Global solar radiation prediction in time by the ELM.

5   Conclusions

In this paper we have developed a methodology for global solar radiation based on the application of an ELM network to satellite data. The study has been made over the radiometric station of Toledo, Spain, where several input variables have been used in the experiments: a clear-sky model solar radiation estimation, the cloud index, and several reflectivity values from Meteosat visible images. The data are available from May of 2013 to April of 2014, although a preprocessing of the data-base has been necessary because of the missing values in the time series. The experiments carried out show how the performance of the ELM is better than the physical-based model, with a Pearson's correlation Coefficient around 87% in the best case against the 76% achieved by the physical-based model. As a future work, it could be interesting to develop more experiments, where the performance of different machine learning techniques are compared.

Acknowledgement. This work has been partially supported by the Spanish Ministry of Economy, through project number TIN2017-85887-C2-2-P.

References

1. Kalogirou, S.A.: Designing and modeling solar energy systems. In: Solar Energy Engineering, 2nd edn, chap. 11, pp. 583–699 (2014)
2. Kannan, N., Vakeesan, D.: Solar energy for future world: - a review. Renew. Sustain. Energy Rev. **62**, 1092–1105 (2016)
3. Khatib, T., Mohamed, A., Sopian, K.: A review of solar energy modeling techniques. Renew. Sustain. Energy Rev. **16**, 2864–2869 (2012)
4. Inman, R.H., Pedro, H.T., Coimbra, C.F.: Solar forecasting methods for renewable energy integration. Prog. Energy Combust. Sci. **39**(6), 535–576 (2013)
5. Mellit, A., Kalogirou, S.A.: Artificial intelligence techniques for photovoltaic applications: a review. Prog. Energy Combust. Sci. **34**(5), 574–632 (2008)
6. Mubiru, J.: Predicting total solar irradiation values using artificial neural networks. Renew. Energy **33**, 2329–2332 (2008)

7. Alharbi, M.A.: Daily global solar radiation forecasting using ANN and extreme learning machines: a case study in Saudi Arabia. Master of Applied Science thesis, Dalhousie University, Halifax, Nova Scotia (2013)

8. Dong, H., Yang, L., Zhang, S., Li, Y.: Improved prediction approach on solar irradiance of photovoltaic power station. TELKOMNIKA Indones. J. Electr. Eng. **12**(3), 1720–1726 (2014)

9. Salcedo-Sanz, S., Casanova-Mateo, C., Pastor-Sánchez, A., Gallo-Marazuela, D., Labajo-Salazar, A., Portilla-Figueras, A.: Direct solar radiation prediction based on soft-computing algorithms including novel predictive atmospheric variables. In: Yin, H., et al. (eds.) IDEAL 2013. LNCS, vol. 8206, pp. 318–325. Springer, Heidelberg (2013). https://doi.org/10.1007/978-3-642-41278-3_39

10. Salcedo-Sanz, S., Casanova-Mateo, C., Pastor-Sánchez, A., Sánchez-Girón, M.: Daily global solar radiation prediction based on a hybrid coral reefs optimization - extreme learning machine approach. Sol. Energy **105**, 91–98 (2014)

11. Senkal, O., Kuleli, T.: Estimation of solar radiation over Turkey using artificial neural network and satellite data. Appl. Energy **86**(7–8), 1222–1228 (2009)

12. Sahin, M., Kaya, Y., Uyar, M., Yidirim, S.: Application of extreme learning machine for estimating solar radiation from satellite data. Int. J. Energy Res. **38**(2), 205–212 (2014)

13. Schmid, J.: The SEVIRI instrument. In: Proceedings of the 2000 EUMETSAT Meteorological Satellite, Data User's Conference, Bologna, Italy, 29 May–2 June 2000, pp. 13–32. EUMETSAT ed., Darmstadt (2000)

14. Aminou, D.M.A.: MSG's SEVIRI instrument. ESA Bull. **111**, 15–17 (2002)

15. Schmetz, J., Pili, P., Tjemkes, S., Just, D., Kerkmann, J., Rota, S., et al.: An introduction to meteosat second generation (MSG). Am. Meteorol. Soc. **83**(7), 977–992 (2002)

16. Harries, J.E.: The geostationary earth radiation budget experiment: status and science. In: Proceedings of the 2000 EUMETSAT Meteorological Satellite Data Users' Conference, Bologna, EUM-P29, pp. 62–71 (2000)

17. Rigollier, C., Lefévre, M., Wald, L.: The method Heliosat-2 for deriving shortwave solar radiation from satellite images. Sol. Energy **77**, 159–169 (2004)

18. Huang, G.B., Zhu, Q.Y.: Extreme learning machine: theory and applications. Neurocomputing **70**, 489–501 (2006)

19. Huang, G.B., Zhou, H., Ding, X., Zhang, R.: Extreme learning machine for regression and multiclass classification. IEEE Trans. Syst. Man Cybern. Part B **42**(2), 513–529 (2012)

20. Huang, G.B.: ELM matlab code. http://www.ntu.edu.sg/home/egbhuang/elm_codes.html

Distribution-Based Discretisation and Ordinal Classification Applied to Wave Height Prediction

David Guijo-Rubio[✉], Antonio M. Durán-Rosal, Antonio M. Gómez-Orellana, Pedro A. Gutiérrez, and César Hervás-Martínez

Department of Computer Science and Numerical Analysis, Universidad de Córdoba, Córdoba, Spain
{dguijo,aduran,am.gomez,pagutierrez,chervas}@uco.es

Abstract. Wave height prediction is an important task for ocean and marine resource management. Traditionally, regression techniques are used for this prediction, but estimating continuous changes in the corresponding time series can be very difficult. With the purpose of simplifying the prediction, wave height can be discretised in consecutive intervals, resulting in a set of ordinal categories. Despite this discretisation could be performed using the criterion of an expert, the prediction could be biased to the opinion of the expert, and the obtained categories could be unrepresentative of the data recorded. In this paper, we propose a novel automated method to categorise the wave height based on selecting the most appropriate distribution from a set of well-suited candidates. Moreover, given that the categories resulting from the discretisation show a clear natural order, we propose to use different ordinal classifiers instead of nominal ones. The methodology is tested in real wave height data collected from two buoys located in the Gulf of Alaska and South Kodiak. We also incorporate reanalysis data in order to increase the accuracy of the predictors. The results confirm that this kind of discretisation is suitable for the time series considered and that the ordinal classifiers achieve outstanding results in comparison with nominal techniques.

Keywords: Wave height prediction · Distribution fitting
Time series discretisation · Autoregressive models
Ordinal classification

This work has been subsidised by the projects with references TIN2017-85887-C2-1-P and TIN2017-90567-REDT of the Spanish Ministry of Economy and Competitiveness (MINECO), FEDER funds, and the project PI15/01570 of the Fundación de Investigación Biomédica de Córdoba (FIBICO). David Guijo-Rubio's and Antonio M. Durán-Rosal's researches have been subsidised by the FPU Predoctoral Program (Spanish Ministry of Education and Science), grant references FPU16/02128 and FPU14/03039, respectively.

© Springer Nature Switzerland AG 2018
H. Yin et al. (Eds.): IDEAL 2018, LNCS 11315, pp. 171–179, 2018.
https://doi.org/10.1007/978-3-030-03496-2_20

1   Introduction

Due to the difficulty of predicting a real valued output, discretisation can be considered in order to transform the original problem into a classification one [7,8], in those cases where the information given by the resulting categories is enough for taking the corresponding decisions. This transformation simplifies the prediction task and can increase the robustness of the obtained models. The discretisation is defined by a set of threshold values, which are usually given by experts. However, this can introduce some bias in the models, and it seems more appropriate to guide the process by using the properties of the data.

Discretisation of time series has been applied in several fields, including wave height prediction, which is a hot topic in renewable and sustainable systems for energy supply [10]. Oceans are becoming a promising source of clean and sustainable energy in many countries. Among other techniques, wave energy conversion is gaining popularity because of its good balance between cost and efficiency. However, wave height prediction becomes necessary for designing and controlling wave energy converters, which are the devices responsible for transforming the energy of waves into electricity using either the vertical oscillation or the linear motion of the waves. The data for characterising waves and performing the prediction is usually obtained from sensors integrated at buoys located in the sea, resulting in a time series. Moreover, reanalysis data can provide further information to increase the accuracy of the predictions.

On the other hand, AutoRegressive models (AR) [6] are one of the most common approaches for time series prediction, where past lagged values of the time series are used as inputs. The main reason behind their use is the high correlation among the lagged events of real-world time series. In the specific field of wave height prediction, previous approaches include the use of artificial neural networks [4], dynamic AR models [8] and soft computing methods [11]. Although some of these works approach a categorical prediction of wave height [7,8], the categories are directly defined by experts.

Specifically, this paper deals with a problem of significant wave height prediction, tackling it as a classification problem. We propose an automated procedure to characterise the time series, using different statistical distributions, such as Generalized Extreme Value (GEV), Weibull, Normal and Logistic distributions. According to these distribution, the quartiles are used to categorise the time series in four wave height categories. We use AR models considering four time series as inputs, obtained from four different reanalysis variables, and the wave height (target variable), which is directly obtained from the buoy. Finally, because of the order of the corresponding categories, ordinal classifiers [9] are used, which are able to incorporate this order during learning. We test five different ordinal classifiers, achieving better performance than their nominal counterparts.

The rest of the paper is organised as follows: Sect. 2 shows the proposed methodology and the main contribution of this paper. Section 3 describes the data considered, the experimental design and the discussion of the results obtained. Finally, Sect. 4 concludes the work.

2   Methodology

The contribution of this paper is twofold: firstly, we determine the best-fitting probabilistic distribution for reducing the information of a time series into categories; and secondly, we apply ordinal classification methods to exploit the order information of the obtained categories.

2.1   Discretization of Wave Height

To discretise the wave height time series, we consider a method based on deciding the best fitting distribution, from a set of candidate ones. Four distributions are considered:

- Generalized Extreme Value (GEV) distribution, whose cumulative distribution is:

$$F(y; k, \mu, \sigma) = \begin{cases} \exp(-(1 + \frac{k(y-\mu)}{\sigma})^{-1/k}) & \text{for } k \neq 0, \\ \exp(-\exp(-\frac{y-\mu}{\sigma})) & \text{for } k = 0, \end{cases} \tag{1}$$

  where k is the shape parameter, σ is the scale parameter and μ is the location parameter.
- The Normal distribution with the following cumulative function:

$$F(y; \mu, \sigma) = \frac{1}{2}\left[1 + \text{erf}\left(\frac{y-\mu}{\sigma\sqrt{2}}\right)\right], \tag{2}$$

  where erf is the Gauss error function.
- The Weibull distribution, defined by:

$$F(y; k, \sigma) = 1 - \exp\left[-\left(\frac{y}{\sigma}\right)^k\right]. \tag{3}$$

- And the Logistic distribution:

$$F(y; \mu, \sigma) = \frac{1}{2} + \frac{1}{2}\tanh\left(\frac{y-\mu}{2\sigma}\right). \tag{4}$$

Using the training data, we apply a Maximum Likelihood Estimator (MLE) procedure [12] to adjust the different parameters of the four distributions. After that, the best distribution is selected based on two objectives criteria:

- The Bayesian Information Criterion (BIC) [13] minimizes the bias between the fitted model and the unknown true model, and it is defined as:

$$\text{BIC} = -2\ln L + n_p \ln N, \tag{5}$$

where L is the likelihood of the fit, N is the number of data points, and n_p the number of parameters of the distribution.

– The Akaike Information Criterion (AIC) [5] searches for the best compromise between bias and variance:

$$\text{AIC} = -2\ln L + 2n_p. \tag{6}$$

Once the best distribution is selected, we use the corresponding 25%, 50% and 75% percentiles as the thresholds (Q_1, Q_2, Q_3, respectively) to discretise the output variable, in training and test sets. Our main hypothesis is that these theoretical distributions, properly adjusted to fit the training data, will provide better robustness in the test set than selecting the thresholds directly from the histograms of the training set.

2.2   Ordinal Classification

As we stated in the previous section, we discretise the wave weight (target variable, y_t) in four different categories. In this way, $y_t \in C_1, C_2, C_3, C_4$, where C_1 ($y_t \leq Q_1$) represents LOW wave height, C_2 ($y_t \in (Q_1, Q_2]$) represents AVERAGE wave height, C_3 ($y_t \in (Q_2, Q_3]$) represents BIG wave height, and, finally, C_4 ($y_t > Q_3$) represents HUGE wave height. Therefore, we have a natural order between these labels. The type of classification in which there is an order relationship between the categories is known as ordinal classification [9]. In this paper, we consider the following ordinal classifiers:

– The Proportional Odds Model (POM) is the first model developed for ordinal classification. POM is a linear model, which extends binary logistic regression for obtaining cumulative probabilities. The model includes a linear projection and a set of thresholds, which divide this linear projection into categories.
– Kernel Discriminant Learning for Ordinal Regression (KDLOR) is a widely used discriminant learning method used for OR. This algorithm minimises a quantity to ensure the order, which measures the distance between the averages of the projection of any two adjacent classes.
– Support Vector Machines (SVMs) have been widely used for OR (SVOR). Different methods have been introduced in the literature:
  • SVOR considering Explicit Constraints (SVOREX) only uses the patterns from adjacent classes to compute the error of a hyperplane, so this algorithms tends to lead to a better performance in terms of accuracy.
  • SVOR considering Implicit Constraints (SVORIM) uses all the patterns to compute the error of a hyperplane, in this way, this algorithm tends to get a better performance in terms of absolute difference between predicted categories.
  • REDuction framework applied to Support Vector Machines (REDSVM), which applies a reduction from ordinal regression to binary support vector classifiers in three steps: extracting extended examples from the original examples, learning a binary classifier on the extended examples, and constructing a ranker from the binary classifier.

Further details about these methods can be found in [9] and references therein.

In this paper, we focus on AR models, which generate a set of input variables based on the previous values of the observed time series, i.e. lagged events are used to predict the current value. Specifically, the m previous events of the time series are used as input. In this way, the dataset will be defined by $\mathbf{D} = (\mathbf{X}, \mathbf{Y}) = \{(\mathbf{x}_t, y_t)\}_{t=1}^n$, where y_t is the target discretised category and $\mathbf{x}_t$ is a set of inputs based on the previous events, $\mathbf{x}_t = \{\mathbf{x}_{t-1}, y_{t-1}, \mathbf{x}_{t-2}, y_{t-2}, \ldots, \mathbf{x}_{t-m}, y_{t-m}\}$. Note that the inputs take into account the independent and dependent variables of the m previous events.

3   Experiments and Results

In this section, we present the dataset and the experimental setting used, and we discuss the results obtained.

3.1   Dataset Used

The presented methodology has been tested on meteorological time series data obtained from two different buoys located at the Gulf of Alaska of the USA. These buoys collect meteorological data hourly using the sensors installed on it. These data are stored, and they can be obtained by downloadable annual text files in the National Oceanic and Atmospheric Administration (NOAA) [3], specifically in the National Data Buoy Center (NDBC), that maintains a network of data collecting for buoys and coastal stations. Specifically, we have selected the following two buoys: 1) Station 46001 (LLNR 984) – Western Gulf of Alaska, geographically located at 56.304N 147.92W (56° 18′ 16″ N 147° 55′ 13″ W). 2) Station 46066 (LLNR 984.1) – South Kodiak, geographically located at coordinates 52.785N 155.047W (52° 47′ 6″ N 155° 2′ 49″ W). The data from Station 46001 covers from 2013 January 1st (0:00) to 2017 December 31st (23:00), while the second one from 2013 August 24th (13:00) to 2017 December 31st (23:00).

From these two buoys, we have considered the wave height as the variable to predict, after discretising it using the theoretical distributions described in Sect. 2.1. On the other hand, we include reanalysis data from the NCEP/NCAR Reanalysis Project web page [1], which maintains sea surface level data around the world in a global grid of resolution $2.5° \times 2.5°$. In order to collect accurate information, we have considered the four points closest to each buoy (north, south, east and west), in a 6-hours time horizon resolution, which is the minimum resolution given by the Earth System Research Laboratory (ESRL). We have used four variables as inputs: pressure, air temperature, the zonal component of the wind and the meridional component of the wind [2]. A matching procedure has been performed every 6 hours between the reanalysis data obtained from ESRL and the wave height measured by the buoys. As there were some missing points, these values were approximated by taking the mean values between the previous three instants and the next three instants. The total number of patterns, N, was 7304 and 6351 for the Station 46001 and 46066, respectively. Finally, for both buoys, the datasets are split in training and test data. The training set is formed by the values collected in 2013, 2014 and 2015; while the test sets are formed by the rest of values (data in 2016 and 2017). The number of values of training/test are 4380/2924 for the buoy 46001, and 3427/2924 for the buoy 46066.

3.2   Experimental Settings

The five ordinal regression models presented in Sect. 2.2 have been compared to the following base-line methods: (1) Support Vector Regression (SVR), which is extensively used due to its good performance for complex regression techniques. We apply this regression technique by mapping ordinal labels to real values. (2) Nominal classification techniques, which can be applied in ordinal classification, by ignoring order information. In our case, we use a Support Vector Classifier (SVC) which is adapted to multiclass classification following two different strategies: SVC1V1, which considers a one-vs-one decomposition, and SVC1VA, which considers one-vs-all. (3) Cost-Sensitive techniques which weight the misclassification errors with different costs. We select the Cost-Sensitive Support Vector Classifier (CSSVC), with one vs all decomposition. The costs of misclassifications are different depending on the distance between the real class and the class consider by the corresponding binary, taking into account the ordinal scale. All these comparisons are focused on showing the necessity of using ordinal classification in this real problem.

Two different performance measures were used to evaluate the predictions obtained, $\hat{y}_i$, against the actual ones, y_i: (1) the accuracy (ACC) is the percentage of correct predictions on individual samples: $ACC = (100/N) \sum_{i=1}^{N} I(\hat{y}_i = y_i)$, where $I(\cdot)$ is the zero-one loss function and n is the number of patterns of the dataset. This measure evaluates a globally accurate prediction. (2) The mean absolute error (MAE) is the average deviation of the predictions: $MAE = (1/N) \sum_{i=1}^{N} |O(\hat{y}_i) - O(y_i)|$, where $O(C_k) = k; k = \{1, \ldots, K\}$, i.e. $O(y_i)$ is the order of class label y_i. K represents the number of categories ($K = 4$, in our case). This measure evaluates how far are the predictions, in average number of categories, from the true targets.

Now we discuss how the parameters are tuned for all the steps of the methodology. To optimize the hyper-parameters of the classification models, a cross-validation method is applied to the training dataset, deciding the most adequate parameter values without checking out the test performance. The validation criterion used for selecting the parameters is the minimum MAE value. As we use AR models, the best number of previous events to be considered is adjusted using the grid $m \in \{1, 2, \ldots, 5\}$. To adjust the kernel width and cost parameter for the SVM-based methods (SVC1V1, SVC1VA, SVR, CSSVC, REDSVM, SVOREX and SVORIM), the range considered is $k \in \{10^{-3}, 10^{-2}, \ldots, 10^3\}$.

The kernel width of KDLOR is optimized using the same range than SVM-based methods, while the KDLOR regularization parameter (for avoiding singularities while inverting matrices) is adjusted in the range $u \in \{10^{-2}, 10^{-3}, \ldots, 10^{-6}\}$, and the cost of KDLOR in range $C \in \{10^{-1}, 10^{0}, 10^{1}\}$. Finally, for SVR, an additional parameter is needed, ϵ, which is adjusted as $\epsilon \in \{10^{0}, 10^{1}, \ldots, 10^{3}\}$. Note that the POM algorithm does not have hyper-parameters to be optimized.

3.3   Results and Discussion

Table 1 shows the BIC and AIC criteria for all fitted distributions in both datasets. As can be seen, the best fitted distribution in the two buoys is the GEV distribution, presenting the lowest values in BIC and AIC criteria. The second best-fitted one is the Weibull distribution. These results show that, for wave height time series, extreme values distributions are more adequate. In this way, GEV distributions are used for the prediction phase.

Table 1. BIC and AIC for the four distributions considered and the two buoys.

Station 46001			Station 46066		
Distribution	$BIC(\downarrow)$	$AIC(\downarrow)$	Distribution	$BIC(\downarrow)$	$AIC(\downarrow)$
GEV	**13901**	**13882**	GEV	**11523**	**11504**
Normal	14809	14796	Normal	12335	12323
Weibull	*14116*	*14104*	Weibull	*11777*	*11765*
Logistic	14741	14728	Logistic	12169	12157

The best result is in bold face and the second one in italics

Once the output variable is discretised, Table 2 shows the results of the prediction for all the classification algorithms compared, including ACC and MAE. In general, very good results are obtained (with values of ACC higher than 75% and MAE lower than 0.25). AR models are shown to provide enough information for this prediction problem, possibly due to the high persistence of the data. Ordinal classifiers obtain better performance than regression techniques, nominal classification methods or cost-sensitive methods, thus justifying the use of ordinal methods. In case of Station 46001, there are two algorithms that achieve the best performance, REDSVM and SVOREX, with $ACC = 77.4624$ and $MAE = 0.2305$. For the Station 46066, REDSVM also obtained the best results with $ACC = 76.8126$ and $MAE = 0.2349$.

Table 2. Results obtained by the different classification algorithms on the two buoys, using the GEV discretisation.

Station 46001			Station 46066		
Algorithm	$ACC(\uparrow)$	$MAE(\downarrow)$	Algorithm	$ACC(\uparrow)$	$MAE(\downarrow)$
SVC1V1	76.4706	0.2411	SVC1V1	76.0260	0.2442
SVC1VA	74.8632	0.2579	SVC1VA	74.7606	0.2579
SVR	75.3420	0.2538	SVR	76.5048	0.2387
CSSVC	75.2736	0.2541	CSSVC	75.1026	0.2421
POM	73.8030	0.2733	POM	73.3584	0.2777
KDLOR	76.9152	0.2350	KDLOR	75.9918	0.2421
REDSVM	**77.4624**	**0.2305**	REDSVM	**76.8126**	**0.2349**
SVOREX	**77.4624**	**0.2305**	SVOREX	76.3338	0.2401
SVORIM	*77.3940*	*0.2538*	SVORIM	*76.7442*	*0.2353*

The best result is in bold face and the second one in italics

4   Conclusions

This paper evaluates the use of four different distributions to reduce the information of a time series in a problem of wave height prediction. The best distribution is selected based on two estimators of their quality and used for discretising wave height in four different classes, using the four quartiles of the distribution. After that, an autoregressive structure is combined with ordinal classification methods to tackle the prediction of the categories. Two real datasets were considered in the experimental validation, and the REDuction applied to Support Vector Machines (REDSVM) was the ordinal classifier achieving the best performance in both. As future work, dynamic windows could be used instead of the fixed ones, which could be able to better exploit the dynamics of the time series.

References

1. NCEP/NCAR: The NCEP/NCAR Reanalysis Project, NOAA/ESRL Physical Sciences Division. https://www.esrl.noaa.gov/psd/data/reanalysis/reanalysis.shtml. Accessed 19 July 2018
2. NCEP/NCAR: The NCEP/NCAR Reanalysis Project Sea Surface Level Variables 6-hourly. https://www.esrl.noaa.gov/psd/data/gridded/data.ncep.reanalysis.surface.html. Accessed 19 July 2018
3. NOAA/NDBC: National Oceanic and Atmospheric Administration (NOAA), National Data Buoy Center (NDBC). http://www.ndbc.noaa.gov. Accessed 19 July 2018
4. Agrawal, J., Deo, M.: Wave parameter estimation using neural networks. Mar. Struct. **17**(7), 536–550 (2004)
5. Akaike, H.: Information theory and an extension of the maximum likelihood principle. In: Parzen, E., Tanabe, K., Kitagawa, G. (eds.) Selected Papers of Hirotugu Akaike. Springer Series in Statistics (Perspectives in Statistics), pp. 199–213. Springer, Heidelberg (1998). https://doi.org/10.1007/978-1-4612-1694-0_15

6. Brockwell, P.J., Davis, R.A.: Time Series: Theory and Methods. Springer, New York (2013). https://doi.org/10.1007/978-1-4899-0004-3
7. Fernández, J.C., Salcedo-Sanz, S., Gutiérrez, P.A., Alexandre, E., Hervás-Martínez, C.: Significant wave height and energy flux range forecast with machine learning classifiers. Eng. Appl. Artif. Intell. **43**, 44–53 (2015)
8. Gutiérrez, P.A., et al.: Energy flux range classification by using a dynamic window autoregressive model. In: Rojas, I., Joya, G., Catala, A. (eds.) IWANN 2015. LNCS, vol. 9095, pp. 92–102. Springer, Cham (2015). https://doi.org/10.1007/978-3-319-19222-2_8
9. Gutiérrez, P.A., Pérez-Ortiz, M., Sánchez-Monedero, J., Fernandez-Navarro, F., Hervás-Martínez, C.: Ordinal regression methods: survey and experimental study. IEEE Trans. Knowl. Data Eng. **28**(1), 127–146 (2016)
10. López, I., Andreu, J., Ceballos, S., de Alegría, I.M., Kortabarria, I.: Review of wave energy technologies and the necessary power-equipment. Renew. Sustain. Energy Rev. **27**, 413–434 (2013)
11. Mahjoobi, J., Etemad-Shahidi, A., Kazeminezhad, M.: Hindcasting of wave parameters using different soft computing methods. Appl. Ocean Res. **30**(1), 28–36 (2008)
12. Mathiesen, M., et al.: Recommended practice for extreme wave analysis. J. Hydraul. Res. **32**(6), 803–814 (1994)
13. Schwarz, G., et al.: Estimating the dimension of a model. Ann. Stat. **6**(2), 461–464 (1978)

Wind Power Ramp Events Ordinal Prediction Using Minimum Complexity Echo State Networks

M. Dorado-Moreno[1]([✉]), P. A. Gutiérrez[1], S. Salcedo-Sanz[2], L. Prieto[3], and C. Hervás-Martínez[1]

[1] Department of Computer Science and Numerical Analysis, University of Cordoba, Córdoba, Spain
manuel.dorado@uco.es
[2] Department of Signal Processing and Communications, University of Alcalá, Alcalá de Henares, Spain
[3] Department of Energy Resource, Iberdrola, Madrid, Spain

Abstract. Renewable energy is the fastest growing source of energy in the last years. In Europe, wind energy is currently the energy source with the highest growing rate and the second largest production capacity, after gas energy. There are some problems that difficult the integration of wind energy into the electric network. These include wind power ramp events, which are sudden differences (increases or decreases) of wind speed in short periods of times. These wind ramps can damage the turbines in the wind farm, increasing the maintenance costs. Currently, the best way to deal with this problem is to predict wind ramps beforehand, in such way that the turbines can be stopped before their occurrence, avoiding any possible damages. In order to perform this prediction, models that take advantage of the temporal information are often used. One of the most well-known models in this sense are recurrent neural networks. In this work, we consider a type of recurrent neural networks which is known as Echo State Networks (ESNs) and has demonstrated good performance when predicting time series. Specifically, we propose to use the Minimum Complexity ESNs in order to approach a wind ramp prediction problem at three wind farms located in the Spanish geography. We compare three different network architectures, depending on how we arrange the connections of the input layer, the reservoir and the output layer. From the results, a single reservoir for wind speed with delay line reservoir and feedback connections is shown to provide the best performance.

This work has been subsidized by the projects with references TIN2017-85887-C2-1-P, TIN2017-85887-C2-2-P and TIN2017-90567-REDT of the Spanish Ministry of Economy and Competitiveness (MINECO) and FEDER funds. Manuel Dorado-Moreno's research has been subsidised by the FPU Predoctoral Program (Spanish Ministry of Education and Science), grant reference FPU15/00647. The authors acknowledge *NVIDIA Corporation* for the grant of computational resources through the *GPU Grant Program*.

H. Yin et al. (Eds.): IDEAL 2018, LNCS 11315, pp. 180–187, 2018.
https://doi.org/10.1007/978-3-030-03496-2_21

Keywords: Echo state networks · Wind energy
Ordinal classification · Wind power ramp events
Recurrent neural networks

1   Introduction

Nature provides us with multiple ways of producing sustainable energy without pollution emissions. These type of energy exploit natural renewable resources and are currently the fastest growing sources worldwide. Among them, the most common are solar, wind and marine energies, as well as their combinations, although there are other alternatives such as biomass or hydropower. Our work focuses on wind energy, specifically, on its production at wind farms, where wind turbines use wind speed to generate energy. One of the main problems in wind farms is known as wind power ramp events (WPREs), defined as large increases or decreases of wind speed in a short period of time. Wind ramps can be positive (increases of the wind speed) or negative (due to a decrease). The effect of positive ramps is mainly the possible damage that can be caused to the turbines, which leads to an increase in the maintenance costs of the wind farm. On the other hand, negative ramps can produce a sudden decrease in the energy production, which can carry energy supply problems if it is not predicted with sufficient advance.

Many problems related to renewable energies have been approached using machine learning techniques, e.g. in solar energy [2], wave energy [8] or wind energy [4–6]. In machine learning, one of the most well known models to deal with time series and perform predictions are recurrent neural networks [11]. Their difference with standard neural networks is the inclusion of cycles among their neurons, i.e. connections from a neuron to itself are allowed, or from a neuron to another neuron located in previous layers. In any case, when we increase the number of layers of a recurrent neural network to increase its computational capacity, it usually suffer from what is known as vanishing gradient problem [11]. This problem causes that, while computing the derivatives among the cycles, these tend to zero and do not contribute to the gradient, hindering the update of the network weights. One of the most widely accepted proposals to overcome this are the echo state networks (ESNs) which have a hidden layer, known as reservoir, which includes all the cycles, and where all the connection weights are randomly initialized. This reservoir is fully connected with the inputs and the outputs, and these last connections are the only ones which are trained. In this way, the vanishing gradient problem is avoided, because the cycle connections within the reservoir are not trained.

One of the difficulties associated to ESNs is their stochastic nature, because part of their performance depends on their random initialization. In order to solve this problem, in this work, we use minimum complexity ESNs proposed in [13], which establish their connections following a given pattern and initialize them in a deterministic way, which can be used to justify their performance. Furthermore, we propose three different architectures in line with a previous work [5], to compare the different ways in which the reservoir affects to the model results, depending on how the inputs are connected to it. It is important to note that, due to the natural order of the different categories to predict (positive ramp, non ramp and negative ramp), the problem is approached from an ordinal regression perspective [9].

Finally, two data sources are used for generating the different input variables. The first source of information corresponds to wind speed measurements, obtained hourly in three wind farms located in Spain, as can be observed in Fig. 1. We derive wind ramp categories as objective values to be predicted, using a ramp function a set of predictive variables. The second source of information, from which we obtain these predictive variables, is the ERA-Interim reanalysis project [7], which stores weather information every 6 h.

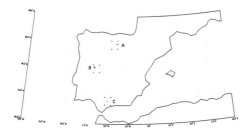

Fig. 1. Location of the three wind farms (A, B and C) and the reanalysis nodes.

The different architectures proposed for the modeling will be introduced in Sect. 2, just before explaining the experimental design in Sect. 3 and discussing the results obtained. Section 4 will conclude this work.

2   Proposed Architectures

In this paper, we propose a modification of the models considered in [5], including an ordinal multiclass prediction of WPREs. Moreover, we modify the reservoir structure based on the different proposals in [13], which reduce the complexity of the reservoir and also remove the randomness in the weights initialization without drastically reducing the model performance. A scheme of the three reservoir structures: Delay line reservoir (DLR), DLR with feedback connections (DLRB) and Simple cycle reservoir (SCR) can be analyzed in Fig. 2.

In the output layer, we use a threshold based ordinal logistic regression model [9,12], which projects the patterns into a one dimensional space and then optimizes a set of thresholds to classify the patterns into different categories.

Now, we describe the different architectures proposed to solve WPRE prediction, which explore different ways of combining the past values of wind speed and the reanalysis data from the ERA-Interim. In the input layer, we include the wind speed (at time t) and 12 reanalysis variables which can be estimated

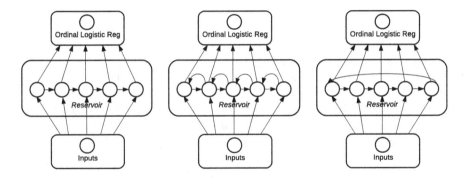

Fig. 2. Reservoir structures considered (DLR, DLRB, SCR) [13]

at time $t + 1$ (more details will be given in Sect. 3.1). We propose three architectures, which can be observed in Fig. 3. The first one (Simple) has a single reservoir directly connected to the past values of wind speed measured in the wind farm, while the reanalysis variables are directly connected to the output layer. The second proposal (Double) has two independent reservoirs, one for wind speed, and the other one for the reanalysis variables. Finally, our third proposal (Shared) has a single reservoir, but, in this case, it receives its inputs both from wind speed measured at the wind farm and the reanalysis variables. With these three architectures, we study and evaluate the computing capacity of the reservoir, as well as the usefulness of each type of variable.

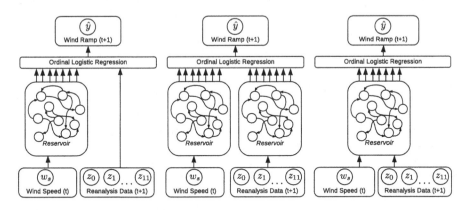

Fig. 3. Network architectures proposed (Simple, Double, Shared)

3 Experiments

In this section, we first describe the data considered, the evaluation metrics used for comparing the different methodologies and the experimental design carried out. Finally, we introduce and discuss the results obtained.

3.1 Dataset Considered

As previously explained, we consider data from three different wind farms in Spain (see Fig. 1). Wind speed is recorded hourly in each wind farm. The ramp function S_t will be used to decide whether a WPRE has happened. Our definition of S_t includes the production of energy (P_t) as criterion to describe the ramp, $S_t = P_t - P_{t-\Delta t_r}$, where Δt_r is the interval of time considered for characterizing the event (6 h in our case, to match the reanalysis data). Using a threshold value for the ramp function (S_0), we transform the regression problem into ordinal classification:

$$y_t = \begin{cases} \mathcal{C}_{\text{NR}}, & \text{if} \quad S_t \leq -S_0, \\ \mathcal{C}_{\text{NoR}}, & \text{if} \quad -S_0 < S_t < S_0, \\ \mathcal{C}_{\text{PR}}, & \text{if} \quad S_t \geq S_0. \end{cases}$$

where $\{\mathcal{C}_{\text{NR}}, \mathcal{C}_{\text{NoR}}, \mathcal{C}_{\text{PR}}\}$ correspond to negative ramp category, no ramp category and positive ramp category, respectively. We fix S_0 as a percentage of the production capacity of the wind farm (in our case, 50%).

The prediction of ramps will be based on past information of the ramp function and reanalysis data ($\mathbf{z}$) from the ERA-Interim reanalysis project [7]. Specifically, we consider 12 variables, including surface temperature, surface pressure, zonal wind component and meridional wind component at 10 m, and temperature, zonal wind component, meridional wind component and vertical wind component at 500 hPa and 850 hPa. These variables are taken from the four closest reanalysis nodes (see Fig. 1), considering a weighted average according to the distance from the wind farm to the reanalysis node. This reanalysis data is computed using physical models, i.e. they do not depend on any sensor which can generate missing data. More details about the data processing and the merge of both sources of information are given in [5].

3.2 Evaluation Metrics

There are many evaluation metrics for ordinal classifiers. The most common ones include accuracy and mean absolute error (MAE) [9], where the second one measures the average absolute deviation (in number of categories of the ordinal scale) of the predicted class with respect to the target one. Given that the problem considered is imbalanced (check Table 1), these metrics have to be complemented [1], giving more importance to minority classes. In this way, we have considered four metrics to evaluate the models (more details can be found in [6]): the minimum accuracy evaluated independently for each class (minimum sensitivity, MS), the geometric mean of these sensitivities (GMS), the average of the MAE values ($AMAE$) from the different classes and the standard accuracy or correctly classified rate (CCR). GM, $AMAE$ and MS are specifically designed for imbalanced datasets, while $AMAE$ is the only metric from these four ones taking into account the ordinal character of the targets.

3.3   Experimental Design

The three wind farms from Fig. 1 have been used in the results comparison of the different proposed structures and architectures. All the dataset covers data from 2002/3/2 to 2012/10/29. To evaluate the results, the three datasets have been divided in the same manner: the last 365 days are used for the test set and the rest of the dataset is used for training purposes. With this partition, the patterns per class of each of the three datasets is shown in Table 1, where we can find the number of patterns in each category (negative ramp, NR, non-ramp, NoR, and positive ramp, PR).

Table 1. Number of patterns per class of each wind farm

Dataset	A			B			C		
	#NR	#NoR	#PR	#NR	#NoR	#PR	#NR	#NoR	#PR
Train	753	12469	886	1161	11804	1074	661	12768	679
Test	67	1288	105	117	1220	123	58	1340	62

The different architectures presented in Sect. 2 have been compared among them, comparing also the different internal structures of the reservoir according to [13]. We want to find the architecture with the best performance and check whether the minimum complexity ESNs are enough to approach our problem.

Due to the imbalance degree of the problem, we perform a preliminary over-sampling using the SMOTE methodology [3] applied to the reservoir outputs (not to the input vectors), as explained and justified in [4]. For both minority classes, a 60% of the number of patterns of the majority class are generated as synthetic patterns.

The regularization parameter of ordinal logistic regression (α) is adjusted using a $5 - fold$ cross-validation over the training set. The grid of values considered is $\alpha \in \{2^{-5}, 2^{-4}, \ldots, 2^{-1}\}$. The selection of the best model is based on the maximum MS. The rest of parameters are configured in the following form: the number of neurons within the reservoir is $M = 50$, assuming that this is a sufficient size to approach this problem without incurring in a too high computational cost. The connection weights in the reservoir are established following an uniform distribution in $[-0.9, 0.9]$, and the matrix of weights is rescaled to fulfill the Echo State Property [10].

3.4   Results

All the results are included in Table 2, for the three architectures proposed, the three structures, the three wind farms and the four evaluation metrics. As can be observed, for the DLR structure, the Simple architecture wins in two out of the three wind farms, the Double wins in one and the Shared obtains the worst results. The high value of CCR should not be confused with good results, because the GMS of the model is really low, meaning that the performance is low for minority classes. On the contrary, for DLRB, the Double architecture wins for two wind farms, while the Single one obtains the second best result. The bad performance of the Shared architecture can also be observed for this reservoir structure. Finally, the results obtained with the SCR structure follow the same direction than the ones obtained with the DLR one: in two of the three wind farms, the Simple model obtains the best results.

Table 2. Results for the three architectures proposed (Double, Shared and Simple, Fig. 3) and the three structures (DLR, DLRB and SCR, Fig. 2). For each structure, the best architecture for each metric in each wind farm is in **bold** face and the second best in *italics*. The best global results for each wind farm is <u><u>double underlined</u></u> while the second best is <u>underlined</u>

Struc.	Metric	Wind farm								
		A			B			C		
		Architecture			Architecture			Architecture		
		Simple	Double	Shared	Simple	Double	Shared	Simple	Double	Shared
DLR	GMS	*0.6607*	**0.6951**	0.3056	**0.6394**	*0.6311*	0.3185	**0.6344**	*0.6227*	0.2443
	$AMAE$	*0.3485*	**0.3060**	0.6207	**0.3850**	*0.3903*	0.5921	**0.3768**	*0.3931*	0.6598
	CCR	0.7212	**0.7411**	*0.7328*	*0.7082*	0.7000	**0.7630**	0.7383	*0.7452*	**0.7636**
	MS	*0.5671*	**0.5820**	0.1791	**0.5811**	*0.5726*	0.0813	*0.5689*	**0.5862**	0.0967
DLRB	GMS	*0.6715*	**0.6971**	0.3630	**0.6397**	*0.6352*	0.1648	*0.6290*	**0.6437**	0.1922
	$AMAE$	0.3389	**0.3057**	0.5484	**0.3847**	*0.3912*	0.6634	*0.3871*	**0.3733**	0.6454
	CCR	0.7294	*0.7397*	**0.7863**	*0.7089*	0.7006	**0.7821**	0.7376	*0.7486*	**0.8445**
	MS	*0.5970*	**0.6268**	0.1343	**0.5726**	*0.5470*	0.0427	*0.5645*	**0.5862**	0.0645
SCR	GMS	*0.6607*	**0.6951**	0.3056	**0.6394**	*0.6311*	0.3185	**0.6344**	*0.6227*	0.2443
	$AMAE$	0.3485	**0.3060**	0.6207	**0.3850**	*0.3903*	0.5921	**0.3768**	*0.3931*	0.6598
	CCR	0.7212	**0.7411**	*0.7328*	*0.7082*	0.7000	**0.7630**	*0.7383*	*0.7383*	**0.7636**
	MS	*0.5671*	**0.5970**	0.1492	**0.5726**	*0.5641*	0.1623	**0.5689**	*0.5517*	0.1290

Comparing the three tables, the reservoir structure that obtains better performance for WPRE prediction is DLRB. Besides, the Double architecture of the network only improves the results for the DLRB structure, but not for the other two. If we consider the increase of complexity that is induced in the training of the ordinal logistic regression (62 inputs in the Single architecture, versus 100 inputs for the Double one), we can affirm that the Simple architecture is the most adequate for this problem.

4   Conclusions

This paper evaluates three different recurrent neural network architectures, combined with three different minimum complexity ESN structures. They are used

to three ordinal wind ramp classes, where a high degree of imbalance is observed (because of which, over-sampling is applied to the reservoir activations). The best architecture and structure is a single reservoir for wind speed with delay line reservoir with feedback connections, although, for a few cases, another reservoir for reanalysis data works better.

References

1. Baccianella, S., Esuli, A., Sebastiani, F.: Evaluation measures for ordinal regression. In: Proceedings of the Ninth International Conference on Intelligent Systems Design and Applications, pp. 283–287 (2009)
2. Basterrech, S., Buriánek, T.: Solar irradiance estimation using the echo state network and the flexible neural tree. In: Pan, J.-S., Snasel, V., Corchado, E.S., Abraham, A., Wang, S.-L. (eds.) Intelligent Data analysis and its Applications, Volume I. AISC, vol. 297, pp. 475–484. Springer, Cham (2014). https://doi.org/10.1007/978-3-319-07776-5_49
3. Chawla, N.V., Bowyer, K.W., Hall, L.O., Kegelmeyer, W.P.: SMOTE: synthetic minority over-sampling technique. J. Artif. Intell. Res. **16**, 321–357 (2002)
4. Dorado-Moreno, M., et al.: Multiclass prediction of wind power ramp events combining reservoir computing and support vector machines. In: Luaces, O., Gámez, J.A., Barrenechea, E., Troncoso, A., Galar, M., Quintián, H., Corchado, E. (eds.) CAEPIA 2016. LNCS (LNAI), vol. 9868, pp. 300–309. Springer, Cham (2016). https://doi.org/10.1007/978-3-319-44636-3_28
5. Dorado-Moreno, M., Cornejo-Bueno, L., Gutiérrez, P.A., Prieto, L., Salcedo-Sanz, S., Hervás-Martínez, C.: Combining reservoir computing and over-sampling for ordinal wind power ramp prediction. In: Rojas, I., Joya, G., Catala, A. (eds.) IWANN 2017. LNCS, vol. 10305, pp. 708–719. Springer, Cham (2017). https://doi.org/10.1007/978-3-319-59153-7_61
6. Dorado-Moreno, M., Cornejo-Bueno, L., Gutiérrez, P.A., Prieto, L., Hervás-Martínez, C., Salcedo-Sanz, S.: Robust estimation of wind power ramp events with reservoir computing. Renew. Energy **111**, 428–437 (2017)
7. Dee, D.P., Uppala, S.M., Simmons, A.J., Berrisford, P., Poli, P.: The ERA-Interim reanalysis: configuration and performance of the data assimilation system. Q. J. R. Meteorol. Soc. **137**, 553–597 (2011)
8. Fernandez, J.C., Salcedo-Sanz, S., Gutiérrez, P.A., Alexandre, E., Hervás-Martínez, C.: Significant wave height and energy flux range forecast with machine learning classifiers. Eng. Appl. Artif. Intell. **43**, 44–53 (2015)
9. Gutiérrez, P.A., Pérez-Ortiz, M., Sánchez-Monedero, J., Fernández-Navarro, F., Hervás-Martínez, C.: Ordinal regression methods: survey and experimental study. IEEE Trans. Knowl. Data Eng. **28**, 127–146 (2016)
10. Jaeger, H.: The 'echo state' approach to analysing and training recurrent neural networks. GMD report 148, German National Research Center for Information Technology, pp. 1–43 (2001)
11. Lukosevicius, M., Jaeger, H.: Reservoir computing approaches to recurrent neural network training. Comput. Sci. Rev. **3**(3), 127–149 (2009)
12. McCullagh, P.: Regression models for ordinal data. J. R. Stat. Soc. **42**(2), 109–142 (1980)
13. Rodan, A., Tiňo, P.: Minimum complexity echo state network. IEEE Trans. Neural Netw. **22**(1), 131–144 (2011)

Special Session on Evolutionary Computing Methods for Data Mining: Theory and Applications

GELAB - A Matlab Toolbox
for Grammatical Evolution

Muhammad Adil Raja$^{(\boxtimes)}$ and Conor Ryan

Department of Computer Science and Information Systems,
University of Limerick, Limerick, Ireland
{adil.raja,conor.ryan}@ul.ie

Abstract. In this paper, we present a Matlab version of libGE. libGE is
a famous library for Grammatical Evolution (GE). GE was proposed initially in [1] as a tool for automatic programming. Ever since then, GE has
been widely successful in innovation and producing human-competitive
results for various types of problems. However, its implementation in
C++ (libGE) was somewhat prohibitive for a wider range of scientists
and engineers. libGE requires several tweaks and integrations before it
can be used by anyone. For anybody who does not have a background
in computer science, its usage could be a bottleneck. This prompted us
to find a way to bring it to Matlab. Matlab, as it is widely known, is
a fourth generation programming language used for numerical computing. Details aside, but it is well known for its user-friendliness in the
wider research community. By bringing GE to Matlab, we hope that
many researchers across the world shall be able to use it, despite their
academic background. We call our implementation of GE as GELAB.
GELAB is currently present online as an open-source software (https://
github.com/adilraja/GELAB). It can be readily used in research and
development.

1 Introduction

Artificial intelligence (AI) has become a buzzword in almost every feat of life
these days. Not only AI enabled gadgets are becoming commonplace, it is
thought that in the near future AI enabled applications and bots shall take
over the whole world [2]. On one hand, AI is supposed to make life easier for
humanity. On the other hand, it is also supposed to make problem-solving easier.
Machine learning (ML), the subfield of AI that is responsible for the contemporary autonomous systems that we see all around us today, is a way to steer
computers to solve problems by themselves. This can sound miraculous. And
indeed, it does sound quite miraculous if one observes closely how ML algorithms learn solutions to problems. A problem is given to an ML algorithm, and
in a short span of time the algorithm churns out a solution [3].

However, the main problem in using the ML algorithms remains their tedious
interfaces. Most algorithms can have an esoteric command line interface (CLI)
through which they obtain data. Gluing algorithms to other applications can

© Springer Nature Switzerland AG 2018
H. Yin et al. (Eds.): IDEAL 2018, LNCS 11315, pp. 191–200, 2018.
https://doi.org/10.1007/978-3-030-03496-2_22

also be problematic. Consider the case where an ML algorithm, such as an evolutionary algorithm (EA), has to be glued to a flight simulator [4,5]. If the EA is implemented in C++, the engineer would have to peek inside the source code and try to figure out as to how data could be exchanged back and forth with the algorithm. If the engineer does not have a decent level of experience in computer programming, he/she will be hampered by the need to develop know-how with the source code. In most of the cases, taking advanced programming courses may be required. This can inhibit the process of innovation.

In [6] it was argued that if domain-specific simulators could be dovetailed with ML algorithms, innovation would follow naturally. As a matter of fact, embracing such a philosophy could help in developing a worldwide innovation culture. However, as discussed above, the process of dovetailing may remain cumbersome in cases where the ML algorithm is implemented in a way that it does not lend itself easily to be integrated with other applications.

In the hope of acquiring user-friendliness and making the process of dovetailing easier, we are proposing a Matlab version of GE. As it will be found later in the paper, the proposed version can be invoked through the Matlab CLI with a single statement. Most of the peripheral code that is responsible for running GE is also written in Matlab. To this end, it makes it easier for an engineer to plug in their own logic or code with the algorithm.

Rest of the paper is organized as follows. In Sect. 2, we briefly describe GE. In Sect. 3 we discuss GELAB. Section 6 concludes the paper with an outlook on the future prospects of GELAB.

2   Grammatical Evolution

GE was first proposed by the Bio-Developmental Systems (BDS) Research Group, CSIS, University of Limerick, Ireland[1]. GE is a type of an EA that inspires from Darwinian evolution. Given a user-specified problem it creates a large population of randomly generated computer programs. Each of the programs is a possible candidate solution to the problem at hand. It eventually evaluates each of the problems and assigns it a fitness score. After that, the genetic operators of selection, crossover, and mutation are applied to produce an offspring generation. This follows by the fitness evaluation of the child population. Replacement is applied as a final step to remove the undesirable solutions and to retain better candidates for the next iteration of the evolutionary process. Evolution commences until the time the desired solution is found.

To this end, the search space on which GE operates is the set of all the possible computer programs. GE is a variant of genetic programming (GP). However, GE differs from GP in certain ways.

GE is not only inspired by biological evolution alone. It is also inspired by genetics. At the heart of this is the idea that a certain genotype gives rise to a certain phenotype. Here, the genotype refers to the genetic makeup of an individual. The phenotype, on the other hand, refers to the physical qualities of an

---

[1] Web: http://bds.ul.ie/libGE/.

individual organism. In order to leverage from this idea, GE consumes geno-types from some source and performs a so-called *genotype to phenotype mapping* step. In this mapping process, GE converts the genome to a corresponding com-puter program. Figure 1(a) exhibits the analogy between genotype to phenotype mapping as it happens in biology as well as in GE.

For all practical purposes, the genome is normally integer coded. The com-puter program it maps to is governed by a grammar specified in a Backus Naur Form (BNF). Figure 1(b) shows the conceptual diagram of the GE's mapping process.

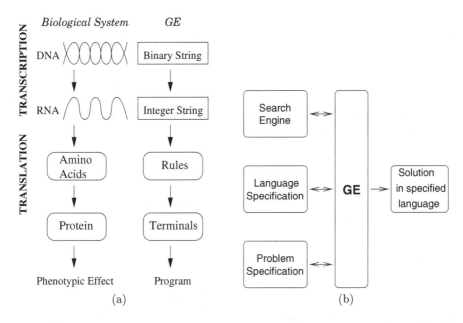

Fig. 1. (a) Genotype to phenotype mapping in biological systems and in GE. (b) Conceptual diagram of GE's mapping process.

Figure 2 shows how an integer-coded genotype is mapped to a corresponding computer program using a grammar. The figure shows the various steps of the mapping process.

In order to yield computer programs, GE requires a source that can generate a large number of genomes. To accomplish this the GE mapper is normally augmented with an integer-coded genetic algorithm (GA). The GA creates a large number of integer-coded genomes at each iteration. The genomes are then mapped to the corresponding genotype using the mapper. Pseudo-code of GE is given in Algorithm 1.

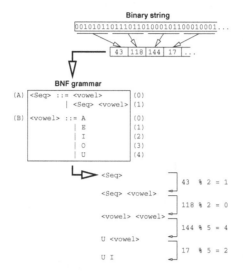

Fig. 2. Mapping of integer-coded genome to a corresponding computer program using a grammar.

Algorithm 1. Pseudo-code of GELAB.

```
parentPop=initPop;
parentPop=genotype2phenotypeMapping(parentPop);
parentPop=evalPop(parentPop).

for(i=1:numGens)
        childPop=selection(parentPop);
        childPop=crossover(parentPop);
        childPop=mutation(parentPop);
        childPop=genotype2phenotypeMapping(childPop);
        childPop=evalPop(childPop);
        parentPop=replacement(parentPop, childPop);
end
```

2.1 libGE

libGE is the original implementation of GE in C++. Its initial version was released around 2003 and has since been used in everything involving GE by the BDS group. It can be integrated with almost any search technique. However, as suggested earlier, mostly it is augmented with a GA. The class diagram of libGE is shown in Fig. 3.

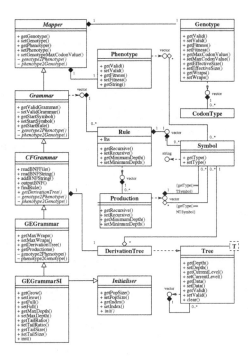

Fig. 3. Class diagram of libGE.

In order to invoke the mapper of libGE, the user has to instantiate an object of type *Grammar* (*GEGrammar* or *GEGrammarSI*). The *Grammar* object extends *Mapper*. The *Mapper* is in turn an *ArrayList* of *Rule* objects. The *Rule* is an *ArrayList* of *Production* objects. And the *Production* is an *ArrayList* of *Symbol* objects. The *Grammar* object parses a *bnf* file containing production rules. In turn, it uses the above-mentioned data structures to hold the production rules in the memory. Given any random integer array, representing genomes, it maps it to the corresponding phenotype representation.

2.2 libGE in Java

In order to make libGE portable to a wider set of computing platforms, we re-implemented it in Java. Another benefit is that Java objects can be readily called into Matlab code. This is opposed to the requirement of creating complicated *MEX* files to call C++ code from within Matlab. The Java version of libGE is a verbatim copy of its C++ counterpart. Its source code is the reflection of the class diagram shown in Fig. 3. The salient feature of the code is that anyone who wishes to use this program should invoke an object of type GEGrammarSI. This object is an ArrayList (i.e. a vector or simply an array) of Rules, which in turn is an ArrayList of Productions, and which in turn is an ArrayList of Symbols. So given a grammar in BNF format, the object first reads it into appropriate data structures mentioned above. After that, whenever an integer-coded genome is provided, the grammar object can perform a mapping step to convert the genotype to its corresponding phenotype.

3   GELAB

Calling Java code in Matlab is simple. Matlab has functions *javaObject* and *javaMethod* that allow Java objects and methods to be called from Matlab respectively. We simply levered from these methods. Once a Java object or a method is called from within Matlab, the later renders it to the Java virtual machine (JVM) for execution. Every installation of Matlab maintains JVM for the purpose of running Java code.

In our implementation of GELAB, we have created two main functions to invoke GE. These are *load_grammar.m* and *genotype2phenotype.m*. The *load_grammar.m* function invokes a GEGrammarSI object. It subsequently reads a BNF file and loads the production rules of the grammar. This object can then be passed to the *genotype2phenotype.m* function along with a genome. As the name suggests, this function maps the genotype to phenotype. However, it leverages from the *genotype2phenotype* method of the GEGrammar object. The *genotype2phenotype* object returns a computer program in the form of a string that can be evaluated using the Matlab's built-in function *eval*. The result returned by the *eval* function can be used for subsequent fitness evaluation of the individual.

In order for the whole algorithm to run successfully in Matlab, we implemented a simple GA. The simple GA performs the typical evolutionary steps as discussed in the previous section. Functions are created for all the genetic operators such as crossover, mutation, and replacement etc. Fitness evaluation is based on mean squared error (MSE). Linear scaling as proposed in [7] is also implemented. Moreover, the nature of the software is such that user-defined schemes can be easily integrated with it.

It is extremely easy to invoke the toolbox. In order to run GELAB, simply run *ge_main.m* or *ge_inaloop.m*. The former runs GELAB for one complete run. Whereas the latter runs GELAB for fifty runs by default. The user can specify parameters such as the number of runs, number of generations and population size. Software invocation does not require even typing anything on the Matlab CLI. It can be done simply by pressing the *Run* button in the Matlab IDE. Doing so will run GELAB with example data. Once some familiarity with GELAB is acquired, data files would need to be supplied for a user-specified problem.

GELAB also accumulates statistics relevant to a typical evolutionary experiment. These statistics are returned at the end of an experiment.

4   Results

We have performed a number of preliminary tests to analyze the performance of GELAB. GELAB was run over a few data sets for almost two months. During this period, the working of GELAB was observed from various perspectives. Our initial concern was to observe any unexpected behaviors of the toolbox. To this end, we analyzed the causes for which the software crashed and rectified those both in the Java and Matlab code. After that we sought to benchmark the software.

To this end, we employed data from the domain of speech quality estimation. Exact description of the data is given in Sect. 3 of [8]. Initially, feature extraction was performed by processing the MOS labeled speech databases discussed using the ITU-T P.563 algorithm [8]. Values of 43 features corresponding to each of the speech files were accumulated as the input domain variables. The corresponding MOS scores formed the target values for training and testing.

An evolutionary experiment comprising of 50 runs was performed using GELAB. Complete details of the experiment are given in Table 1.

Table 1. Parameters of the GE experiment

Parameter	Value
Runs	50
Generations	100
Population size	1,000
Selection	Tournament
Tournament size	2
Genetic operators	Crossover, mutation
Fitness function	Scaled MSE
Survival	Elitist
Function set	$+, -, *, /$, sin, cos, log_{10}, log_2, log_e, power, lt, gt
Terminal set	Random numbers P.563 features

Figure 4(a) shows the average fitness of all the individuals at each generation during evolution. Results are plotted for five runs. Similarly, Fig. 4(b) shows the average time (in seconds) it took for the creation, evaluation and replacement of each generation during a run.

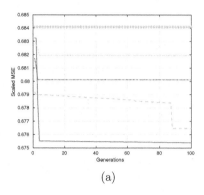
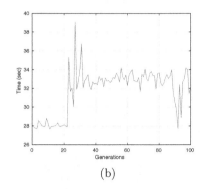

(a)   (b)

Fig. 4. (a) Mean fitness history of five runs using GELAB. (b) Time taken by GELAB at each generation.

5   Additional Features of GELAB

We have implemented some additional features and integrated them with GELAB. The purpose of these features is to speed up GELAB, reduce its memory requirements and to make it possible to solve a wider range of problems. These features are as follows:

5.1   GELAB and the Compact Genetic Algorithm (cGA)

We have implemented an integer-valued version of the cGA and integrated it with GELAB. The cGA [9] works by evolving a probability distribution (PDF) that describes the distribution of solutions in a hypothetical population of individual chromosomes. The PDF is maintained with the help of a couple of probability vectors (PVs). The ith element of both the PVs store the means of the ith genes of chromosomes of the whole population. The traditional cGA works on binary-coded chromosomes [9]. In [10] a cGA was proposed for real-valued chromosomes. The motivation was to run cGA on a floating-point micro-controller. The binary-coded cGA would have consumed considerable additional resources for doing binary to floating-point conversions and vice versa. A cGA that worked directly on floating point numbers addressed this problem. They assume that the distribution of the genes can be approximated with the Gaussian PDF. Since contemporary GE employs integer-valued variable length GA, we have created an integer-valued compact genetic algorithm (icGA). Our implementation is similar to [9] and [10].

5.2   Caching

In order to reduce the computational requirements, we have implemented a genotype cache to store the results of the pre-evaluated individuals. As the individuals are evaluated, they are looked up in the cache if their evaluation is already there. If so, the evaluations (such as fitness values and results) are used. If not, the individual is evaluated and stored in the cache. In any subsequent computations, if the algorithm produces the same individual again, its evaluation from the cache is used. To this end, our scheme is similar to the one proposed by Keijzer in [11]. However, instead of subtree caching, we maintain a cache based on the genotypes.

5.3   GELAB and Multiple Input Multiple Output (MIMO) Systems

Certain systems are inevitably of MIMO nature. Controllers for unmanned aerial vehicles (UAVs), driverless cars are MIMO. They accept multiple inputs and are expected to generate multiple outputs simultaneously, such as speed, steering etc. Certain regression problems are also MIMO. To address such problems we have implemented a capability in GELAB to have multiple trees per individual. Each of the trees produces an output for the desired MIMO system. Together, all the trees provide output values for the whole MIMO system.

6   Conclusions and Future Work

GELAB is a convenient way to use GE in research and development. It is far easier to use than libGE. The user only needs to specify the data for which the software should be run. More keen users can even tweak the code with a lot more ease. It is easy to integrate with third-party software too. The inherent abilities of Matlab to gather data and plotting make it a viable choice for easy, useful and repeatable research. Currently, we are benchmarking GELAB on Matlab R2018a and R2017b.

In the future we expect GELAB to be used by a wider research community. The expected users are of course who would be dealing with optimization problems in their work. Invention of autonomous systems can be accelerated with GELAB.

References

1. O'Neill, M., Ryan, C.: Grammatical evolution. IEEE Trans. Evol. Comput. **5**, 349–358 (2001)
2. Müller, V.C., Bostrom, N.: Future progress in artificial intelligence: a survey of expert opinion. In: Müller, V.C. (ed.) Fundamental Issues of Artificial Intelligence. SL, vol. 376, pp. 553–570. Springer, Cham (2016). https://doi.org/10.1007/978-3-319-26485-1_33
3. Mitchell, T.: Machine Learning. McGraw Hill, New York (1997)
4. Raja, M.A., Rahman, S.U.: A tutorial on simulating unmanned aerial vehicles. In: 2017 International Multi-topic Conference (INMIC), pp. 1–6 (2017)
5. Habib, S., Malik, M., Rahman, S.U., Raja, M.A.: NUAV - a testbed for developing autonomous unmanned aerial vehicles. In: 2017 International Conference on Communication, Computing and Digital Systems (C-CODE), pp. 185–192 (2017)

6. Raja, M.A., Ali, S., Mahmood, A.: Simulators as drivers of cutting edge research. In: 2016 7th International Conference on Intelligent Systems, Modelling and Simulation (ISMS), pp. 114–119 (2016)

7. Keijzer, M.: Scaled symbolic regression. Genet. Program. Evolvable Mach. **5**, 259–269 (2004)

8. Raja, A., Flanagan, C.: Real-time, non-intrusive speech quality estimation: a signal-based model. In: O'Neill, M., et al. (eds.) EuroGP 2008. LNCS, vol. 4971, pp. 37–48. Springer, Heidelberg (2008). https://doi.org/10.1007/978-3-540-78671-9_4

9. Harik, G.R., Lobo, F.G., Goldberg, D.E.: The compact genetic algorithm. IEEE Trans. Evol. Comput. **3**, 287–297 (1999)

10. Mininno, E., Cupertino, F., Naso, D.: Real-valued compact genetic algorithms for embedded microcontroller optimization. IEEE Trans. Evol. Comput. **12**, 203–219 (2008)

11. Keijzer, M.: Alternatives in subtree caching for genetic programming. In: Keijzer, M., O'Reilly, U.-M., Lucas, S., Costa, E., Soule, T. (eds.) EuroGP 2004. LNCS, vol. 3003, pp. 328–337. Springer, Heidelberg (2004). https://doi.org/10.1007/978-3-540-24650-3_31

Bat Algorithm Swarm Robotics Approach for Dual Non-cooperative Search with Self-centered Mode

Patricia Suárez[1], Akemi Gálvez[1,2], Iztok Fister[3], Iztok Fister Jr.[3],
Eneko Osaba[4], Javier Del Ser[4,5,6], and Andrés Iglesias[1,2(✉)]

[1] University of Cantabria, Avenida de los Castros s/n, 39005 Santander, Spain
iglesias@unican.es
[2] Toho University, 2-2-1 Miyama, Funabashi 274-8510, Japan
[3] University of Maribor, Smetanova ulica 17, 2000 Maribor, Slovenia
[4] TECNALIA, Derio, Spain
[5] University of the Basque Country (UPV/EHU), Bilbao, Spain
[6] Basque Center for Applied Mathematics (BCAM), Bilbao, Spain

Abstract. This paper presents a swarm robotics approach for dual non-cooperative search, where two robotic swarms are deployed within a map with the goal to find their own target point, placed at an unknown location of the map. We consider the self-centered mode, in which each swarm tries to solve its own goals with no consideration to any other factor external to the swarm. This problem, barely studied so far in the literature, is solved by applying a popular swarm intelligence method called bat algorithm, adapted to this problem. Five videos show some of the behavioral patterns found in our computational experiments.

1 Introduction

Swarm robotics is attracting a lot of interest because of its potential advantages for several tasks [1]. For instance, robotic swarms are well suited for navigation in indoor environments, as a swarm of simple interconnected mobile robots has greater exploratory capacity than a single sophisticated robot. Additional benefits of the swarm are greater flexibility, adaptability, and robustness. Also, swarm robotics methods are relatively simple to understand and implement, and are quite affordable in terms of the required budget and computing resources.

The most common case of swarm robotics is the cooperative mode, with several robotic units working together to accomplish a common task. Much less attention is given so far to the non-cooperative mode. In this paper we are particularly interested in a non-cooperative scheme that we call *egotist* or *self-centered* mode. Under this regime, each swarm tries to solve its own goals with little (or none at all) consideration to any other factor external to the swarm. In this work, we consider two swarms of robotic units deployed within the same spatial environment. Each swarm is assigned the task to reach its own target point, placed in an unknown location of a complex labyrinth with narrow corridors and several

© Springer Nature Switzerland AG 2018
H. Yin et al. (Eds.): IDEAL 2018, LNCS 11315, pp. 201–209, 2018.
https://doi.org/10.1007/978-3-030-03496-2_23

dead ends, forcing the robots to turn around to escape from these blind alleys. It is assumed that the geometry of the environment is completely unknown to the robots. Finding these unknown target points requires the robots of both swarms to perform exploration of the environment, hence interfering each other during motion owing to potential intra- and inter-swarm collisions. Under these conditions, the robots have to navigate in a highly dynamic environment where all moving robots are additional obstacles to avoid. This brings the possibility of potentially conflicting goals for the swarms, for instance when the robots are forced to move on the same corridors but in opposite directions. In our approach, there is no centralized control of the swarm, so the robot decision-making is completely autonomous and each robot takes decisions by itself without any external order from a central server of any other robot.

Similar to [2], in this paper we consider robotic units based on ultrasound sensors and whose internal functioning is based on a popular swarm intelligence method: the *bat algorithm*. To this aim, we developed a computational simulation framework that replicates very accurately all features and functionalities of the real robots and the real environment, including the physics of the process (gravity, friction, motion, collisions), visual appearance (cameras, texturing, materials), sensors (ultrasound, spatial orientation, cycle-based time clock, computing unit emulation, bluetooth), allowing both graphical and textual output and providing support for additional components and accurate interaction between the robots and with the environment.

The structure of this paper is as follows: firstly, we describe the bat algorithm, its basic rules and its pseudocode. Then, our approach for the dual non-cooperative search problem for the self-centered mode posed in this paper is described. Some experimental results are then briefly discussed. The paper closes with the main conclusions and some plans for future work in the field.

2   The Bat Algorithm

The *bat algorithm* is a bio-inspired swarm intelligence algorithm originally proposed by Xin-She Yang in 2010 to solve optimization problems [4–6]. The algorithm is based on the echolocation behavior of microbats, which use a type of sonar called *echolocation*. The idealization of this method is as follows:

1. Bats use echolocation to sense distance and distinguish between food, prey and background barriers.
2. Each virtual bat flies randomly with a velocity $\mathbf{v}_i$ at position (solution) $\mathbf{x}_i$ with a fixed frequency f_{min}, varying wavelength λ and loudness A_0 to search for prey. As it searches and finds its prey, it changes wavelength (or frequency) of their emitted pulses and adjust the rate of pulse emission r, depending on the proximity of the target.
3. It is assumed that the loudness will vary from an (initially large and positive) value A_0 to a minimum constant value A_{min}.

Require: (Initial Parameters)
    Population size: $\mathcal{P}$; Maximum number of generations: $\mathcal{G}_{max}$; Loudness: $\mathcal{A}$
    Pulse rate: r ; Maximum frequency: f_{max} ; Dimension of the problem: d
    Objective function: $\phi(\mathbf{x})$, with $\mathbf{x} = (x_1, \ldots, x_d)^T$; Random number: $\theta \in U(0, 1)$
1: $g \leftarrow 0$
2: Initialize the bat population $\mathbf{x}_i$ and $\mathbf{v}_i$, $(i = 1, \ldots, n)$
3: Define pulse frequency f_i at $\mathbf{x}_i$
4: Initialize pulse rates r_i and loudness $\mathcal{A}_i$
5: **while** $g < \mathcal{G}_{max}$ **do**
6:    **for** $i = 1$ **to** $\mathcal{P}$ **do**
7:       Generate new solutions by using eqns. (1)-(3)
8:       **if** $\theta > r_i$ **then**
9:          $\mathbf{s}^{best} \leftarrow \mathbf{s}^g$ //select the best current solution
10:          $\mathbf{ls}^{best} \leftarrow \mathbf{ls}^g$ //generate a local solution around $\mathbf{s}^{best}$
11:       **end if**
12:       Generate a new solution by local random walk
13:       **if** $\theta < \mathcal{A}_i$ *and* $\phi(\mathbf{x_i}) < \phi(\mathbf{x}^*)$ **then**
14:          Accept new solutions, increase r_i and decrease $\mathcal{A}_i$
15:       **end if**
16:    **end for**
17:    $g \leftarrow g + 1$
18: **end while**
19: Rank the bats and find current best $\mathbf{x}^*$
20: **return** $\mathbf{x}^*$

Algorithm 1. Bat algorithm pseudocode

Some additional assumptions are advisable for further efficiency. For instance, we assume that the frequency f evolves on a bounded interval $[f_{min}, f_{max}]$. This means that the wavelength λ is also bounded, because f and λ are related to each other by the fact that the product $\lambda.f$ is constant. For practical reasons, it is also convenient that the largest wavelength is chosen such that it is comparable to the size of the domain of interest (the search space for optimization problems). For simplicity, we can assume that $f_{min} = 0$, so $f \in [0, f_{max}]$. The rate of pulse can simply be in the range $r \in [0, 1]$, where 0 means no pulses at all, and 1 means the maximum rate of pulse emission. With these idealized rules indicated above, the basic pseudo-code of the bat algorithm is shown in Algorithm 1. Basically, the algorithm considers an initial population of $\mathcal{P}$ individuals (bats). Each bat, representing a potential solution of the optimization problem, has a location $\mathbf{x}_i$ and velocity $\mathbf{v}_i$. The algorithm initializes these variables with random values within the search space. Then, the pulse frequency, pulse rate, and loudness are computed for each individual bat. Then, the swarm evolves in a discrete way over generations, like time instances until the maximum number of generations, $\mathcal{G}_{max}$, is reached. For each generation g and each bat, new frequency, location and velocity are computed according to the following evolution equations:

$$f_i^g = f_{min}^g + \beta(f_{max}^g - f_{min}^g) \tag{1}$$

$$\mathbf{v}_i^g = \mathbf{v}_i^{g-1} + [\mathbf{x}_i^{g-1} - \mathbf{x}^*] f_i^g \tag{2}$$

$$\mathbf{x}_i^g = \mathbf{x}_i^{g-1} + \mathbf{v}_i^g \tag{3}$$

where $\beta \in [0,1]$ follows the random uniform distribution, and $\mathbf{x}^*$ represents the current global best location (solution), which is obtained through evaluation of the objective function at all bats and ranking of their fitness values. The superscript $(.)^g$ is used to denote the current generation g. The best current solution and a local solution around it are probabilistically selected according to some given criteria. Then, search is intensified by a local random walk. For this local search, once a solution is selected among the current best solutions, it is perturbed locally through a random walk of the form: $\mathbf{x}_{new} = \mathbf{x}_{old} + \epsilon \mathcal{A}^g$, where ϵ is a uniform random number on $[-1,1]$ and $\mathcal{A}^g = <\mathcal{A}_i^g>$, is the average loudness of all the bats at generation g. If the new solution achieved is better than the previous best one, it is probabilistically accepted depending on the value of the loudness. In that case, the algorithm increases the pulse rate and decreases the loudness (lines 13–16). This process is repeated for the given number of generations. In general, the loudness decreases once a bat finds its prey (in our analogy, once a new best solution is found), while the rate of pulse emission decreases. For simplicity, the following values are commonly used: $\mathcal{A}_0 = 1$ and $\mathcal{A}_{min} = 0$, assuming that this latter value means that a bat has found the prey and temporarily stop emitting any sound. The evolution rules for loudness and pulse rate are as: $\mathcal{A}_i^{g+1} = \alpha \mathcal{A}_i^g$ and $r_i^{g+1} = r_i^0[1 - exp(-\gamma g)]$ where α and γ are constants. Note that for any $0 < \alpha < 1$ and any $\gamma > 0$ we have: $\mathcal{A}_i^g \rightarrow 0, r_i^g \rightarrow r_i^0$ as $g \rightarrow \infty$. Generally, each bat should have different values for loudness and pulse emission rate, which can be achieved by randomization. To this aim, we can take an initial loudness $\mathcal{A}_i^0 \in (0,2)$ while r_i^0 can be any value in the interval $[0,1]$. Loudness and emission rates will be updated only if the new solutions are improved, an indication that the bats are moving towards the optimal solution.

3   Bat Algorithm Method for Robotic Swarms

In this work we consider two robotic swarms $\mathcal{S}_1$ and $\mathcal{S}_2$ comprised by a set of μ and ν robotic units $\mathcal{S}_1 = \{\mathcal{R}_1^i\}_{i=1,...,\mu}$, $\mathcal{S}_2 = \{\mathcal{R}_2^j\}_{j=1,...,\nu}$, respectively. For simplicity, we assume that $\mu = \nu$ and that all robotic units are functionally identical, i.e., $\mathcal{R}_1^i = \mathcal{R}_2^j, \forall i,j$. For visualization purposes, the robots in $\mathcal{S}_1$ and $\mathcal{S}_2$ are displayed graphically in red and yellow, respectively. They are deployed within a 3D synthetic labyrinth, $\mathbf{\Omega} \subset \mathbb{R}^3$ with a complex geometrical configuration (unknown to the robots and shown in Fig. 1) to perform dynamic exploration. The figure is split into two parts for better visualization, corresponding to the side view (left) and top view (right). As the reader can see, the scene consists of a large collection of cardboard boxes arranged in a grid-like structure and forming challenging structures for the robots such as corridors, dead ends, bifurcations and T-junctions to simulate the walls and corridors of a labyrinth. Figure 1 also shows a set of 10 robotic units for each swarm, scattered throughout the environment. The goal of each swarm $\mathcal{S}_k$ is to find a static target point $\mathbf{\Phi}_k$ ($k = 1, 2$), placed in a certain (unknown) location of the environment. We assume that $\|\mathbf{\Phi}_1 - \mathbf{\Phi}_2\| > \delta$ for a certain threshold value $\delta > 0$, meaning that both target points are not very close to each other so as to broaden the spectrum of possible interactions between the swarms. They are represented in Fig. 1 by two spherical-shaped points of light in red and yellow for $\mathcal{S}_1$ and $\mathcal{S}_2$, respectively. Target $\mathbf{\Phi}_1$ is located inside a room at a relatively accessible location, while target $\mathbf{\Phi}_2$ is placed at the deepest part of the labyrinth (the upper left corner in top view). The scene also includes many dead ends to create a challenging environment for the robotic swarms.

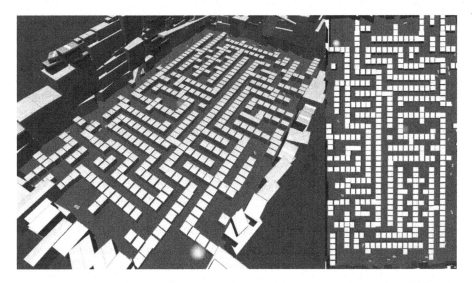

Fig. 1. Graphical representation of the labyrinth: side view (left); top view (right). The image corresponds to the initialization step of our method, when two robotics swarms are deployed at random positions in the outermost parts of the map. (Color figure online)

In our approach, each robot moves autonomously, according to the current values of its fitness function and its own parameters. To this aim, each virtual robot $\mathcal{R}_k^i$ is mathematically described by a vector $\mathbf{\Xi}_k^{i,j} = \left\{ \varphi_k^{i,j}, \mathbf{x}_k^{i,j}, \mathbf{v}_k^{i,j} \right\}$, where $\varphi_k^{i,j}$, $\mathbf{x}_k^{i,j} = (x_k^{i,j}, y_k^{i,j})$ and $\mathbf{v}_k^{i,j} = (v_{k,x}^{i,j}, v_{k,y}^{i,j})$ represent the fitness value, position, and velocity at time instance j, respectively. Note here that, although the environment is a 3D world, we consider the case of mobile walking robots, therefore moving on a two-dimensional map $\mathcal{M} = \mathbf{\Omega}|_{z=0}$. The robots are deployed at initial random positions $\mathbf{x}_k^{i,0}$ and with random but bounded velocities $\mathbf{v}_k^{i,0}$ so that the robots move strictly within the map $\mathcal{M}$. Moreover, the robots are initialized in the outermost parts of the map to avoid falling down very near to the target.

For the robot motion, we assume that the two-dimensional map $\mathcal{M}$ is described by a tessellation of convex polygons $\mathcal{T}_{\mathcal{M}}$. Then, we consider the set $\mathcal{N}_{\mathcal{M}} \subset \mathcal{T}_{\mathcal{M}}$ (called the *navigation mesh*) comprised by all polygons that are fully traversable by the robots. At time j the fitness function $\varphi_k^{i,j}$ can be defined as the distance between the current position $\mathbf{x}_k^{i,j}$ and the target point $\mathbf{\Phi}_k$, measured on $\mathcal{N}_{\mathcal{M}}$ as $\varphi_k^{i,j} = ||\mathbf{x}_k^{i,j} - \mathbf{\Phi}_k||_{\mathcal{N}_{\mathcal{M}}}$. In this case, our 2D robot navigation can be seen as an optimization problem, that of minimizing the value of $\varphi_k^{i,j}$, $\forall i, j, k$, which represents the distance from the current location to the target point. This problem is solved through the bat algorithm described above. About the parameter tuning, our choice has been fully empirical, based on computer simulations for different parameter values. We consider a population size of 10 robots for each swarm, as larger values make the labyrinth too populated and increase the number of collisions among robots. Initial and minimum loudness and parameter α are set to 0.5, 0, and 0.6, respectively. We also set the initial pulse rate and parameter γ to 0.5 and 0.4, respectively. However, our results do not change significantly when varying these values. All executions are performed until all robots reach their target point. Our method is implemented in *Unity 5* on a 3.8 GHz quad-core Intel Core i5, with 16 GB of DDR3 memory, and a graphical card AMD RX580 with 8 GB VRAM. All programming code in this paper has been created in *JavaScript* using the *Visual Studio* programming framework.

4   Experimental Results

The proposed method has been tested through several computational experiments. Five of them are recorded in five MPEG-4 videos (labelled as *Video1* to *Video5*) (generated as accompanying material of this paper and stored as a single 18.8 MB ZIP file publicly available at [3]), selected to show different behavioral patterns obtained with our method and corresponding to as many random initial locations of the robots.

Video 1 shows that the robots are initially wandering to explore the environment, searching for their target points. As explained above, at every iteration each robot receives the value of its fitness as well as the position of the best member of the swarm. After a few iterations some robots are able to find a path that improves their fitness (this fact is clearly visible for the red robots in lower right part of top view) and then they move to follow it. One of the members of the red swarm is successfully approaching its target so, at a certain iteration, it becomes the best of the swarm. At this point, the other members of the swarm try to follow it according to Eqs. (1)–(3). This behavioral pattern is not general, however, as other robots exhibit different behaviors. For instance, we can also see an example of a grouping pattern, illustrated by several yellow robots gathering in the upper part of the map in top view. Also, we find a case of collision between robots in central area of the scene, where two red robots trying to move to the south to approach its best, and two yellow robots trying to gather with other members of their swarm collide each other, getting trapped in a narrow corridor for a while. Note also that, although the yellow robot on the right of this group does not visually collapse with any other when moving ahead and hence might advance, it keeps idle because its ultrasound sensor detects the neighbor robots even if they are not exactly in front. When a possible collision is detected, the robots involved stop moving and start yawing slightly to left and right trying to escape from the potentially colliding area. But since the yawing motion angle is small, this situation can last for a while, as it actually happens in this video for the four robots in this group, until one of the yellow robots is able to turn around and moves in the opposite direction to its original trajectory. In its movement, this yellow robot attracts its fellow teammate. Once freed, the two red robots follow their original plan and move to the south as well. In the meanwhile, all other red robots reached their target point and all other yellow robots gathered in north part of the map and try to find a passage to advance towards their target point. Eventually, all robots find a path to their target points, and the simulation ends successfully.

Some other behavioral patterns can be seen in the other videos. For instance, *Video 2* shows an interesting example of the ability of the robots to escape from dead ends. In the video, several yellow robots gather in the upper part of the map in top view but in their exploration of the environment they move to a different corridor, getting trapped in a dead end. Unable to advance, the robots in the rear turn around and after some iterations all robots get rid of the alley, and start moving to the previous corridor in the north for further exploration. However, other robots of the group follow the opposite direction getting closer to the target point. As their fitness value improves, they attract the other members of the swarm in their way to the target point. This kind of wandering behavior where the robots move back and forth apparently in erratic fashion, also shown in *Video 3* to *Video 5*, can be explained by the fact that the robots do not have any information about the environment, so they explore it according to their fitness value at each iteration and the potential collisions with the environment and other robots of the other and their own swarm. *Video 2* also shows some interesting strategies for collision avoidance. For example, how some robots entering a corridor stop their motion to allow other robots to enter first. This behavioral pattern is even more evident in the initial part of *Video 4*, where a red robot in the upper part trying to enter into a corridor to move to the south finds four yellow robots in front trying to move to the north and forcing the red robot to back off and wait until they all pass first. This video also shows some of the technical problems we found in this work. At the middle part of the simulation, one red robot gets trapped at a corner, unable to move because the geometric shape of the corner makes the ultrasound signal very different to left and right. Actually, the robot was unable to move until other robots arrived to the area, modifying its ultrasound signal and allow it to move. A different illustrative example appears in *Video 5*, where a red robot in the central part of the map stays in front of two yellow robots moving in a small square with an obstacle in the middle. Instead of avoiding the yellow robots by moving around the obstacle in the opposite direction to them, the red robot stays idle for a while waiting for the other robots to pass first. We also found these unexpected situations with the real robots, a clear indication that our simulations reflect the actual behavior of the robots in real life very accurately. Of course, these videos are only a few examples of some particular behaviors, and additional behavioral patterns can be obtained from other executions. We hope however that they allow the reader to get a good insight about the kind of behavioral patterns that can be obtained from the application of our method to this problem.

5   Conclusions and Future Work

This paper develops a swarm robotics approach to solve the problem of dual non-cooperative search with self-centered mode, a problem barely studied so far in the literature. In our setting, two robotic swarms are deployed within the same environment and assigned the goal to find their own target point (one for each swarm) at an unknown location of the map. This search must be performed under the self-centered regime, a particular case of non-cooperative search in which each swarm tries to solve its own goals with little (or none at all) consideration to any other factor external to the swarm. To tackle this issue, we apply a popular swarm intelligence method called bat algorithm, which has been adapted to this particular problem. Five illustrative videos generated as supplementary material show some of the behavioral patterns found in our computational experiments.

As discussed above, the bat algorithm allows the robotic swarms to find their targets in reasonable time. In fact, after hundreds of simulations we did not find any execution where the robots cannot get a way to the targets. We also remark the ability of the robots to escape from dead ends and other challenging configurations, as well as to avoid static and dynamic obstacles. From these observations, we conclude that our method performs very well for this problem.

There also some limitations in our approach. As shown in the videos, some configurations are still tricky (even impossible) to overcome for individual robots. Although the problem is solved for the case of swarms, it still requires further research for individual robots. We also plan to analyze all behavioral patterns that emerge from our experiments, the addition of dynamic obstacles and moving targets and larger and more complex scenarios.

Acknowledgements. Research supported by project PDE-GIR of the European Union's Horizon 2020 (Marie Sklodowska-Curie grant agreement No. 778035), grant #TIN2017-89275-R (Spanish Ministry of Economy and Competitiveness, Computer Science National Program, AEI/FEDER, UE), grant #JU12 (SODERCAN and European Funds FEDER UE) and project EMAITEK (Basque Government).

References

1. Bonabeau, E., Dorigo, M., Theraulaz, G.: Swarm Intelligence: From Natural to Artificial Systems. Oxford University Press, New York (1999)
2. Suárez, P., Iglesias, A., Gálvez, A.: Make robots be bats: specializing robotic swarms to the bat algorithm. Swarm and Evolutionary Computation (in press)
3. https://goo.gl/JUbBYw, (password: RobotsIDEAL2018)
4. Yang, X.S.: A new metaheuristic bat-inspired algorithm. In: González, J.R., Pelta, D.A., Cruz, C., Terrazas, G., Krasnogor, N. (eds.) NICSO 2010. SCI, vol. 284, pp. 65–74. Springer, Berlin (2010). https://doi.org/10.1007/978-3-642-12538-6_6
5. Yang, X.S., Gandomi, A.H.: Bat algorithm: a novel approach for global engineering optimization. Eng. Comput. 29(5), 464–483 (2012)
6. Yang, X.S., He, X.: Bat algorithm: literature review and applications. Int. J. Bio-Inspired Comput. 5(3), 141–149 (2013)

Hospital Admission and Risk Assessment Associated to Exposure of Fungal Bioaerosols at a Municipal Landfill Using Statistical Models

W. B. Morgado Gamero[1(✉)], Dayana Agudelo-Castañeda[2],
Margarita Castillo Ramirez[5], Martha Mendoza Hernandez[2],
Heidy Posso Mendoza[3], Alexander Parody[4], and Amelec Viloria[6]

[1] Deparment of Exact and Natural Sciences, Universidad de la Costa,
Calle 58#55-66, Barranquilla, Colombia
wmorgado1@cuc.edu.co

[2] Department of Civil and Environmental Engineering, Universidad del Norte,
Km 5 Vía Puerto Colombia, Barranquilla, Colombia
mdagudelo@uninorte.edu.co, Marticamh15@hotmail.com

[3] Department of Bacteriology, Universidad Metropolitana, Calle 76 No. 42-78,
Barranquilla, Colombia
Heidy_posso@unimetro.edu.co

[4] Engineering Faculty, Universidad Libre Barranquilla, Carrera 46 No. 48-170,
Barranquilla, Colombia
alexandere.parodym@unilibre.edu.co

[5] Barranquilla Air Quality Monitoring Network, EPA – Barranquilla Verde,
Barranquilla, Colombia, Carrera 60 # 72-07, Barranquilla, Atlántico, Colombia
Margaritacastillo87@gmail.com

[6] Department of Industrial, Agro-Industrial and Operations Management,
Calle 58#55-66, Barranquilla, Colombia
aviloria7@cuc.edu.co

Abstract. The object of this research to determine the statistical relationship and degree of association between variables: hospital admission days and diagnostic (disease) potentially associated to fungal bioaerosols exposure. Admissions included acute respiratory infections, atopic dermatitis, pharyngitis and otitis. Statistical analysis was done using Statgraphics Centurion XVI software. In addition, was estimated the occupational exposure to fungal aerosols in stages of a landfill using BIOGAVAL method and represented by Golden Surfer XVI program. Biological risk assessment with sentinel microorganism A. fumigatus and Penicillium sp, indicated that occupational exposure to fungal aerosols is Biological action level. Preventive measures should be taken to reduce the risk of acquiring acute respiratory infections, dermatitis or other skin infections.

Keywords: Fungal aerosols · Biological risk assessment · Hospital admission Respiratory infections · Landfill

© Springer Nature Switzerland AG 2018
H. Yin et al. (Eds.): IDEAL 2018, LNCS 11315, pp. 210–218, 2018.
https://doi.org/10.1007/978-3-030-03496-2_24

1 Introduction

Some activities, there exist no deliberate intention to manipulate biological agents, but these ones are associated to the presence and exposure to infectious, allergic or toxic biological agents in air [1]. Bioaerosols consist of aerosols originated biologically such as metabolites, toxins, microorganisms or fragments of insects and plants that are present ubiquitously in the environment [2]. Bioaerosols play a vital role in the Earth system, particularly in the interactions between atmosphere, biosphere, climate and public health [3]. Studies suggest adverse health effects from exposure to bioaerosols in the environment, especially in workplaces. However, there is still a lack of specific environmental-health studies, diversity of employed measuring methods for microorganisms and bioaerosol-emitting facilities, and insufficient exposure assessment [4, 5]. Bioaerosols exposure does not have threshold limits to assess health impact/toxic effects; reasons include: complexity in their composition, variations in human response to their exposure and difficulties in recovering microorganisms that can pose hazard during routine sampling [4, 7]. Occupational exposure to bioaerosols containing high concentrations of bacteria and fungi, e.g., in agriculture, composting and waste management workplaces or facilities, may cause respiratory diseases, such as allergies and infections [3]. Also, there is no international consensus on the acceptable exposure limits of bioaerosol concentration, too [4].

More research is needed to properly assess their potential health hazards including inter-individual susceptibility, interactions with non-biological agents, and many proven/unproven health effects (e.g., atopy and atopic diseases) [2]. Consequently, the aim of this research was to evaluate if the exposure to fungal bioaerosols becomes a risk factor that increases the number of landfill operator's hospital admission.

2 Materials and Methods

2.1 Site Selection and Fungi Aerosol Collection

Bioaerosol sampling is the first step toward characterizing bioaerosol exposure risks [8]. Samples were collected for 12 months (April 2015–April 2016) in a municipal landfill located near Barranquilla, Colombia. Landfill has a waste discharge zone, where the waste deposited is compacted (active cell), there are some terraces with cells are no longer in operation (passive cells) and a leachate treatment system divided into three treatment steps: one pre-sedimentator, two leach sedimentation ponds and biological treatment pond. Sampling stations were located in the passive cell 1, the passive cell 2, the leachate pool and the active cell. In each sampling station, samples were collected once a month by triplicate in two journeys, morning (7:00 to 11:00) and afternoon (12:30 to 18:00). Fungi aerosol collection procedures and methodology are described in researches recently published [9].

2.2   Analysis Data

Chi-square analysis was performed to determine the significative statistical relationship and degree of association between variables: hospital admission days and diagnostic (disease) with 95% confidence ($p < 0.05$) [10] using Statgraphics Centurion XVI software. Hospital admission reported diseases were acute diarrheal disease, acute respiratory infections, atopic dermatitis, pharyngitis, otitis and tropical diseases. Analysed period was January 2015–July 2016. The landfill has 90 workers, while 50 are operators [11]. So, the research was done just with the operators.

2.3   Risk Assessment: Estimation of the Occupational Exposure to Fungal Aerosols in a Landfill

Technical guide for the evaluation and prevention of risks related to exposure to biological agents [7] and the Practical Manual for the evaluation of biological risk in various work activities BIOGAVAL were used to the estimation of the occupational risk of non-intentional exposure to fungal aerosols. Calculation of the level of biological risk (R) was done with the following Eq. (1):

$$R = (D \times V) + T + I + F \tag{1}$$

Where:
R is the level of biological risk, D is Damage, D* is Damage after reduction with the value obtained from the hygienic measures, V is Vaccination, T is Transmission way, T* is transmission way (having subtracted the value of the hygienic measures), I is Incidence rate and F is Frequency of risk activities.

For the interpretation of biological risk levels, after validation, two levels were considered: Biological action level (BAL) and Biological exposure limit (BEL). BAL: from this value preventive measures must be taken to try to reduce the exposure. Although this exposure is not considered dangerous for the operators, it constitutes a situation that can be clearly improved, from which the appropriate recommendations will be derived. (BAL) = 12, higher values require the adoption of preventive measures to reduce exposure. BEL: It must not be exceeded. BEL = 17, higher values represent situations of intolerable risk that require immediate corrective actions. To establish the distribution of the risk in the landfill, risk level map was made using the Golden Surfer 11 program.

2.4   Operator Type vs Exposure Time

Exposure operator time in active cell and leachate pool corresponded to 12 h for 5 days, except the mechanical technician.

3 Results and Discussion

3.1 Sentinel Microorganism

Results of air samples showed more prevalence of *Aspergillus*. Species reported were *A. fumigatus, A. versicolor, A. niger and A. nidulans*. The highest concentration corresponds to *A. fumigatus* during study period [9] microorganism associated to toxins production with cytotoxic properties. *A. fumigatus* has been reported as allergic and toxic microorganism in working environments [1, 6]. Other taxa reported in this study, although in lower concentration during the sampling period, was *Penicillium sp*, associated with dermatitis and respiratory conditions [1, 13, 14]. Airborne fungi causing respiratory infections and allergic reactions include *Penicillium, Aspergillus, Acremonium, Paecilomyces, Mucor* and *Cladosporium* [15]. Most infections, specifically Aspergillosis can occur in immune compromised hosts or as a secondary infection, which is caused due to inhalation of fungal spores or the toxins produced by *Aspergillus* fungus [16]. In addition, for the sentinel microorganism exposure, *Aspergillus fumigatus* and *Penicillium* sp are contemplated in the Technical Guide for the evaluation and prevention of risks related to exposure to biological agents in Appendix 14. Biological Risk in Waste Disposal Units [7].

3.2 Risk Assessment

Table 1 presents the damage quantification data, according to the Manual of Optimal Times of Work Disability [12]. The damage of acute respiratory infections and atopic dermatitis corresponds to temporary disability less than 30 days but that may have sequels about the patient.

Table 1. Damage rating

Sentinel microorganism	Manual of optimal times of work incapacity	Damage	Score
Respiratory infections, bronchitis, pharyngitis, or other. *A. fumigatus*	10 Days	Days of absence <30 days, sequels	3
Atopic dermatitis, allergic urticaria, *Penicillium sp*	14 Days	Days of absence <30 days, sequels	3

A. fumigatus and *Penicillium* sp have the highest score according to the transmission route due to their aerial dispersion. In addition, the health threats from bioaerosol exposure can be also greatly enhanced by airborne transmission of infectious agents breathing via [8]. The risk assessment was applied between April 2015 and April 2016, corresponding to the months of monitoring analyzed. For atopic dermatitis and urticaria, the data to calculate the incidence rate correspond to the number of disability cases in relation to the number of operators according to the previous year, the study period was March 2014 to March 2015, and the organization provided information (Table 2).

Table 2. Results incidence rate in the population

Sentinel microorganism	Incidence rate	Incidence/100.000 habitant	Score
Respiratory infections, bronchitis, pharyngitis, or other. *A. fumigatus*	16.000	≥ 1000	5
Atopic dermatitis, allergic urticaria, or other *Penicillium sp*	12.000	≥ 1000	5

Percentage of time in which operators are in contact with the different biological agents under analysis was calculated, ignoring the total of the working day (12 h) and the time spent on breaks (1 h); obtaining a grade of 5 for presenting a percentage of habitually >80% of the time. Vaccination variable was classified as 5 since there is no completely effective vaccine for the conditions evaluated. Hygienic Measures Adopted was used as a corrective value of −2, according to 83.2% of affirmative responses, as results of the survey hygienic measures adopted.

3.3   Biological Risk Level

The following is the calculation of the risk level of exposure to sentinel microorganisms in operational area, personnel who work in Active Cell and leachate pool, or personnel who works 12 h per day (Table 3).

Table 3. Biological risk level of operator

Sentinel microorganism	D	T	Corrective value	D*	V	T*	I	F	R
A. fumigatus	3	3	2	1	5	1	5	5	**16**
Penicillium sp	3	3	2	1	5	1	5	5	**16**

According to Biogaval method, the biological risk assessment with sentinel microorganism *A. fumigatus* and *Penicillium* sp, indicates that occupational exposure to fungal aerosols is in Biological action level (BAL). Preventive measures should be taken to reduce the risk of acquiring pharyngitis, bronchitis or other acute respiratory infections; preventive measures should also be taken to reduce the risk of dermatitis or other skin infections. Figure 1 shows the risk level map according to biological risk level in stages of the landfill and the time exposure; active cell is the stage of the landfill that presents the greatest risk of exposure to bioaerosols fungi, followed by the Leachate pool.

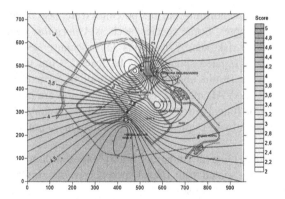

Fig. 1. Biological risk level map – exposure to fungal aerosols in the landfill.

3.4   Chi-Square Analysis

Chi-square analysis are shown in Table 4. Chi-squared statistic established a p-value less than 0.05, means statistical significance relationship between the hospital admission days and the diagnosed diseases (95% confidence). Figure 2 establish the sense of the relationship between the operator days and diagnostic hospital admissions. Acute respiratory diseases contribution had representation when the hospital admissions day was one, two or three days (Fig. 2). Moreover, otitis generated a significant number of days, whereas pharyngitis and dermatitis had a lower contribution of days (2). Acute diarrheal disease was represented on (1), (2) and (3) day of admission.

Table 4. Chi-squared analysis

Test	Statistic	Gl	P Value
Chi-square	56,089	15	**0,0000**

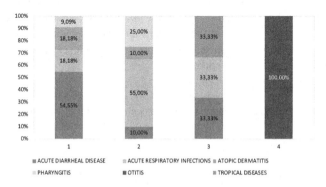

Fig. 2. Chi-square contribution for each number of hospital admission days.

Non-hazardous waste landfilling has the potential to release biological agents into the air, notably mould spores. Some species, such as *Aspergillus fumigatus*, may be a cause of concern for at-risk nearby residents [17] because aerodynamic diameter (AED) of single and aggregated spores could be from 1.9 μm to 2.7 μm [18]. Chi-square results support the results obtained in the sampling carried out [9] and the risk analysis based on BIOGAVAL method. Hospital admission are potentially associated to operator exposure of fungal aerosols by topic and respiratory via. Other hospital admission disease reported was tropical disease (chikungunya) not associated to fungal aerosols.

According to epidemiological studies of highly exposed populations, diarrhoea is one of the symptoms associated with fungal spore exposure, similar association with endotoxins (acute diarrheal disease) [18]. Endotoxin associated with ambient PM (particulate matter) has been linked to adverse respiratory symptoms, but there have been few studies of ambient endotoxin and its association with co-pollutants and inflammation [19]. The highest concentration corresponds to *A. fumigatus*, but the reported value is lower than those reported by other studies who reported geometrical averages of 9300 CFU /m^3 [11]. But it exceeds the values by The Health and Welfare Department in Canada [2, 20, 21].

4   Conclusion

The concentrations of bioaerosols did not show a major difference at a reference distance of 200 m, stating that this distance was not enough to reduce the microorganisms to background levels [22]. Environment Agency recommends a limit distance of at least 250 m, to ensure that composting plants do not have any adverse impact on the health of people living in the area of influence [23]. However, other authors have measured higher than background concentrations, at a distance of 550 m or more of composting sites [20], showing that the radius of action for the decay of the concentration in a natural way may require a wide distance [22].

Statistical analysis of the disease and days hospital admission provided objective information on the potential effect of the operator's bioaerosols exposure indicated changes in working hours urgently, according to risk assessment and results of the sampler campaigns [9]. Looking at the concerns and risks associated with bioaerosols, the area demands a substantial research culminate personal exposure to bioaerosol, formation, distribution, and its validation.

Acknowledgments. This research was supported by grants from GESSA Research Group, Universidad de la Costa. Special mention to environmental engineering students: Alfonso Lopez, Ericka Arbelaez, Andres De la Cruz, Wilson Guerrero, Andres Reales and Carlos Guerra.

References

1. Llorca, J., Soto, P., Roberto, G., Laborda, R., Benavent, S.: Practical manual for the assessment of biological risk in various labor activities *BIOGAVAL*. INVASSAT (2013)
2. Kim, K., Kabir, E., Jahan, S.: Airborne bioaerosols and their impact on human health. J. Environ. Sci. (2017). https://doi.org/10.1016/j.jes.2017.08.027
3. Fröhlich-Nowoisky, J., et al.: Bioaerosols in the earth system: climate, health, and ecosystem interactions. Atmos. Res. **182**, 346–376 (2016)
4. Walser, S.M., et al.: Evaluation of exposure–response relationships for health effects of microbial bioaerosols – a systematic review. Int. J. Hyg. Environ. Health **218**, 577–589 (2015). https://doi.org/10.1016/j.ijheh.2015.07.004
5. Haig, C.W., Mackay, W.G., Walker, J.T., Williams, C.: Bioaerosol sampling: sampling mechanisms, bioefficiency and field studies. J. Hosp. Infect. (2016). https://doi.org/10.1016/j.jhin.2016.03.017. W. B. Saunders Ltd.
6. Macher, J., Ammann, H.A., Burge, H.A., Milton, D.K., Morey, P.R. (Eds): Chapter 1. Bioaerosols: Assessment and Control. In: American Conference of Governmental Industrial Hygienists, pp 1–5. ACGIH, Cincinnati (1999)
7. Technical guide for the evaluation and prevention of risks related to exposure to biological agents. INSHT, Madrid, Spain (2014)
8. Yao, M.: Bioaerosol: a bridge and opportunity for many scientific research fields. J. Aerosol Sci. **115**, 108–112 (2018). https://doi.org/10.1016/j.jaerosci.2017.07.010
9. Morgado Gamero, W.B., Ramírez, M.C., Parody, A., Viloria, A., López, M.H.A., Kamatkar, S.J.: Concentrations and size distributions of fungal bioaerosols in a municipal landfill. In: Tan, Y., Shi, Y., Tang, Q. (eds.) Data Mining and Big Data. LNCS, vol. 10943, pp. 244–253. Springer, Cham (2018). https://doi.org/10.1007/978-3-319-93803-5_23
10. Landim, A.A., Teixeira, E.C., Agudelo-Castañeda, D., Kumar, P.: Spatio-temporal variations of sulfur dioxide concentrations in industrial and urban area via a new statistical approach (2018)
11. Morgado Gamero, W.B.: Evaluación de bioaeroroles fungí asociados a un relleno sanitario ubicado en el municipio de Tubara, departamento del Atlántico. Universidad de Manizales, Manizales, Colombia (2017)
12. Manual of optimal times of temporary disability. INSS (2013)
13. Torres-Rodríguez, J., Pulido-Marrero, Z., Vera-García, Y.: Respiratory allergy to fungi in Barcelona, Spain: clinical aspects, diagnosis and specific treatment in a general allergy unit. Allergol. Immunopathol. **40**(5), 295–300 (2012)
14. Alter, S., McDonald, M., Schloemer, J., Simon, R., Trevino, J.: Common child and adolescent cutaneous infestations and fungal infections. Curr. Probl. Pediatri. Adolesc. Health Care **48**(1), 3–25 (2018)
15. Kanaani, H., Hargreaves, M., Ristovski, Z., Morawska, L.: Deposition rate of fungal spores in indoor environments, factors effecting them and comparison with nonbiological aerosols. Atmos. Environ. **42**, 7141–7154 (2008)
16. Swan, J.R., Crook, B., Gilbert, E.J.: Microbial emission from composting sites. Issues Environ. Sci. Technol. **18**, 23–85 (2002)
17. Schlosser, O., et al.: *Aspergillus fumigatus* and mesophilic moulds in air in the surrounding environment downwind of non-hazardous waste landfill sites. Int. J. Hyg. Environ. Health (2016). http://dx.doi.org/10.1016/j.ijheh.2016.02.003
18. Eduard, W.: Fungal spores: a critical review of the toxicological and epidemiological evidence as a basis for occupational exposure limit setting. Crit. Rev. Toxicol. **39**(10), 799–864. https://doi.org/10.3109/10408440903307333. ISSN 1040-8444

19. Mahapatra, P.S., Jain, S., Shrestha, S., Senapati, S., Puppala, S.P.: Ambient endotoxin in PM10 and association with inflammatory activity, air pollutants, and meteorology, in Chitwan, Nepal. Sci. Total Environ. **618**, 1331–1342 (2018). https://doi.org/10.1016/J. SCITOTENV.2017.09.249

20. WHO (World Health Organization): Indoor Air Quality: 1094 Biological Contaminants. World Health Organization, 1095 European Series no. 31, Copenhagen, Denmark (1988) http://www.euro.who.int/__data/assets/pdf_file/0005/156146/1097WA754ES.pdf. 1096

21. Douglas, P., et al.: Sensitivity of predicted bioaerosol exposure from open windrow composting facilities to ADMS dispersion model parameters. J. Environ. Manag. **184**, 448–455 (2016). https://doi.org/10.1016/j.jenvman.2016.10.003

22. Sánchez-Monedero, M.A., Roig, A., Cayuela, M.L., Stentiford, E.I.: Emisión de bioaerosoles asociada a la gestión de residuos sólidos. Revista Ingeniería **10**(1), 39–47 (2006)

23. Environment Agency: Technical guidance on composting operations (2002). www. environment-agency.gov.uk/commondata/105385/compostin.pdf

Special Session on Data Selection in Machine Learning

Novelty Detection Using Elliptical Fuzzy Clustering in a Reproducing Kernel Hilbert Space

Maria Kazachuk, Mikhail Petrovskiy[⊠], Igor Mashechkin,
and Oleg Gorohov

Computer Science Department, Lomonosov Moscow State University,
Vorobjovy Gory, Moscow 119899, Russia
kazachuk@mlab.cs.msu.su, {michael,mash}@cs.msu.su,
owlman995@gmail.com

Abstract. Nowadays novelty detection methods based on one-class classification are widely used for many important applications associated with computer and information security. In these areas, there is a need to detect anomalies in complex high-dimensional data. An effective approach for analyzing such data uses kernels that map the input feature space into a reproducing kernel Hilbert space (RKHS) for further outlier detection. The most popular methods of this type are support vector clustering (SVC) and kernel principle component analysis (KPCA). However, they have some drawbacks related to the shape and the position of contours they build in the RKHS. To overcome the disadvantages a new algorithm based on fuzzy clustering with Mahalanobis distance in the RKHS is proposed in this paper. Unlike SVC and KPCA it simultaneously builds elliptic contours and finds optimal center in the RKHS. The proposed method outperforms SVC and KPCA in such important security related problems as user authentication based on keystroke dynamics and detecting online extremist information on web forums.

Keywords: Outlier and novelty detection · Kernel methods · Fuzzy clustering Mahalanobis distance · Keystroke dynamics · Online extremism discovering

1 Introduction

Novelty detection methods play an important role in solving many applied problems, primarily related to information and computer security. In such tasks, as a rule, only training data from one (legitimate) class is available. A one-class classification model is developed without using samples of other (illegitimate) target class. However, it may be assumed that a training sample contains a small percentage of observations from the target illegitimate class that are not labeled. Such tasks arise in computer [1], financial [2] and public security [3] areas. An anomaly (or outlier) is an observation in a sample whose features or their combinations do not correspond to the characteristics of the other observations in the sample. Traditional outlier detection methods use probabilistic (statistical) or distance-based (metric) approaches. In the statistical approach, the outlier is defined as an unlikely event and probabilistic interpretation is used. The metric

© Springer Nature Switzerland AG 2018
H. Yin et al. (Eds.): IDEAL 2018, LNCS 11315, pp. 221–232, 2018.
https://doi.org/10.1007/978-3-030-03496-2_25

approach uses a geometric interpretation: an outlier is an observation that is distant from the most part of observations in the sample. Methods that do not use probabilistic and geometric interpretations of the concept of the outlier are called deviation based.

It is worth to note that security-related tasks usually deal with a high dimensional input feature space. Therefore, many features are irrelevant to target (novelty) class and many features are interrelated and dependent. This fact makes difficult to apply traditional probabilistic and metric approaches that suffer from «the curse of dimensionality». Kernel methods represent a promising approach in this situation. The most popular kernel based methods are support vector clustering machine (or SVC) [4] and kernel principal components (KPCA) method [5]. But they have several drawbacks. In particular, SVC finds the optimal center of the training observations images in the RKHS and bounds them by the hypersphere of minimum radius, considering observations with images out of the hypersphere as outliers. The obvious disadvantage of the method is a spherical form of the non-outliers region in the RKHS. Dependences and correlations among input features may also lead to dependences between induced features in the RKHS. Therefore elliptic form is more suitable than spherical. KPCA, that is a kernel version of well-known PCA method, builds elliptic regions in the RKHS containing the most part of observations images. It provides effective approximation of areas for correlated features in the RKHS. Nevertheless, KPCA fixes the distribution center of the ellipse in the RKHS and does not recalculate it. That means that the position of the center in the RKHS is affected by outliers and may be biased.

To overcome these disadvantages a new method of anomaly detection is proposed in the paper. The input feature space is mapped into infinite dimensional feature space (using RBF kernel), and then fuzzy clustering algorithm with single cluster and the Mahalanobis distance metric is applied in the RKHS. Thus the elliptical single fuzzy cluster (ESFC) with iteratively adopted center is constructed in the RKHS, and the image of each observation has its own membership degree in this cluster. Parameters of the clustering algorithm (cluster radius regularization parameter and the degree of fuzziness) are set so that the degree of membership for the «core part» of observations images in the cluster would be high enough (above a predefined threshold, e.g. 0.5). Observations with calculated membership degrees smaller than a predefined threshold are marked as outliers. The Mahalanobis distance is calculated by projecting the input data into the space of eigenvectors of the covariance matrix in the RKHS. This allows taking into account the variance and the correlation between the features in the RKHS. Thus, the resulting cluster has an ellipsoidal form and allows to describe the area containing images of the most part of the legitimate training samples more precisely and, as a result, to build more accurate one-class model.

This paper is organized as follows. Section 2 provides an overview of popular kernel-based novelty detection methods. Section 3 describes the proposed approach that develops the idea of existing kernel-based methods and try to overcome their drawbacks. Section 4 is devoted to the experimental study of the proposed approach for anomaly detection in two typical applied problems related to computer and information security. The first problem is a user authentication based on keystroke dynamics. The second one is discovering extremism information in users' messages on the Internet forums. Conclusions are presented in the Sect. 5.

2   Review of Existing Kernel-Based Methods for Novelty Detection

Well-known methods of anomaly and novelty detection in a RKHS are support vector clustering (SVC) [4], Single Class SVM [6], Kernel PCA [5], and kernel based fuzzy outlier detection method (Fuzzy) [7]. The main idea behind of all these methods is in mapping of the input feature space X into a high (or infinite) dimensional feature space H called kernel reproducing kernel Hilbert space (RKHS), where images $\varphi(x)$ and $\varphi(y)$ are associated with the original observations x and y as follows:

$$K(x,y) = \langle \varphi(x), \varphi(y) \rangle_H, \tag{1}$$

where K is some Mercer kernel function. The Euclidian distance d between images of x and y in new induced space is calculated as follows:

$$d(x,y) = \sqrt{K(x,x) - 2K(x,y) + K(y,y)}. \tag{2}$$

The choice of the kernel function depends on the particular problem. Gaussian radial basis function (RBF) is frequently used in outlier detection tasks:

$$K(x,y) = e^{\frac{-\|x-y\|^2}{2\sigma^2}}, \tag{3}$$

where σ is a kernel width parameter.

The purpose of the mapping is the ability to use simple geometric structures in the RKHS to describe complex relations existing in the input space.

Consider these methods in more details. In SVC method [4] observations from the input space are implicitly mapped by a kernel function into a high dimensional Hilbert space, where a hypersphere of minimum radius R containing «the main part» of images is found. Observations whose images are outside of the hypersphere are marked as outliers. Thus, the following optimization problem is solved in this method:

$$\min_{\zeta \in \mathbb{R}^N, R \in \mathbb{R}, a \in H} \left[R^2 + \frac{1}{vN} \sum\nolimits_{i=1}^N \xi_i \right], \tag{4}$$

$$\|\varphi(x_i) - a\|_H^2 \leq R^2 + \xi_i, \forall i \in [1, N], \tag{5}$$

where R is a radius of the hypersphere; a is a center of the hypersphere in the RKHS; N is the number of samples in the training set, $0 < v \leq 1$ is a predefined expected percent of outliers; ξ_i are slack variables. Solving this problem by the Lagrange multipliers method, one comes to the binary decision function of the following form:

$$f(z) = sgn\left(R^2 - \sum\nolimits_{i,j=1}^N \beta_i \beta_j K(x_i, x_j) + 2 \sum\nolimits_{i=1}^N \beta_i K(x_i, z) - K(z, z)\right), \tag{6}$$

where $\beta_i\, i \in [1, N]$ are Lagrange multipliers, $\beta_i = \frac{1}{vN}$ for outliers, $0 < \beta_i < \frac{1}{vN}$ for margin (boundary) observations and $\beta_i = 0$ for others, z is a tested observation.

Single Class SVM [6] is similar to SVC. It finds the hyperplane that separates «the main part» of the observations images from the origin. Observations with images located closer to the origin on the «wrong» site of the hyperplane are marked as outliers. The results of the Single Class SVM and SVC methods are very similar if the RBF kernel is used. The main disadvantage of these methods is an excessively simple form of the boundaries for identifying outliers in the RKHS. That does not allow taking into account dependencies among features in the RKHS. Another disadvantage has any support vector machine method – the resulting model depends only on support vectors. These methods do not provide a correct estimation of the anomaly degree for observations on the «correct» site of the boundary.

The latter problem can be solved by kernel based fuzzy outlier detection method [7], which is a modification of the SVC method. In this approach single fuzzy cluster containing all observations images is built in the RKHS instead of a hypersphere. The fuzzy cluster membership degree for the «core» of observations should be high enough. The membership degree is interpreted as a measure of «typicalness». The observations with typicalness less than a predefined threshold are marked as outliers. Thus, the problem is reduced to the following optimization problem:

$$\min_{U,a,\eta} J(U,a,\eta) = \sum\nolimits_{i=1}^{N} u_i^m (\varphi(x_i) - a)^2 + \eta \sum\nolimits_{i=1}^{N} (1 - u_i)^m, \qquad (7)$$

where a is a center of fuzzy cluster in the RKHS; N is a number of observations; U is a vector of fuzzy membership degrees; $u_i \in [0, 1]$ is a measure of typicalness of the i-th observation; $m > 1$ is a degree of fuzziness (a parameter that determines the rate of decreasing membership function depending on the distance to the center of the cluster) and $\eta > 0$ is a regularization parameter referred to the radius of the fuzzy cluster in the RKHS. Practically η is a distance from the cluster center a in the RKHS, where the degree of typicalness is exactly equal to 0.5. The observations that images are closer to the center have higher typicalness, and others have smaller. Minimizing $J(U,a,\eta)$ by the iterative procedure [7], one can find a measure of typicalness $u(z)$ for any observation z in the following form:

$$u(z) = \left[1 + \left(\frac{\sum_{j=1}^{N} u_j^m \sum_{i=1}^{N} u_i^m K(x_i, x_j)}{\eta \left(\sum_{i=1}^{N} u_i^m \right)^2} - 2 \frac{\sum_{i=1}^{N} u_i^m K(z, x_i)}{\eta \sum_{i=1}^{N} u_i^m} + \frac{K(z,z)}{\eta} \right)^{\frac{1}{m-1}} \right]^{-1}.$$

$$(8)$$

Unlike SVC, Fuzzy method assigns a certain weight (degree of typicalness) to each observation, that reduces the influence of less significant (less typical) observations in X and increases the influence of more significant observations.

It is worth noting that the training data may contain correlations and other types of interrelations among features remained in the RKHS after mapping. Thus, using of hypersphere, spherical fuzzy cluster or separating hyperplane in the RKHS seems to be not optimal. In this case, the training sample boundary can be better described using an ellipsoid in the RKHS. The use of elliptical structures instead of a hypersphere allows taking into account the scale variation in each direction and features interrelations. As a result, more accurate one class model can be built.

Kernel PCA method [5] exploits this idea. This is a kernel-based version of a traditional PCA method. It projects images of training samples to selected top principal components in the RKHS. As a result, the hyperellipsoid (not the hypersphere) contour containing the most part of images of samples is constructed in the RKHS. The principal components correspond to the eigenvectors of the covariance matrix with the largest eigenvalues. The number of selected principal components is a key parameter set a priori.

To find outliers the KPCA method introduces the concept of reconstruction error as a measure of novelty. For each sample x_i this value is calculated as follows:

$$\|\tilde{\varphi}(x_i) - P\tilde{\varphi}(x_i)\|_H^2 = (\tilde{\varphi}(x_i), \tilde{\varphi}(x_i)) - 2(\tilde{\varphi}(x_i), P\tilde{\varphi}(x_i)) + (P\tilde{\varphi}(x_i), P\tilde{\varphi}(x_i)), \quad (9)$$

where $\tilde{\varphi}(x_i)$ is a centered image of this observation in the centered RKHS, $\tilde{\varphi}(x_i) = \varphi(x_i) - \frac{1}{N}\sum_{i=1}^{N} \varphi(x_i), P\tilde{\varphi}(x_i)$ is a projection of $\tilde{\varphi}(x_i)$ to the subspace with maximum variance of the data.

Thus, the level of anomaly $p(z)$ for a tested vector z is calculated as follows:

$$p(z) = K(z, z) - \frac{2}{N}\sum_{i=1}^{N} K(z, x_i) + \frac{1}{N^2}\sum_{i,j=1}^{N} K(x_i, x_j)$$
$$- \sum_{l=1}^{q}\left(\sum_{i=1}^{N} a_i^l(K(z, x_i) - \frac{1}{N}\sum_{r=1}^{N} K(x_i, x_r) - \frac{1}{N}\sum_{r=1}^{N} K(z, x_r)\right.$$
$$\left. + \frac{1}{N^2}\sum_{r,s=1}^{N} K(x_r, x_s))\right)^2, \quad (10)$$

where N is a number of observations; a_i^l are weight coefficients, calculated by the method; q is a number of used principal components. Samples with the highest reconstruction error are marked as outliers. Although the model constructs elliptical contours in the RKHS (see (9), (10)), their center is fixed by centralized kernel matrix even before the algorithms starts. That means the center is not optimally recalculated as it is done in SVC and Fuzzy methods. The center of the KPCA ellipsoid is placed in the center of mass of the distribution in the RKHS and, as a result, it is strongly affected by outliers. In addition, the model significantly dependents on the parameter q, which determines the number of used principal components.

3 Proposed Approach

The proposed novelty detection method called Elliptical Single Fuzzy Clustering (ESFC) in a RKHS is a modification of the kernel-based Fuzzy outlier detection method [7]. It combines advantages of Fuzzy, SVC and Kernel PCA methods. A key feature of the ESFC is using of Mahalanobis distance metric for calculating distances between images of samples and the cluster center in the RKHS. The Mahalanobis distance takes into account a heterogeneous variance and correlations of the data in the RKHS. It is calculated by projecting the data images to the space specified by the eigenvectors of the covariance matrix in the RKHS. As a result, in contrast to the cluster in Fuzzy method the constructed cluster is not spherical but elliptical (thus, it takes the advantage of the KPCA approach).

The optimal values of typicalness membership degree of observations are found from solving the following optimization problem:

$$\min_{U,a,\eta} E(U,a,\eta) = \sum_{i=1}^{N} u_i^m \|a - \varphi(x_i)\|_C^2 + \eta \sum_{i=1}^{N} (1 - u_i)^m, \quad (11)$$

that is similar to (7), but Mahalanobis kernel-based distance is used instead of traditional kernel-based distance (2). Here a is a fuzzy cluster center in the RKHS; N is a number of samples; U is a vector of fuzzy measures of typicalness, $u_i \in [0, 1]$ is a component of U – the measure of typicalness for the i-th sample; $m > 1$ is a fuzzy degree parameter and $\eta > 0$ is a regularization parameter responsible to the size of the cluster. The latter one can be either fixed as a constant or can be iteratively calculated at each iteration, for example as the distance to the image of the q-th most distant image from a; $\left\| a - \varphi(x_i)_C \right\|^2$ is a square of Mahalanobis distance in the RKHS between the image $\varphi(x_i)$ and the cluster center a:

$$\|a - \varphi(x_i)\|_C^2 = (a - \varphi(x_i))^T C^{-1}(a - \varphi(x_i)) \quad (12)$$

with covariance matrix C. The following iterative block coordinate descent algorithm can be applied to minimize (11):

Step 0. Initialization.

$$u_i^{(0)} = random([0, 1]), \; \eta^{(0)} = 0.5. \quad (13)$$

Step 1. *l-th* iteration
The center of the cluster $a^{(l)} = argmin E\left(U^{(l-1)}, a^{(l-1)}, \eta^{(l-1)}\right)$ is found as:

$$a^{(l)} = \sum_{i=1}^{N} a_i^{(l)} \varphi(x_i), \text{ where } a_i^{(l)} = \frac{\left(u_i^{(l-1)}\right)^m}{\sum_{i=1}^{N} \left(u_i^{(l-1)}\right)^m}. \quad (14)$$

Covariance matrix approximation C is recalculated at each iteration because it depends on the degrees of typicalness and the center of the cluster a:

$$C_{ij}^{(l)} = \left(\varphi(x_i) - a^{(l)}\right)\left(\varphi(x_j) - a^{(l)}\right) =$$

$$\left(\varphi(x_i) - \frac{\sum_{r=1}^{N}\left(u_r^{(l-1)}\right)^m \varphi(x_r)}{\sum_{r=1}^{N}\left(u_r^{(l-1)}\right)^m}\right)\left(\varphi(x_j) - \frac{\sum_{s=1}^{N}\left(u_s^{(l-1)}\right)^m \varphi(x_s)}{\sum_{s=1}^{N}\left(u_s^{(l-1)}\right)^m}\right) =$$

$$K_{ij} - \frac{\sum_{r=1}^{N}\left(u_r^{(l-1)}\right)^m \left(K_{ir} + K_{rj}\right)}{\sum_{r=1}^{N}\left(u_r^{(l-1)}\right)^m} + \frac{\sum_{r,s=1}^{N}\left(u_r^{(l-1)}\right)^m \left(u_s^{(l-1)}\right)^m K_{rs}}{\left(\sum_{r=1}^{N}\left(u_r^{(l-1)}\right)^m\right)^2}, \tag{15}$$

where $K_{ij} = K(x_i, x_j)$. Inverse matrix $M = C^{-1}$ is also found.

Step 2. The distance $D_k\left(a^{(l)}\right)$ from the cluster center to the image of the k-th sample x_k on the l-th iteration is calculated for each training sample as follows:

$$D_k\left(a^{(l)}\right) = \left\|a^{(l)} - \varphi(x_k)\right\|_c^2 = \left(a^{(l)} - \varphi(x_k)\right)^T M\left(a^{(l)} - \varphi(x_k)\right) =$$

$$\sum_{j=1}^{N}\sum_{i=1}^{N}\left(M_{ij}^{(l)}\frac{\sum_{r,s=1}^{N}\left(u_r^{(l-1)}\right)^m \left(u_s^{(l-1)}\right)^m K_{rs}}{\left(\sum_{r=1}^{N}\left(u_r^{(l-1)}\right)^m\right)^2} - 2M_{ij}^{(l)}\frac{\sum_{r=1}^{N}\left(u_r^{(l-1)}\right)^m K_{rk}}{\sum_{r=1}^{N}\left(u_r^{(l-1)}\right)^m} + M_{ij}^{(l)} M_{kk}\right) =$$

$$\sum_{j=1}^{N}\sum_{i=1}^{N} M_{ij}^{(l)} M_{kk}^{(l)}. \tag{16}$$

If η is not fixed, it is calculated as the distance to the image of the $(N - q)$-th most distant from a sample, as a result q samples are atypical (their measure of typicalness is less than 0.5). Parameter q (expected number of outliers) must be specified a priori and it is fixed during all iterations.

$$\eta^{(l)} = D_{i:q}^2\left(a^{(l)}\right), D_{i:1}^2\left(a^{(l)}\right) \geq D_{i:2}^2\left(a^{(l)}\right) \geq \ldots \geq D_{i:N}^2\left(a^{(l)}\right). \tag{17}$$

Step 3. New degrees of typicalness $U^{(l)} = argminE\left(U^{(l)}, a^{(l)}, \eta^{(l)}\right)$ are found:

$$u_i^{(l)} = \left[1 + \left(\frac{D_i\left(a^{(l)}\right)}{\eta^{(l)}}\right)^{\frac{1}{m-1}}\right]^{-1}. \tag{18}$$

If Frobenius norm $\left\|U^l - U^{l+1}\right\|_F > \varepsilon$ go to **Step 1**, otherwise the iterative procedure stops. The convergence of the algorithm is proved but the formal proof is not included in this paper due to the size limitation.

The decision function for the typicalness degree $u(z)$ for an arbitrary observation z is calculated as follows:

$$u(z) = \left[1 + \left(\frac{D_z(a)}{\eta} \right)^{\frac{1}{m-1}} \right]^{-1},\qquad(19)$$

where the Mahalanobis distance from the image $\varphi(z)$ of the analyzing observation z to the fuzzy cluster center a defined by (14) in the RKHS is calculated as following:

$$D_z(a) = \|a - \varphi(z)\|_C^2 = (a - \varphi(z))^T M(a - \varphi(z)) =$$
$$\sum_{j=1}^{N} \sum_{i=1}^{N} \left(M_{ij} \frac{\sum_{r,s=1}^{N} u_r^m u_s^m K_{rs}}{\left(\sum_{r=1}^{N} u_r^m \right)^2} - 2M_{ij} \frac{\sum_{r=1}^{N} u_r^m K(x_r, z)}{\sum_{r=1}^{N} u_r^m} + M_{ij} K(z, z) \right).\qquad(20)$$

Typicalness degree u_r assigned to the r-th object from the training set (18), inverse covariance matrix M (15) and the parameter η (17) are found by the proposed algorithm ESFC, and $K(x_r, z)$ is a kernel function calculated for the tested observation z and the r-th sample from the training set X.

The toy example of application of SVC, KPCA and ESFC algorithms for simple simulated data is presented on figures below. They show sample data points (dark solid circles) and projected decision contours of each algorithm for linear kernel on Fig. 1 and for RBF kernel (with a kernel width of 0.01) on Fig. 2.

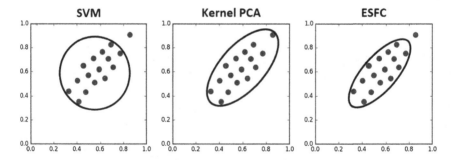

Fig. 1. Decision contours with linear kernel.

The figures show that the proposed algorithm allocates more precise contours with smaller area and that is why it better describes the simulated data distribution. The reasons are elliptical shape (unlike SVC) and adopted center (unlike KPCA) of the contour. For all algorithms with linear kernel the shape of the contours are smooth and analytically described by ellipsoid or circle equations. For RBF kernel the contours have more complex shape because they are back projected from infinite dimensional RKHS into the input space. For both types of kernels the decision boundary of the proposed algorithm looks more precise than boundaries produced by competitors.

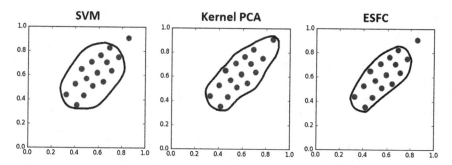

Fig. 2. Decision contours with RBF kernel.

4 Experiments

To estimate the performance of the proposed Elliptical Single Fuzzy Clustering algorithm we applied it to two security-related real-world problems and compared the results with SVC, KPCA and Fuzzy. The first problem is related to the computer security. It is a user authentication based on keystroke dynamics. The second problem is related to the information and public security. It is online detection of radical content (extremism) on web forums in the Internet.

In the first problem dataset Villani [8, 9] is used for experiments. It contains data of keystroke dynamics of 144 users, accumulated in background during their work on computers. These users wrote essays on given topics and produced 1345 samples overall. The following data was collected for each user: platform (desktop or laptop), gender, age group, handedness and awareness of data collection. Pressed key code, type of event (press/release) and time stamp are also used in our calculations. The data for each user is divided into two equal parts. The novelty detection model for each user is built on the first part of the data. The second part of the user's data is used for testing as positive examples and data of all other users is used for testing as negative examples. The same was repeated for each user. The feature vectors are constructed in the following way. The sequence is divided into subsequences (windows) of events, if the number of events exceeds the threshold (minimum window size – 300 events; maximum window size – 500 events), or if there is a pause (of more than 40 s) between the sequential events. Features describing time delays between keyup and keydown for separate keystrokes and digraphs (two sequentially pressed keys) were calculated. Obtained sparse feature vectors consist of about 250 features, many of them are correlated. Quintile based discretization is applied to all features to avoid problems with multi-modal distributions of press-release times. This dataset and its preparation procedures are described in details in [1].

The second problem is devoted to the online discovering of extremist's content on Internet web forums. In this problem, we assume that users had some «clean» legitimate history of publishing forum messages. The novelty detection model is built to discover the moment of appearing illegitimate radical content. Examples of radical messages were taken from jihadist «private» (by invitation only) Ansar1 web forum, collected in the Dark Web Project by the University of Arizona research group [10].

As legitimate messages we took 20 Newsgroups dataset [11]. To make the data more realistic we took messages from the following discussion groups: *talk.politics.misc, talk.politics.guns, talk.politics.mideast, talk.religion.misc, alt.atheism, soc.religion. christian*. Similar topics were discussed on the jihadist forum Ansar1. The novelty detection model was trained on 2525 legitimate messages (from 20 newsgroups dataset). Than the model was tested on a mixture of 2524 legitimate messages from 20 newsgroups dataset (as positive examples) and on 500 extremist messages from Ansar1 dataset (as negative examples – novelties). Stemming was applied as a preprocessing step and lexeme based features were extracted with tf-idf weights. Words that appeared less than in five messages were not included in the vocabulary. As a result, sparse feature vectors with many correlated features were constructed.

To set a priori metaparameters for the algorithms we used the following procedure. Setting metaparameters for one-class machine learning methods is a non-trivial problem, because usually only data from a legitimate class are available. Examples of illegitimate class are absent or they are «hidden» (that means they are not labeled in the training set). Therefore standard methods of parameters tuning with validation set, containing labeled examples of both classes (as it is for supervised learning methods), cannot be used. In this paper, we adopted the approach of tuning parameters with validation dataset for unsupervised learning case in the following way. All data of each legitimate class are divided randomly into training and holdout-subsamples in the ratio of one to one. Each of the considered kernel based anomaly detection methods has the important fixed metaparameter, which can be described as "expected ratio of outliers to the number of all observations" in the considered sample (or expected percentage of outliers). For example, SVC has a parameter v (4) that is the ratio of the expected number of outliers. In Fuzzy and ESFC methods such parameter has the form of the relation $\frac{q}{N}$, where q is the number of non-typical observations (with membership degree less than 0.5), N is the size of the training sample (17). For KPCA this parameter denotes the part of objects with the largest anomaly level, threshold of the value is calculated for the training set (10) and is used for classification. The remaining parameters of the algorithms are the following: kernel width for all considered algorithms, fuzzy degree for Fuzzy and ESFC, and the number of principle components for Kernel PCA. Our idea is to fix the «expected ratio of outliers» parameter for each method and vary other parameters to determine their best combination such that the value of «expected ratio of outliers» on the training and holdout (validation) datasets would be as close as possible. If they are close, it means that our key assumption about the "expected ratio of outliers" in the training dataset is confirmed on the validation dataset. In this case found combination of other parameters produces stable results in the context of "expected proportion of outliers". It should be noted that both training and validation datasets are still not labeled in this approach. In particular, for SVC we are looking such kernel width that the proportion of observations with negative values of the decision function is the same for training and validation datasets. Kernel width and degree of fuzziness for ESFC and Fuzzy are chosen to provide equal proportions of observations with typicalness degree less than 0.5 for training and validation datasets and so on. The following values of the expected proportion of outliers were considered: 5%, 10%, 15%. The optimal values of other parameters of algorithms are obtained by greed search with fixed «expected proportion of outliers» using described validation method. Results are represented in Table 1.

Table 1. Optimal values of parameters for anomaly detection methods with RBF kernel obtained using proposed holdout approach.

	Parameter name	Keystroke dataset	Extremism dataset
SVC	Kernel width	0.01	0.5
	Proportion of outliers	10%	10%
KPCA	Kernel width	0.01	0.5
	Proportion of outliers	10%	5%
	Number of principle components	30	40
Fuzzy	Kernel width	0.008	0.7
	Proportion of outliers	10%	15%
	Degree of fuzziness	1.5	1.2
ESFC	Kernel width	0.006	0.7
	Proportion of outliers	5%	5%
	Degree of fuzziness	1.5	1.2

Performance of algorithms is estimated using the area under ROC-curve (ROC–AUC) measure. This measure is tolerant to the ratio of positive and negative examples in the test sample. For each of two problems several series of experiments with different randomly generated subsets of the same size for training samples were run and finally the median and interquartile range (to estimate the variance) of AUC were calculated. Experimental results for the given settings of anomaly detection methods are represented in Table 2. Experimental results confirm that the proposed method is more precise and more stable in the same time.

Table 2. Experimental results: median $\pm$ interquartile range of AUC.

	Keystroke dataset	Extremism dataset
SVC	0.8849 $\pm$ 0.1884	0.7035 $\pm$ 0.092
KPCA	0.8871 $\pm$ 0.1863	0.7131 $\pm$ 0.071
Fuzzy	0.9031 $\pm$ 0.1705	0.7513 $\pm$ 0.074
ESFC	**0.9138 $\pm$ 0.1475**	**0.7710 $\pm$ 0.062**

5 Conclusions

Kernel based approach is an effective way to deal with high-dimensional correlated data structures in novelty detection tasks. However, existing popular methods such as one-class support vectors machines and kernel principal components analysis have several drawbacks related to the shape and the center position of bounding contours in the RKHS that they use to isolate outliers. In particular, SVC finds optimally adopted center with spherical contour. However, images of samples with correlated features in the RKHS are better approximated by elliptical contours. On the other hand, KPCA uses elliptical contours with the predefined fixed center. This center is estimated even before the algorithm starts, thus, it is affected by outliers. To avoid these drawbacks we develop Elliptical Single Fuzzy Clustering algorithm, which builds the elliptical contour with optimal center in the RKHS. It combines the advantages of both KPCA and SVC approaches. To estimate the practical utility of the proposed approach we use it for two important applied problems. The former is a user authentication based on keystroke dynamics and the latter is an online extremist content detection on web forums. Experimental study on real-world data confirms that the proposed method is more precise and, in the same time, more stable than SVC, KPCA and Fuzzy for both practical problems.

Future research directions for our approach include its application to other novelty detection problems that arise in the field of computer, information and public security; comparison the performance to other modern novelty detection methods.

Acknowledgements. The research is supported by RFFI Grant 16-29-09555.

References

1. Kazachuk, M., et al.: One-class models for continuous authentication based on keystroke dynamics. In: Yin, H., et al. (eds.) IDEAL 2016. LNCS, vol. 9937, pp. 416–425. Springer, Cham (2016). https://doi.org/10.1007/978-3-319-46257-8_45
2. Ngai, E.W., Hu, Y., Wong, Y.H., Chen, Y., Sun, X.: The application of data mining techniques in financial fraud detection: a classification framework and an academic review of literature. Decis. Support Syst. **50**(3), 559–569 (2011)
3. Petrovskiy, M., Tsarev, D., Pospelova, I.: Pattern based information retrieval approach to discover extremist information on the internet. In: Ghosh, A., Pal, R., Prasath, R. (eds.) MIKE 2017. LNCS (LNAI), vol. 10682, pp. 240–249. Springer, Cham (2017). https://doi.org/10.1007/978-3-319-71928-3_24
4. Ben-Hur, A., Horn, D., Siegelmann, H.T., Vapnik, V.: Support vector clustering. J. Mach. Learn. Res. **2**(Dec), 125–137 (2001)
5. Hoffmann, H.: Kernel PCA for novelty detection. Pattern Recogn. **40**(3), 863–874 (2007)
6. Scholkopf, B., Williamson, R.C., Smola, A.J., Shawe-Taylor, J., Platt, J.C.: Support vector method for novelty detection. In: Advances in Neural Information Processing Systems, pp. 582–588 (2000)
7. Petrovskiy, M.: A fuzzy kernel-based method for real-time network intrusion detection. In: Böhme, T., Heyer, G., Unger, H. (eds.) IICS 2003. LNCS, vol. 2877, pp. 189–200. Springer, Heidelberg (2003). https://doi.org/10.1007/978-3-540-39884-4_16
8. Monaco, J.V., Bakelman, N., Cha, S.H., Tappert, C.C.: Developing a keystroke biometric system for continual authentication of computer users. In: 2012 European Intelligence and Security Informatics Conference (EISIC), pp. 210–216. IEEE (2012)
9. Tappert, C.C., Cha, S., Villani, M., Zack, R.S.: Keystroke biometric identification and authentication on long-text input. Int. J. Inf. Secur. Priv. (IJISP) **4**, 32–60 (2010)
10. Zhang, Y., Zeng, S., Fan, L., Dang, Y., Larson, C.A., Chen, H.: Dark web forums portal: searching and analyzing jihadist forums. In: IEEE International Conference on Intelligence and Security Informatics, pp. 71–76. IEEE (2009)
11. The 20 Newsgroups data set. http://people.csail.mit.edu/jrennie/20Newsgroups/. Accessed 18 Aug 2017

Semi-supervised Learning to Reduce Data Needs of Indoor Positioning Models

Maciej Grzenda[(✉)]

Faculty of Mathematics and Information Science, Warsaw University of Technology,
ul. Koszykowa 75, 00-662 Warszawa, Poland
M.Grzenda@mini.pw.edu.pl
http://www.mini.pw.edu.pl/~grzendam

Abstract. Indoor positioning systems answer the need for ubiquitous localisation systems. Frequently, indoor positioning relies on machine learning models developed based on the training data composed of WiFi received signal strength (RSS) vectors observed in different indoor locations. However, this requires expensive collection of RSS vectors in precisely measured locations. In this study, we propose a semi-supervised method, which can reduce the volume of expensive labelled training data and exploit the availability of unlabelled signal strength measurements. The method relies, *inter alia*, on the measures of similarity among nearest neighbours of unlabelled vectors. Tests performed with a number of testbed areas confirm that the method improves the accuracy of random forest models used to estimate indoor location of mobile terminals.

Keywords: Semi-supervised learning · Fingerprinting
Indoor positioning

1 Introduction

Estimating the location of Internet of Things (IoT) objects is fundamental for many IoT applications. In case indoor location has to be estimated, indoor localisation [4] providing basis for Indoor Positioning Services (IPS) [2] can be used. Such services can be applied to estimate the location of both the objects and the persons using them for purposes such as increasing safety at construction sites [5]. Many IPS solutions follow *fingerprinting* paradigm [2,4,7] i.e. rely on a *radio map* to estimate the location of mobile objects in a building. The radio map is a set of RSS vectors collected during off-line phase in a building of interest. Importantly, these measurements are linked to known locations. In terminal-centric approach, vectors of signal strengths emitted by network infrastructure such as WiFi access points (APs) [1] are collected in a number of reference points in a building. The collection of such data requires a grid of hundreds or thousands of points to be planned and measurements of signal strengths made in each of these points. Hence, the collection of a radio map is a time consuming task [4,8].

© Springer Nature Switzerland AG 2018
H. Yin et al. (Eds.): IDEAL 2018, LNCS 11315, pp. 233–240, 2018.
https://doi.org/10.1007/978-3-030-03496-2_26

Once radio map is available, based on a signal strength vector observed at a mobile object, the indoor location of this object can be estimated. In the simplest case, techniques such as Nearest Neighbour (NN) estimating unknown location to be this location in the building, in which the most similar RSS vector present in the radio map has been observed, can be used. The radio map provides the training data, based on which a classification model for floor detection or regression models for x and y coordinate estimation can be developed.

Typically, the higher the density of the measurement grid, the higher the accuracy of location estimates, as the distribution of RSS values in $(x, y, floor)$ space can be more precisely sampled. Still, this increases the cost of the off-line phase. A major part of this cost is measuring horizontal coordinates (x, y) of each measurement point. The RSS part of the data itself could be easily collected during random or approximately planned walks [6] made in the area of a building covered with radio map. This illustrates a frequent problem of the cost of obtaining labelled data. In the analysed case, the labels are known locations i.e. $(x, y, floor)$. For this reason, an attempt to apply semi-supervised approach i.e. to use both labelled radio map data linked to precisely known locations and unlabelled RSS data missing location labels is worth considering. Importantly, the unlabelled data could be collected also after the IPS system is deployed, possibly raising its accuracy with time. However, the use of semi-supervised techniques for fingerprinting-based IPS is still limited. Existing attempts are largely focused on the use of manifold learning including the use of Isomap method suggested in [8] and locally linear embedding [3]. A proposal of the method integrating semi-supervised manifold regularisation and extreme learning machines has recently been made in [4]. The latter study includes also a recent survey of semi-supervised approaches to indoor localisation, including methods involving the use of Laplacian regularization, but also deep learning.

In this study, we propose a different approach i.e. providing unlabelled RSS instances with propagated labels i.e. predicted location labels in order to include them in the extended training data. An important new aspect that is addressed in this study is whether to select the subset of unlabelled RSS vectors to be provided with propagated labels based on the estimated confidence of these labels. We show how the consistency of the locations of instance neighbours can be used to estimate such confidence and guide the development of a training data set.

The remainder of this work is as follows. First, the reference data is introduced in Sect. 2. Next, a proposal for semi-supervised methods is made. This is followed by the analysis of experimental results in Sect. 4. Finally, conclusions are made and future works are outlined in Sect. 5.

2   Reference Data

The radio map data used in this study come from measurement sites located in the reference building of the Warsaw University of Technology, referred to as testbeds below. On each of six floors of the building, one testbed was planned in which WiFi RSS vectors linked to known (x, y, f) coordinates were collected.

The dimensions of the testbeds were larger than typically used, e.g. compared to indoor environments used in [4], which has an impact on both spatial dispersion of similar RSS vectors and positioning errors. In every point measurements were made with a terminal rotated into one of four different directions. For every measurement point and terminal direction, measurements were repeated typically 10 times, providing each time one RSS vector associated with location label. There were two groups of measurement points i.e. Reference Points (RP), providing training data and Testing Points (TP) providing testing data.

Let $s_{\text{RP}}^{r,d,t}$ denote an RSS vector collected in r-th RP, for terminal orientation d and measurement series t. Based on r, the coordinates of RP $c(r) = (x, y, f)$ can be determined. Similarly, let $s_{\text{TP}}^{r,d,t}$ denote an RSS vector collected in r-th TP, in direction d and within t-th series. Every training set R_i is composed of measurements $s_{\text{RP}}^{r,d,t} \in R_i$ made in RPs, and every testing set N_i contains measurements $s_{\text{TP}}^{r,d,t} \in N_i$ made in TPs. There were 1401 reference points and 1461 testing points altogether. Further details on the reference data can be found in our work on discretisation of RSS data [2].

In this study, the set of testing RSS in $i - th$ testbed was randomly divided into two subsets, each including the data from approximately equal number of TPs. Out of every pair of subsets $U_i, T_i : U_i \cup T_i = N_i$, U_i was used to provide unlabelled data for semi-supervised approach, while the role of T_i was to provide labelled testing data used to calculate the accuracy of location estimates. Let $\xi(A)$ denote the set of (x, y, f) points, the data contained in A was collected in. All divisions satisfied $\xi(R_i) \cap \xi(U_i) = \phi \wedge \xi(R_i) \cap \xi(T_i) = \phi \wedge \xi(U_i) \cap \xi(T_i) = \phi$.

3   Semi-supervised Methods for Fingerprinting-Based IPS

In this work, the extension of an original training set R to $\tilde{R} : R \subset \tilde{R}$ by adding new RSS vectors with propagated location labels is considered. We will focus on 2D positioning. Hence, by a location label, a tuple (x, y) is meant. The key aspect of the process is to assign location labels to RSS vectors which were collected during random or approximate walks. We expect that a positioning model $\tilde{M}()$ built with extended data $\tilde{R}$ will yield higher accuracy of location estimates than a model $M()$ built with the original reference data R.

To objectively assess the contribution of semi-supervised method, a comparison of errors attained by models developed with R_i data set with the errors of models trained with $\tilde{R}_i$ will be made on testing data T_i coming from the same area of the building. Moreover, let us propose to include in $\tilde{R}_i$ only these instances from U_i that can be assigned labels with high confidence. In the analysed case, it is possible that an RSS vector $\mathbf{v} \in U$ can be observed in distant parts of a building. Importantly, when RSS vectors $\mathbf{v} = [v_1, \ldots, v_k]$ are collected with devices such as smartphones, v_i comes from a limited cardinality data set i.e. $v_i \in Z : card(Z) < 100$. Hence, similar or even identical RSS vectors can be observed in different parts of a building. When such similarity of $\mathbf{v} \in U$ to spatially dispersed vectors in R occurs, a decision may be made not to generate a location label for such a $\mathbf{v}$ vector and not to include the $(\mathbf{v}, l_{\text{x}}(\mathbf{v}), l_{\text{y}}(\mathbf{v}))$ tuple in $\tilde{R}$, where $l_{\text{x}}, l_{\text{y}}$ denote functions assigning propagated location to $\mathbf{v} \in U$. For this reason, methods allowing to set a proportion $\omega \in [0,1]$ of vectors $\mathbf{v} \in U$ to be labelled and placed in $\tilde{R}$ are proposed in this study.

The process of label propagation and the evaluation of its impact taking into account the aforementioned assumptions is proposed in Algorithm 1. The algorithm starts with building localisation models M_x, M_y predicting the values of x and y coordinates, respectively. Next, we propose unlabelled data U to be sorted in the growing sequence of *diversity measures*. By diversity measure we mean a function $d : U \longrightarrow \mathbb{R}_{\geq 0}$ that assigns a non-negative number, describing how consistent location estimates for this RSS vector are, to every RSS vector. Out of the instances in U having the lowest values of diversity measures, the extended training data set $\tilde{R}$ is developed. Importantly, $\tilde{R}$ can include propagated labels provided by various labelling functions l_x, l_y. At the same time, the data sets $\widehat{R}$ and the models $\widehat{M_x}, \widehat{M_y}$ are developed. These are based on true labels rather than propagated labels and provide a baseline to evaluate the performance of semi-supervised approach. It is important to note, that $\widehat{R}$ and the models $\widehat{M_x}, \widehat{M_y}$ are used for comparison purposes only and are not supposed to be used in production. Finally, in Table 1, we propose a number of semi-supervised strategies, each composed of a diversity measure $d()$ and a label function $l()$. In every run of Algorithm 1, one pair $(d(), l())$ will be evaluated.

Input: $\{R_i, U_i, T_i\}_{i=1,\ldots,N_{\max}}$ - a set of testbeds; $R_i \subset D\mathrm{x}\mathbb{R}^2$, $T_i \subset D\mathrm{x}\mathbb{R}^2$ - radio map data and test data, respectively, $U_i \subset D\mathrm{x}\mathbb{R}^2$ - data to be provided with propagated labels, $s(U_i) \subset D$ - the RSS part of U_i, $D \subset \mathbb{Z}^n$ - the set of RSS vectors that can be observed in a building, $\omega \in [0,1]$ - a proportion on unlabelled instances to be included in the training data

Data: $\tilde{U}$ - a list of candidate RSS vectors sorted by growing values of diversity measures

Result: $(\tilde{M}_x, \tilde{M}_y)$ - positioning models developed for individual testing areas, $\tilde{\mathbf{E}}$ - a vector of averaged accuracy indicators of the positioning models built with semi-supervised approach

begin

    for $i \leftarrow 1$ to $N_{\max}$ by 1 do

        (M_x, M_y)=DevelopModel(R_i);

        $\tilde{U} = sort(\{(\mathbf{u_j}, d_x(\mathbf{u_j}, R_i, M_x) + d_y(\mathbf{u_j}, R_i, M_y)) : \mathbf{u_j} \in s(U_i), j = 1, \ldots, card(U_i)\})$;

        $\tilde{R} = R_i \cup \{(\mathbf{u_j}, l_x(\mathbf{u_j}, R_i, M_x), l_y(\mathbf{u_j}, R_i, M_y)) : \mathbf{u_j} \in \tilde{U}, j = 1, \ldots, card(U_i) \times \omega\}$;

        $\widehat{R} = R_i \cup \{(\mathbf{u_j}, x_j, y_j) : \mathbf{u_j} \in \tilde{U}, j = 1, \ldots, card(U_i) \times \omega \wedge (\mathbf{u_j}, x_j, y_j) \in U_i\}$;

        $(\tilde{M}_x, \tilde{M}_y)$=DevelopModel$(\tilde{R})$; $(\widehat{M_x}, \widehat{M_y})$=DevelopModel$(\widehat{R})$;

        $\tilde{\mathbf{E}}_i$=AvgIndicators$(\tilde{M}_x, \tilde{M}_y, T_i)$; $\widehat{\mathbf{E}}_i$=AvgIndicators$(\widehat{M_x}, \widehat{M_y}, T_i)$;

    end

    $\tilde{\mathbf{E}} = mean_{i=1,\ldots,N_{\max}}(\tilde{\mathbf{E}}_i)$; $\widehat{\mathbf{E}} = mean_{i=1,\ldots,N_{\max}}(\widehat{\mathbf{E}}_i)$;

end

Algorithm 1. Semi-supervised development of positioning models

Table 1. Summary of semi-supervised strategies proposed in this study

Strategy	Diversity measure $d()$	Propagated label $l()$
kNN/MM	The difference between maximum and minimum coordinates among k NN in R	The average coordinate of k nearest neighbours in R
RF/MM	The difference between maximum and minimum coordinates suggested by the trees comprising on a random forest M_x or M_y	The coordinate predicted by random forest M_x or M_y
kNN/SD	Standard deviation of coordinates of k nearest neighbours in R	The average coordinate of k nearest neighbours in R
RF/SD	Standard deviation of coordinates suggested by the trees comprising on a random forest M_x or M_y	The coordinate predicted by random forest M_x or M_y

4   Results

To analyse different semi-supervised strategies proposed in this study, extensive calculations were performed for every floor treated as a separate testbed. In each case $\omega \in \{0, 0.2, 0.4, \ldots, 1\}$ was used. In particular, $\omega = 0$ means that $\tilde{R} = R$, whereas $\omega = 1$ represents the case of labelling and adding all data from U into R. Furthermore, in all experiments, a random forest composed of 50 trees was used to develop positioning models $M_x(), M_y()$ and provide basis for RF/MM and RF/SD strategies. Similarly, random forest has also been used to develop $\tilde{M}$ and $\widehat{M}$ models. In the case of kNN/SD and kNN/MM, diversity measures and location labels were calculated based on the coordinates of $k = 20$ nearest neighbours of a vector $\mathbf{v} \in U$ in R. Errors reported below are mean and maximum absolute positioning errors i.e. mean and maximum lengths of the vectors linking true (x, y) and predicted (x_p, y_p) location, respectively.

First of all, let us analyse the way accuracy of the models changes under ideal case scenario of extending training data with instances provided with true location labels. Figure 1 shows averaged mean absolute positioning errors (MAE) of $\widehat{M}$ models developed with different $\widehat{R}$ data sets, produced by different strategies under varied ω settings. These mean error values are denoted by $\widehat{E}(\omega)$ thereafter. Not surprisingly, Fig. 1 shows that the lowest positioning errors are observed when $\omega = 1$ i.e. when all U instances are provided with their true labels and added to the training data. Irrespective of the strategy used, the same $\widehat{R}$ set is used for $\omega = 1$, which directly follows from Algorithm 1. However, it is interesting to note that under kNN/MM strategy the addition of just $\omega = 0.2$ of all U instances yields major mean error reduction. $\widehat{E}(\omega)$ values for all strategies are also reported in Table 2. It follows from the table that under kNN/MM strategy $\widehat{E}(0.2)=2.63$ m, which is less than $\widehat{E}(0.8)=2.65$ m attained by RF/MM. In other words, by adding carefully selected 20% of instances present in U, better impact on the training process and ultimate positioning error reduction can be observed than in the case of addition of 80% of instances based on another diversity measure. The same trend showing superiority of kNN/MM and kNN/SD methods can be observed for maximum errors, denoted by $\widehat{E}_{\mathrm{mx}}(\omega)$.

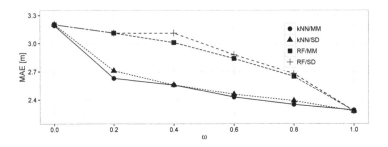

Fig. 1. The impact of extending training data with instances containing true location

Table 2. Mean and maximum positioning errors when U instances are provided with ground truth labels before being included in the training data $\widehat{R}$

ω	RF/MM		RF/SD		kNN/MM		kNN/SD	
	$\widehat{E}(\omega)$	$\widehat{E}_{\mathrm{mx}}(\omega)$	$\widehat{E}(\omega)$	$\widehat{E}_{\mathrm{mx}}(\omega)$	$\widehat{E}(\omega)$	$\widehat{E}_{\mathrm{mx}}(\omega)$	$\widehat{E}(\omega)$	$\widehat{E}_{\mathrm{mx}}(\omega)$
0.00	3.20	35.40	3.20	34.75	3.19	34.34	3.20	33.38
0.20	3.11	34.86	3.11	34.71	2.63	28.49	2.71	29.52
0.40	3.01	33.95	3.11	35.02	2.56	28.93	2.56	26.55
0.60	2.84	33.29	2.88	34.42	2.43	24.51	2.46	25.06
0.80	2.65	33.98	2.68	35.05	2.35	21.32	2.39	21.98
1.00	2.28	20.37	2.28	23.45	2.29	22.09	2.28	20.79

What is even more interesting is whether unlabelled RSS records provided in an automated manner with predicted labels could increase the accuracy of the positioning models. Table 3 shows the mean error rates $\widetilde{E}(\omega)$ and maximum error rates $\widetilde{E}_{\mathrm{mx}}(\omega)$ observed on the testing data, when propagated labels are used to develop $\widetilde{R}$, as defined in Algorithm 1. These values are accompanied by $\xi(\widetilde{R})$, which denotes mean number of distinct TPs, from which the RSS data was selected to be provided with propagated labels. It follows from the table that when propagated labels are used under kNN/MM strategy, a less significant than in the case of the use of true labels, yet important reduction from $\widetilde{E}(0) = 3.18\,\mathrm{m}$ to $\widetilde{E}(0.6){=}2.80\,\mathrm{m}$ is observed. What should be emphasised, an attempt to label, and add to the training data, all available RSS vectors contained in U is not advisable. Figure 2 shows that the inclusion of some instances from U rather than all of them yields the lowest errors. For kNN/MM, $\widetilde{E}(1) = 2.89\,\mathrm{m}$. The lowest mean errors are attained by kNN/MM and kNN/SD strategies and in both cases they occur for $\omega < 1$. As far as maximum errors are concerned, the results contained in Table 3 suggest that data created with semi-supervised approach may reduce also maximum error compared to the models developed with true, yet more limited original training data R.

Table 3. Mean and maximum positioning errors when U instances are provided with propagated location labels before being included in the training data $\tilde{R}$

ω	RF/MM			RF/SD			kNN/MM			kNN/SD		
	$\tilde{E}(\omega)$	$\tilde{E}_{mx}(\omega)$	$\xi(\tilde{R})$	$\tilde{E}(\omega)$	$\tilde{E}_{mx}(\omega)$	$\xi(\tilde{R})$	$\tilde{E}(\omega)$	$\tilde{E}_{mx}(\omega)$	$\xi(\tilde{R})$	$\tilde{E}(\omega)$	$\tilde{E}_{mx}(\omega)$	$\xi(\tilde{R})$
0.00	3.24	34.66	233.50	3.22	34.97	233.50	3.18	33.13	233.50	3.20	34.93	233.50
0.20	3.25	34.70	299.42	3.20	34.69	300.58	2.81	28.26	321.67	2.83	27.90	307.17
0.40	3.17	33.83	323.58	3.13	34.28	323.67	2.82	25.73	342.83	2.85	28.93	341.00
0.60	3.14	34.40	341.00	3.10	34.57	339.00	2.80	27.28	352.50	2.81	26.32	350.83
0.80	3.18	35.37	351.50	3.16	33.90	350.83	2.81	26.77	354.50	2.79	24.61	354.50
1.00	3.15	27.96	355.33	3.13	24.35	355.33	2.89	33.95	355.33	2.88	30.48	355.33

Results for RF/MM and RF/SD strategies show that these strategies yield little reduction of average error $\tilde{E}(\omega)$, $\omega > 0$ compared to the use of R data only. In the case of $\omega = 1$, since all vectors $\mathbf{v} \in U$ are assigned propagated labels, there is no impact of diversity measures on $\tilde{E}(1)$ errors. Hence, all differences in $\tilde{E}(\omega)$ are due to different label functions $l()$. Thus, limited error reduction $\tilde{E}(1) - \tilde{E}(0)$ for RF/* strategies is because of the fact that a random forest is used to generate propagated labels in this case. The use of kNN to generate propagated labels turns out to be more beneficial. More precisely, a semi-supervised approach, in which propagated labels are provided by kNN, and the ultimate models $\tilde{M}_x, \tilde{M}_y$ relying on them are built with random forest provides the largest error reduction, out of the strategies analysed in this study. Furthermore, results attained with kNN/SD closely resemble results for kNN/MM i.e. the fact that kNN is used for $l()$ is more important than how $d()$ for these strategies was calculated.

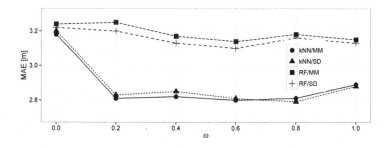

Fig. 2. The impact of extending training data with instances with propagated location.

5   Conclusions

A proposal of semi-supervised techniques matching the noisiness of RSS data has been made. Out of four methods proposed in this study, the methods based on kNN yield substantial reduction of mean errors of the positioning models based on training data sets including originally partly unlabelled data. Importantly, when reduction of mean errors is an objective, the labelling of some, but not all of the available unlabeled records is optimal. This confirms the need for the methods enabling the selection of unlabelled instances to be provided with propagated labels, such as these proposed in this study. Such methods may provide major reduction of expensive labelled data needed to develop positioning models. In the future, further works on diversity measures and label functions are planned, also in combination with other machine learning techniques.

Acknowledgments. This research was partly supported by the National Centre for Research and Development, grant No PBS2/B3/24/2014, app. no. 208921.

References

1. Górak, R., Luckner, M.: Malfunction immune Wi–Fi localisation method. In: Núñez, M., Nguyen, N.T., Camacho, D., Trawiński, B. (eds.) ICCCI 2015. LNCS (LNAI), vol. 9329, pp. 328–337. Springer, Cham (2015). https://doi.org/10.1007/978-3-319-24069-5_31
2. Grzenda, M.: Reduction of signal strength data for fingerprinting-based indoor positioning. In: Jackowski, K., Burduk, R., Walkowiak, K., Woźniak, M., Yin, H. (eds.) IDEAL 2015. LNCS, vol. 9375, pp. 387–394. Springer, Cham (2015). https://doi.org/10.1007/978-3-319-24834-9_45
3. Jain, V.K., Tapaswi, S., Shukla, A.: RSS fingerprints based distributed semi-supervised locally linear embedding (DSSLLE) location estimation system for indoor WLAN. Wirel. Pers. Commun. **71**(2), 1175–1192 (2013)
4. Jiang, X., Chen, Y., Liu, J., Gu, Y., Hu, L.: FSELM: fusion semi-supervised extreme learning machine for indoor localization with Wi-Fi and bluetooth fingerprints. Soft Comput. **22**(11), 3621–3635 (2018)
5. Khoury, H.M., Kamat, V.R.: Evaluation of position tracking technologies for user localization in indoor construction environments. Autom. Constr. **18**, 444–457 (2009)
6. Kim, Y., Chon, Y., Cha, H.: Smartphone-based collaborative and autonomous radio fingerprinting. IEEE Trans. Syst. Man Cybern. - Part C: Appl. Rev. **42**(1), 112–122 (2012)
7. Kjargaard, M.B.: Indoor location fingerprinting with heterogeneous clients. Pervasive Mob. Comput. **7**, 31–43 (2011)
8. Pulkkinen, T., Roos, T., Myllymäki, P.: Semi-supervised learning for WLAN positioning. In: Honkela, T., Duch, W., Girolami, M., Kaski, S. (eds.) ICANN 2011. LNCS, vol. 6791, pp. 355–362. Springer, Heidelberg (2011). https://doi.org/10.1007/978-3-642-21735-7_44

Different Approaches of Data and Attribute Selection on Headache Disorder

Svetlana Simić[1], Zorana Banković[2], Dragan Simić[3(✉)],
and Svetislav D. Simić[3]

[1] Faculty of Medicine, University of Novi Sad, Hajduk Veljkova 1–9,
21000 Novi Sad, Serbia
svetlana.simic@mf.uns.ac.rs
[2] Frontiers Media SA, Pozuelo de Alarcón sn, Madrid, Spain
zbankovic@gmail.com
[3] Faculty of Technical Sciences, University of Novi Sad,
Trg Dositeja Obradovića 6, 21000 Novi Sad, Serbia
dsimic@eunet.rs, {dsimic,simicsvetislav}@uns.ac.rs

Abstract. Half of the general population experiences a headache during any given year. Medical data and information in turn provide knowledge on which physicians base their decisions and actions but, in general, it is not easy to manage them. It becomes increasingly necessary to extract useful knowledge and make scientific decisions for diagnosis and treatment of this disease from the database. This paper presents comparison of data and attribute selected features by automatic machine learning methods and algorithms, and by diagnostic tools and expert physicians, almost all from the last decade.

Keywords: Data selection · Attribute selection · Headache · Diagnosis
Machine learning · Decision support

1 Introduction

Over the last decades, information technology in general, and artificial intelligence, in particular, have gradually stepped into every area of life, starting from industry, business, weather forecasting and media, but the most significant development has taken place in the field of healthcare. Healthcare organizations are continually endeavoring to improve patient care and provide better services. It is expected that by introducing information technology into healthcare system, the information technology will enable more efficient and effective healthcare services.

Headache disorders are the most prevalent of all the neurological conditions and are among the most frequent medical complains seen in a general practice. More than 90% of the general population report experiencing a headache during any given year, which is a lifetime history of head pain [1]. The common practice is to pay attention to the health of working population which is the carrier and the backbone of every society when it comes to the risky occupations, and to perform regular medical checkups. The care about employees who suffer from a primary headache begins only when they themselves seek a physician's help [2].

© Springer Nature Switzerland AG 2018
H. Yin et al. (Eds.): IDEAL 2018, LNCS 11315, pp. 241–249, 2018.
https://doi.org/10.1007/978-3-030-03496-2_27

In recent times, data is the raw material from all information. This data and information in turn provide knowledge through modeling, data mining, analysis, interpretation, and visualization. It is not easy for a decision maker in decision-making processes to handle too much data, information and knowledge. Healthcare organizations are complex and making changes in order to improve general health can therefore be a complex business as well. It is necessary to gain solid evidence through data selection and even transformation measure in order to learn the important data for the forthcoming prediction of unseen data, and to support decision making and necessary actions. It becomes increasingly necessary to extract useful knowledge and make scientific decisions for diagnosis and treatment of a disease. Data selection is usually divided into two phases. First, the experts – physicians select the variables from the enterprise database according to their experience, and in the second phase, the optimal number of variables is sought with the use of attribute selection method [3]. Medical field is primarily directed at patient care activity and only secondarily as research resource. The only justification for collecting medical data is to benefit the individual patient.

This paper presents some of useful techniques and algorithms for data and attribute selection which are almost all developed in the last decade. Also, this paper continues the authors' previous research in medical domain, especially in computer-assisted diagnosis and machine learning methods of primary headaches presented in research papers [4–7].

The rest of the paper is organized in the following way: Sect. 2 provides an overview of the basic idea on primary headache classification. Section 3 presents primary headache clinical features. The comparison between data and attribute selected features by automatic machine learning methods and algorithms, and by diagnostic tools and expert physicians, are presented in Sect. 4. Section 5 provides conclusions and some points for future work.

2 Primary Headache Classification

The International Classification of Headache Disorders – The Third Edition (ICHD-3) established the uniform terminology and consistent operational diagnostic criteria for a wide range of the headache disorders around the world [8]. The ICHD-3 provides a hierarchy of diagnoses with varying degrees of specificity. Headache disorders are identified with three or sometimes five-digit codes which is, in details, presented in [8] but short identification for just two important digit codes is presented in Table 1. All headache disorders are classified into two major groups: (**A**) *Primary headaches from ICHD-3 code 1. to 4.* and (**B**) *Secondary headaches ICHD-3 code from 5. to 12.* The first digit specifies the major diagnostic categories (i.e. *Migraine*). The second digit indicates a disorder within the category (i.e. *Migraine without aura*). Each category is then subdivided into groups, types, subtypes and sub-forms. Subsequent digits permit more specific diagnosis for some headache types.

The starting point of an accurate diagnosis is differentiating primary headaches, without organic cause, from secondary headaches, where etiological cause can be

Table 1. The international classification of headache disorders – the third edition

	ICHD-3 code		Diagnosis
	1.		**Migraine**
		1.1	Migraine without aura
		1.2	Migraine with aura
		1.3	Chronic migraine
		1.4	Complications of migraine
		1.5	Probable migraine
		1.6	Episodic syndromes that may be associated with migraine
	2.		**Tension-type headache (TTH)**
		2.1	Infrequent episodic tension-type headache
		2.2	Frequent episodic tension-type headache
A		2.3	Chronic tension-type headache
		2.4	Probable tension-type headache
	3.		**Trigeminal autonomic cephalalgias (TACs)**
		3.1	Cluster headache (CH)
		3.2	Paroxysmal hemicrania
		3.3	Short-lasting unilateral neuralgiform headache attacks
		3.4	Hemicrania continua
		3.5	Probable trigeminal autonomic cephalalgia
	4.		**Other primary headache disorders**
		4.1	Primary cough headache
		⋮	
		4.10	New daily persistent headache
B	**5.**		
	⋮		**Secondary headache disorders**
	12.		

determined. Headache diagnosing is usually based on: anamnesis, clinical examination and additional examinations.

Everyday clinical reality is such that there are two types of physicians in all social systems and all medical branches:

1. The ones who have tendency to base their diagnosis on anamnesis and clinical examinations, and afterwards on additional examinations – mislabeled "the old-.school physicians".
2. Those who show the tendency to base their diagnosis primarily on additional examinations – mislabeled as "the contemporary physicians".

When first meeting a patient, physicians who are more concerned with the detailed anamnesis and clinical examinations, apply ICHD-3 criteria and can easily establish the primary headache diagnosis. If the criteria are not satisfied, the physicians will have to suggest an additional examination to a patient.

3   Primary Headache Clinical Features

Primary headaches may share certain features; pain is severe for migraine and cluster headache (CH) as an example. However, CH varies from migraine primarily in its pattern of occurrence. CH is in briefer episodic over a period of weeks or months. Migraine usually does not follow this type of pattern. Most of the migraine's features explicitly differentiate this type of headache from tension-type headache (TTH) and therefore help in a precise diagnosis. However, while TTH is commonly generalized, migraine pain is mostly unilateral; and while migraine has a pulsating quality with moderate to severe pain, TTH presents as a mild to moderate in intensity and a dull ache or feeling of a tight band around the head [9].

Table 2. Comparison of primary headache clinical features

	Migraine	Tension-type headache	Cluster headache
Gender ratio (M:F)	1:3	5:4	3:1
Age of onset	15–55 years	25–30 years	28–30 years
Prevalence	18% F–6% M	30 up to 78%	0.9%
Quality	Throbbing	Non-throbbing	Stabbing, sharp
Intensity	Moderate to severe	Mild to moderate	Severe to very severe
Location	Unilateral	Bilateral	Unilateral
Duration of attack	4–72 h	30 min to 7 days	15–180 min
Symptoms	Nausea, vomiting, photophobia, phonophobia	Photophobia, phonophobia	Autonomic dysfunction
Triggers	Physical activity	Stress	Laying down or sleep

A comparison of the main types of primary headache based on previous researches in the world and prevalence is presented in Table 2. The distinct as well as overlapping signs and symptoms of migraine, TTH and cluster headache are illustrated there.

4   Comparison Different Approaches of Attribute Selection

This study analyzes different studies which are all based on IHS recommendations, and therefore comparable to each other. These studies deal with different approaches to data and attribute selection based on automatic methods, expert systems, knowledge–base systems and physicians' expert knowledge as well, as shown in Table 3.

In general, attribute selection – feature selection – feature reduction approaches are widely used in data mining and machine learning. On the other hand, feature selection could be divided on *Stochastic* and *no-Stochastic Feature Selection* methodology, a

Table 3. Comparison of a selection attribute for primary headache based on IHS diagnostic criteria: 1. Consistency measure filter, 2. ReliefF Greedy, 3. ReliefF top10, 4. Genetic algorithm wrapper, 5. Ant Colony Optimization based classification algorithm, 6. Rule Based Fuzzy Logic System, 7. Physician's expert choice, 8. column RES – final decision for important attributes selection

Attributes	1	2	3	4	5	6	7	RES
1. Sex								
2. 1. How old were you when the headache occurred for the first time?								
3. 2. How often do you have headache attacks?	+							
4. 3. How long do the headache attacks last?	+	+	+	+		+	+	+
5. 4. Where is headache located?	+	+		+	+	+	+	+
6. 5. How intense is the pain?	+	+	+	+	+	+	+	+
7. 6. What is the quality of the pain you experience	+	+	+	+	+	+	+	+
8. 7. Do your headaches worsen after physical activities such as walking?	+	+	+	+		+	+	+
9. 8. Do you avoid routine physical activities because you are afraid they might trigger your headache?		+	+			+		
10. 9.a) Are the headaches accompanied by? a) Nausea	+	+	+	+		+	+	+
11. 9.b) Are the headaches accompanied by? b) Vomiting		+	+			+	+	
12. 9.c) Are the headaches accompanied by? c) Photophobia	+	+	+	+	+	+	+	+
13. 9.d) Are the headaches accompanied by? d) Phonophobia		+	+		+	+	+	
14. 10. Do you have temporary visual, sensory or speech disturbance?								
15. 11. Do you, during a headache attack, have tension and/or heightened tenderness of head or neck muscles?	+	+	+	+		+	TTH	
16. 12. Do you have any body numbness or weakness?								
17. 13. Do you have any indications of oncoming headache?	+	+		+				
18. 14. Headache is usually triggered by: Menstrual periods	+							
19. 15. In the half or my visual field, lasting 5 minutes to an hour, along with the headache attack or an hour before.								
20. 16. Along with the headache attack or an hour before one I have sensory symptoms.								

refinement of an initial stochastic feature selection task with a no-stochastic method to reduce a bit more the subset of features to be retained [10]. Therefore, feature selection plays a critical role in many domains for creating the model.

Consistency measure filter is based on the idea that the selected subset of features should be consistent and self-contained, and there should be no conflicts between the objects described by similar features [11]. A dataset, described by a subset of features, is considered inconsistent if there exist at least two objects belonging to it such that they are match – similar except their class labels.

In general, **Relief algorithm** is proposed in [12]. In this method statistical method instead of heuristic search approach is used. Relief requires linear time in the number of given features and number of training instances regardless of the target concept to be learned. It selects the statistically relevant features and usually it is within 95% confidence interval. Two types of **ReliefF algorithms** were tested in this study – with greedy stepwise stop criterion and with heuristic selection of top ten features [13]. There are many ways of describing this code length in the literature and for more detailed exposition of the minimum description length (MDL) principle is presented in [14], and the Akaike information criterion (AIC) suitably modified for use in small samples is shown [15].

Genetic algorithm wrapper is a nature-inspired algorithm, which uses an evolutionary optimization for finding the best subset of features [16]. Feature space is encoded binary, where 1 stands for a selected feature and 0 stands for a discarded feature. The overall classification error is selected as a criterion. This approach starts with generating some random population representing the solution space – feature subsets – and through operations of cross-over – responsible for exchanging information between population members – and mutation – responsible for introducing random diversity into population – searches for an optimal solution. A tournament scheme was applied for selection of population members.

The **Ant Colony Optimization** (ACO) based classification algorithm process is basically discovering of classification rules adjusted in the form of rules like "IF <condition> THEN <class>". The ACO classification algorithm generates rules by simulating ants' foraging. Ants make a trail along the path between the nest and food by leaving a chemical residue called pheromone on the ground. This trail leads other ants to the food by following the path. During this travel the pheromone level decreases if the path is long. The ants that found the shortest way leave more pheromone on the route. Finally, ants are forced to use the shortest way with the guidance of pheromone level [17]. In the research study presented in [18] 850 patients participated in the study; 609 (71.65%) patients suffer from migraine, 185 (21.76%) patients suffer from TTH, and 56 (6.59%) patients suffer from CH. There are some other researches which discussed implementation of ACO as search method in feature subset selection based on correlation or consistency measures [19].

Rule Based Fuzzy Logic (RBFL) model based on *knowledge mining*, means extracting information in the form of "*if - then*" statements made by questioned people, in our research – patients [2]. It is very important to emphasize that establishing the rules in our research was facilitated by our knowledge and the application of the ICHD-3 for headache types and represent an input for our RBFL model. Output of the system is the definition of the specific type of primary headache. 80 patients participated in this study; 10 (12.5%) patients suffer from migraine, 38 (47.5%) patients suffer from TTH, and 32 (40%) patients suffer from other primary headaches.

In this research study [5] methods, such as: consistency measure filter, reliefF algorithms, and genetic algorithm wrapper were used; 579 people suffering from headaches participated: 169 subjects (29.18%) suffered from migraine, 224 subjects (38.60%) suffered from TTH, and 186 subjects presented (32.12%) with other headache types.

The columns from 1 to 7, in Table 3 present different techniques and/or algorithms used in the observed research: (1) Consistency measure filter [5]; (2) ReliefF Greedy [5]; (3) ReliefF top10 [5]; (4) Genetic algorithm wrapper [5]; (5) Ant colony optimization based classification algorithm [17]; (6) Rule Based Fuzzy Logic System [2]; (7) Physician's expert choice; and, finally, column *RES* - shows final decision for data and attributes selection.

Simple plus (+) denotes important selected attribute – feature by the considered method. **Bold red** pluses (+) indicate a feature denoted as a very important (major) by the expert and a simple plus (+) indicates features that are used as an additional support in harder cases (minor). Additionally, one feature according to the physician is important for recognizing TTH. In the column *RES* the attributes marked by **blue bold pluses** (+) represent final decision for important data and attributes selection. This decision is based on the fact that at least six of seven observed studies consider the observed attribute as significant for deciding upon important attributes that can help physicians–neurologists to decide on a type of primary headache. The assessment "at least six of seven" presents more than 85% of observed sample and can, therefore, be considered as a reliable percentage for the observed empirical approximation.

According to experimental results presented in the previously mentioned studies it can be concluded that most important attributes for a decision on the type of a primary headache are: (1) *Length of a headache attack*; (2) *Headache location*; (3) *The intensity of the pain*; (4) *The quality of the pain*; (5) *Are the headaches accompanied by **nausea**?*; (6) *Are the headaches accompanied by **photophobia**?*.

Additionally, it is important to mention that attributes: 13. Are the headaches accompanied by *phonophobia*; and, 15. Do you, during a headache attack, have tension and/or heightened tenderness of head or neck muscles?, can be considered as very important because at least 5 out of 7 observed studies, which is more than 70%, consider them to be important.

5   Conclusion and Future Work

The aim of this paper is to present comparison of different approaches to data and attribute selection based on implementation of automatic methods and algorithms, expert systems, knowledge–base systems and the expert physicians'–neurologists' choice. The diagnostic criterion used in this research is developed according to the IHS recommendations, as are all the analyzed studies, which means that they are comparable to each other. They are all very important for development and implementation as automatic systems in headache decision-making systems which help neurologists, to make diagnoses of primary headache disorders and give their patients better treatment.

Preliminary experimental results of this research compare seven different studies and select data and attributes which are important in headache diagnostic. The experimental results encourage the authors to do further research because 85% of the studies confirm six attributes to be important for the decision on the type of a primary headache. Our future research will focus on creating new model with combined heuristics and meta-heuristics techniques which will efficiently help neurologist in defining a type of a primary headache. The new model and implemented system will be tested with real-world data sets from the Clinical centre of Vojvodina in Serbia.

References

1. Hagen, K., Zwart, J.-A., Vatten, L., Stovner, L.J., Bovin, G.: Prevalence of migraine and non-migrainous headache – head-HUNT, a large population-based study. Cephalalgia **20** (10), 900–906 (2000)
2. Simić, S., Simić, D., Slankamenac, P., Simić-Ivkov, M.: Computer-assisted diagnosis of primary headaches. In: Corchado, E., Abraham, A., Pedrycz, W. (eds.) HAIS 2008. LNCS (LNAI), vol. 5271, pp. 314–321. Springer, Heidelberg (2008). https://doi.org/10.1007/978-3-540-87656-4_39
3. Relich, M., Bzdyra, K.: Knowledge discovery in enterprise databases for forecasting new product success. In: Jackowski, K., Burduk, R., Walkowiak, K., Woźniak, M., Yin, H. (eds.) IDEAL 2015. LNCS, vol. 9375, pp. 121–129. Springer, Cham (2015). https://doi.org/10.1007/978-3-319-24834-9_15
4. Simić, S., Simić, D., Slankamenac, P., Simić-Ivkov, M.: Rule-based fuzzy logic system for diagnosing migraine. In: Darzentas, J., Vouros, George A., Vosinakis, S., Arnellos, A. (eds.) SETN 2008. LNCS (LNAI), vol. 5138, pp. 383–388. Springer, Heidelberg (2008). https://doi.org/10.1007/978-3-540-87881-0_37
5. Krawczyk, B., Simić, D., Simić, S., Woźniak, M.: Automatic diagnosis of primary headaches by machine learning methods. Open Med. **8**(2), 157–165 (2013)
6. Jackowski, K., Jankowski, D., Ksieniewic, P., Simić, D., Simić, S., Wozniak, M.: Ensemble classifier systems for headache diagnosis. In: Piętka, E., Kawa, J., Wieclawek, W. (eds.) Information Technologies in Biomedicine. AISC, vol. 284, pp. 273–284. Springer, Cham (2014). https://doi.org/10.1007/978-3-319-06596-0_25
7. Simić, S., Banković, Z., Simić, D., Simić, S.D.: A hybrid clustering approach for diagnosing medical diseases. In: de Cos Juez, F., et al. (eds.) Hybrid Artificial Intelligent Systems. LNCS, vol. 10870, pp. 741–752. Springer, Cham (2018). https://doi.org/10.1007/978-3-319-92639-1_62
8. The International Classification of Headache Disorders. 3rd edn. https://www.ichd-3.org/
9. Arendt-Nielsen, L.: Headache: muscle tension, trigger points and referred pain. Int. J. Clin. Pract. **69**(Suppl. 182), 8–12 (2015)
10. Tallón-Ballesteros, A.J., Correia, L., Cho, S.-B.: Stochastic and non-stochastic feature selection. In: Yin, H., et al. (eds.) IDEAL 2017. LNCS, vol. 10585, pp. 592–598. Springer, Cham (2017). https://doi.org/10.1007/978-3-319-68935-7_64
11. Arauzo-Azofra, A., Benitez, J.M., Castro, J.L.: Consistency measures for feature selection. J. Intell. Inf. Syst. **30**(3), 273–292 (2008)
12. Kira, K., Rendell, L. A.: A practical approach to feature selection. In: Ninth International Workshop on Machine Learning, pp. 249–256 (1992)
13. Rosario, F.S., Thangadurai, K.: Relief: feature selection approach. Int. J. Innov. Res. Dev. **4** (11), 218–224 (2015)
14. Hansen, M.H., Yu, B.: Model selection and the principle of minimum description length. J. Am. Stat. Assoc. **96**(454), 746–774 (2001)
15. Sugiura, N.: Further analysis of the data by Akaike's information criterion and the finite corrections. Commun. Stat. **7**(454), 13–26 (1978)
16. Guyon, I., Gunn, S., Nikravesh, M., Zadeh, L.: Feature Extraction, Foundations and Applications. Springer, Heidelberg (2006). https://doi.org/10.1007/978-3-540-35488-8

17. Michelakos, I., Mallios, N., Papageorgiou, E., Vassilakopoulos, M.: Ant colony optimization and data mining. In: Bessis, N., Xhafa, F. (eds.) Next Generation Data Technologies for Collective Computational Intelligence. SCI, vol. 352, pp. 31–60. Springer, Heidelberg (2011). https://doi.org/10.1007/978-3-642-20344-2_2

18. Celik, U., Yurtay, N.: An ant colony optimization algorithm-based classification for the diagnosis of primary headaches using a website questionnaire expert system. Turk. J. Electr. Eng. Comput. Sci. **25**(5), 4200–4210 (2017)

19. Tallón-Ballesteros, A.J., Riquelme, J.C.: Tackling ant colony optimization meta-heuristic as search method in feature subset selection based on correlation or consistency measures. In: Corchado, E., Lozano, J.A., Quintián, H., Yin, H. (eds.) IDEAL 2014. LNCS, vol. 8669, pp. 386–393. Springer, Cham (2014). https://doi.org/10.1007/978-3-319-10840-7_47

A Study of Fuzzy Clustering
to Archetypal Analysis

Gonçalo Sousa Mendes and Susana Nascimento[(✉)] [iD]

Computer Science Department and NOVA LINCS,
Faculdade de Ciências e Tecnologia, Universidade Nova de Lisboa, Lisbon, Portugal
`snt@fct.unl.pt`

Abstract. This paper presents a comparative study between a method
for fuzzy clustering which retrieves *pure* individual types from data, the
fuzzy clustering with proportional membership (FCPM), and an archetypal analysis algorithm based on Furthest-Sum approach (FS-AA). A simulation study comprising 82 data sets is conducted with a proper data
generator, FCPM-DG, whose goal is twofold: first, to analyse the ability
of archetypal clustering algorithm to recover Archetypes from data of
distinct dimensionality; second, to analyse robustness of FCPM and FS-
AA algorithms to outliers. The effectiveness of these algorithms are yet
compared on clustering 12 diverse benchmark data sets from machine
learning. The evaluation conducted with five primer unsupervised validation indices shows the good quality of the clustering solutions.

Keywords: Archetypal analysis · Fuzzy clustering
Synthetic multidimensional data · Fuzzy validation indices

1 Introduction

There are domains of application where it is useful to find the representatives
of a group not by an element that expresses the central properties of the group,
but by some sort of *pure* individual type, an extreme point, synthesizing all the
individuals in the groups by their most discriminating properties. Archetypal
Analysis [5] is a statistical method that synthesizes a set of multivariate data
points through a few representatives, called archetypes, in such a way that the
data points are a convex combination of the archetypes, and the archetypes are
also a convex combination of the data points, making those representatives to lie
on the convex hull of the data set. Archetypal Analysis (AA) has been applied
in ecology, benchmarking for performance analysis, talent analysis in sports and
in education, profiling of scientific activities, to name but a few [4,6,8,19].

In the field of fuzzy cluster analysis [2] the Fuzzy Clustering with Proportional Membership (FCPM) [15,16,18] assumes the existence of some prototypes
serving as *ideal* patterns underlying the data. To relate the prototypes to the
observed data, the FCPM model considers that the observations share parts of
the prototypes, such that an entity may bear 80% of a prototype $C1$ and 20%

© Springer Nature Switzerland AG 2018
H. Yin et al. (Eds.): IDEAL 2018, LNCS 11315, pp. 250–261, 2018.
https://doi.org/10.1007/978-3-030-03496-2_28

of prototype $C2$, which simultaneously expresses the entity's membership to the respective clusters. The underlying structure of this model can be described by a fuzzy c-partition such that any entity may independently relate to any prototype, provided the condition that the memberships for any entity must sum to unity. The baseline clustering criterion of the model, FCPM-0, leads to cluster structures with central prototypes, like the popular fuzzy c-means [2], while its smoother version, FCPM-2, leads to cluster structures with extremal prototypes like archetypes. The latter model had succeeded in grouping real data of psychiatric disorders characterized by patterns of extreme and opposite psychosomatic features defining a *syndrome*. Running the FCPM-2 algorithm on the original data augmented with clinical cases bearing less severe syndromes, contradicting the extreme tendency of the data, the prototypes found did not change towards more moderate characteristics of the data; on the contrary, they maintain the extreme prototypes almost unchanged. This highlights the suitability of FCPM to model the concept of *ideal* type, being useful to real world applications following the notion of *interesting* features for cluster analysis as in [10,12,13], and core features in supervised machine learning problems [20].

In the framework of unsupervised machine learning problems, an effective AA algorithm was recently developed [14] to guarantee a fast convergence of AA to find a pre-defined number of archetypes. This alternating optimization (AO) algorithm is initialized with the c data points furthest way from the centre of the data set by the so-called Furthest-Sum, and uses a simple gradient projection method. The Furthest-Sum (FS) method resembles the Anomalous Pattern used to initialize the Fuzzy c-means (AP-FCM) which has been successfully applied to the problem of unsupervised segmentation of Sea Surface Temperature (SST) images [17]. The main goal of this work is to experimentally compare the FS-AA and the FCPM algorithms taking advantage of the Furthest Sum strategy.

The remainder of the paper is organized as follows. The next section briefly describes the AA and FCPM models as well as the corresponding AO algorithms. In Sect. 3 we discuss the analysis of the archetypes' recovery of FS-AA algorithm with multidimensional data properly generated with respect to the FCPM original model. Section 4 presents a systematic study with proper unsupervised validation fuzzy clustering indices to access the quality of the found fuzzy partitions honoring the number of clusters with benchmark real data. Section 5 analyses the sensitivity of FS-AA and FCPM when augmenting the multidimensional synthetic data with outliers. Section 6 concludes the paper.

2   Archetypal Analysis vs Fuzzy Clustering with Proportional Membership

The goal of AA is to find a suitable choice of c archetypes $Z = z_1, z_2, ..., z_c$, given a data set of n observations, $X = x_1, x_2, ..., x_n$, where each observation is a vector of p attributes/features. The c archetypes that better fits the data are found by minimizing the residual sum of squares (RSS) criterion:

$$RSS(p) = \min_{a,b} \sum_{i=1}^{n} ||x_i - \sum_{j=1}^{c} z_j \cdot a_{ji}||^2 = \sum_{i=1}^{n} ||x_i - \sum_{j=1}^{c} \sum_{k=1}^{n} x_k \cdot b_{kj} \cdot a_{ji}||^2, \quad (1)$$

where the data points are convex mixtures of the archetypes,

$$a_{ji} \geq 0, \qquad \sum_{j=1}^{c} a_{ji} = 1, \qquad (2)$$

and the archetypes are convex mixtures of the data points,

$$b_{ij} \geq 0, \qquad \sum_{i=1}^{n} b_{ij} = 1. \qquad (3)$$

As shown in [5], for $p > 1$, the archetypes that minimize RSS criterion (1) fall on the convex hull of the data, making them extreme data-values, while for $p = 1$, the sample mean minimizes the RSS. To minimize the RSS criterion (1), it is then needed to find both matrices $A = [a_{ji}]$ and $B = [a_{ij}]$, which requires an AO algorithm. Several extensions of the original AO algorithm have been developed to optimize the AA clustering criterion (1) (*e.g.* [4,7,14]).

In this work we explore a recent AA algorithm introduced in [14] in the framework of machine learning and data mining. To guarantee a fast convergence of the algorithm on finding a pre-defined number of archetypes lying in the convex hull of the data, the authors take advantage of a simple gradient project method and initialize the algorithm with those c data points furthest away from the center of the data set, called Furthest-Sum (FS) method. In what follows this algorithm is designated as FS-AA algorithm.

The model of Fuzzy Clustering with Proportional Membership (FCPM) [16] is based on the assumption that the data are generated according to the cluster structure:

$$observed\ data\ =\ model\ data\ +\ noise. \qquad (4)$$

Here, the existence of some prototypes which serve as *ideal* patterns to data entities, is assumed. The meaning of *ideal* pattern is something that the researcher needs to explore as a *pure* representative that typifies the characteristics of a cluster.

Consider an entity-to-feature data matrix Y, preprocessed from X by shifting the origin to the center of the data, and rescaling the features by their ranges. The generic proportional membership model instantiates the data recovery model (4) in such a way that the membership value u_{ik} is not just a weight, but an expression of the proportion of prototype v_i which is present in the observation y_k. Formally, the FCPM model is defined as:

$$y_{kh} = u_{ik}v_{ih} + e_{ikh}, \qquad (5)$$

with e_{ikh} as the residual values.

To minimize the residual values e_{ikh} a square-error clustering criterion is defined by fitting each data point sharing a proportion of each of the prototypes, represented by the degree of membership, as follows:

$$E_0(U,V) = \sum_{k=1}^{n} \sum_{i=1}^{c} \sum_{h=1}^{p} (y_{kh} - u_{ik} v_{ih})^2,$$ (6)

with regard to the fuzzy constraints:

$$0 \leq u_{ik} \leq 1, \ for \ all \ i = 1, ..., c; \ k = 1, ..., n,$$ (7)

and

$$\sum_{i=1}^{c} u_{ik} = 1, \ for \ all \ k = 1, ..., n.$$ (8)

However, as this criterion is too strong and sometimes unrealistic [16], an adaptation of the squared error (6) was made, creating a smooth version. The idea is that only meaningful proportions, those with high membership values, were to be taken into account in model (4). Therefore, to smooth the influence of high residual values at small memberships u_{ik}, the squared residuals (6) are weighted by a power m ($m = 0, 1, 2$) of corresponding u_{ik}:

$$E_m(U,V) = \sum_{i=1}^{c} \sum_{k=1}^{n} \sum_{h=1}^{p} u_{ik}^m (y_{kh} - u_{ik} v_{ih})^2,$$ (9)

subject to the fuzzy constraints (7) and (8).

In what follows we only consider the models with parameter $m = 0$ and $m = 2$ designated as FCPM-0 and FCPM-2, respectively [16].

The minimization of clustering criterion E_m (9) considers an AO algorithm as follows: (i) initialize the set of prototypes V with random values selected from the data space; (ii) update the fuzzy membership matrix U from this initial V; (iii) alternate between minimizing the weights, U, given the centroids, $\hat{V}$, and minimizing V, given the updated $\hat{U}$; (iv) stop when algorithm converges. The prototypes feature values are derived by the first order condition of minimizing the clustering criterion (9) as:

$$v_{ih}^{(t)} = \frac{\left\langle \left(u_i^{(t)}\right)^{m+1}, y_h \right\rangle}{\left\langle \left(u_i^{(t)}\right)^{m+1}, u_i^{(t)} \right\rangle}.$$ (10)

To find the fuzzy memberships u_{ih} from (10) over the fuzzy constraints (7) and (8), the process is not as simple, since it is not analytically derivable. This leads to the adaptation of the Gradient Projection Method (GPM) [1] by one of the authors, to calculate the optimal fuzzy memberships in an nested iterative process. The FCPM AO algorithm consists of major iteration to find $V's$ and a nested minor iteration to find $U's$. A detailed description of the FCPM AO algorithm and its foundation can be found in [15, 18].

Extensive study with proper simulated data showed that: (i) the FCPM-2 is able to recover the original ideal types; (ii) the FCPM-0 (in low/intermediate space dimensions) and FCPM-2 (in high space dimension) may act as indicators of the number of clusters present in the data; (iii) the FCPM proportional membership leads to much more clear-cut partitions than the FCM distance membership. Finally, FCPM finds ideal types, even with data set augmented with data cases opposing the extreme tendency cases [16].

3   Testing FCPM and FS-AA on Real Data

This study is conducted with well known benchmark data sets. We consider 10 data sets from the UCI Machine Learning Repository with reference to number of entities, number of features, and number of classes in the form (n, p, L) [21]: Banknote (1372, 4, 2), Wisconsin Breast Cancer (683, 9, 2), Wisconsin Diagnostic Breast Cancer (569, 30, 2), Glass (214, 9, 6), Indian Liver Patient (579, 10, 2), Iris (150, 4, 2), Pima Indians Diabetes (768, 8, 2), Seeds Kernel (210, 7, 3), Vehicle silhouettes (793, 18, 4), and Wine Recognition (178, 13, 3). Two other data sets from the field of psychiatry are taken from [16]: Mental Disorders (44, 17, 4) and Augmented Mental Disorders (80, 17, 4). The attribute class was removed from all the data sets and for the data sets Indian Liver Patient and Wisconsin Breast Cancer the data cases with missing values were removed (4 and 16, respectively).

For each data set, the FS-AA, FCPM-0, and FCPM-2 algorithms were run, looking for $k = L - 1, L, L + 1$ prototypes. Each algorithm was run five distinct times, starting from the same initial k seeds calculated with the *Furthest-Sum* algorithm. To assess the quality of the found FS-AA partitions with the corresponding ones of FCPM we choose five popular internal validation indices for fuzzy clustering analysis: Partition Entropy (PE($\downarrow$)), Partition Coefficient (PC ($\uparrow$)) [2] and its normalized version (MPC ($\uparrow$)), Xie-Beni's index (XB($\downarrow$)), and Fuzzy Silhouette Index (FSI($\uparrow$)) [3]. The arrows ($\downarrow$/$\uparrow$) indicate if the optimal value of the index is its minimum/maximum value, respectively. A comprehensive analysis of most of these indices is in [11]. These indices have been run with the R language toolbox [9]. On the toolbox, the PE index was normalized for the values to be confined to the interval $[0, 1]$ as proposed in [2].

Table 1 presents the mean values for the five validation indices (in columns) for the five runs of the algorithm for each data set (in rows). The best value of each index, for a data set, across the three algorithms, is highlighted in boldface. In case of the FCPM-0 algorithm, when computing the FSI value a division by zero sometimes occurs (noted as NA). This situation happens when the FCPM-0 shifts one or more prototypes outside the data space. The last row of Table 1 shows the proportion of winner index values in the columns.

By majority voting counting it is easy to see that the XB index is the more adequate for the FS-AA algorithm, the FSI for the FCPM-2 algorithm, and the PE, PC, MPC for the FCPM-0. This finding fits the clustering criterion of each algorithm, taking into account that: (i) the XB index uses the fuzzy cluster's center of each cluster as the representative for that cluster, when evaluating the inter-cluster separation; (ii) the FSI uses the average minimum pairwise distance between objects in each fuzzy cluster as the separation measure; and (iii) the PE, PC, MPC are validation indices exclusively based on the membership values, benefiting the clustering partitions which are less fuzzy, *i.e.* with memberships values tending to 1 or 0). Moreover, for FCPM-0 and FCPM-2 this analysis is concordant with the one presented in [16] with Backer's validation index.

Table 1. Validation indices values for the real world data and their counts

Data set	k	AA					FCPM-0					FCPM-2				
		PE(↓)	PC(↑)	MPC(↑)	FSI(↑)	XB(↓)	PE(↓)	PC(↑)	MPC(↑)	FSI(↑)	XB(↓)	PE(↓)	PC(↑)	MPC(↑)	FSI(↑)	XB(↓)
Bank	2	0.701	0.670	0.340	0.603	0.120	**0.192**	**0.912**	**0.823**	0.543	0.323	0.739	0.649	0.299	0.615	0.150
note	3	0.678	0.535	0.302	0.620	**0.105**	0.197	0.869	0.803	NA	0.453	0.765	0.507	0.260	**0.627**	0.113
Wisconsin-BC	2	0.398	0.825	0.649	0.850	**0.066**	0.092	0.958	0.916	0.795	0.121	0.488	0.785	0.570	0.851	0.075
	3	0.356	0.766	0.650	0.795	0.122	**0.021**	**0.985**	**0.978**	0.602	8.213	0.571	0.644	0.466	**0.861**	0.945
Wisconsin-DBC	2	0.615	0.720	0.440	**0.706**	**0.097**	0.183	**0.916**	0.833	0.644	0.252	0.751	0.653	0.306	0.676	0.143
Wisconsin	3	0.597	0.604	0.406	0.685	0.112	**0.129**	0.908	**0.862**	0.523	0.656	0.811	0.477	0.216	0.616	0.208
	5	0.461	0.648	0.560	0.650	0.197	0.438	0.635	0.544	0.277	191	0.886	0.401	0.251	**0.772**	**0.094**
Glass	6	0.422	0.628	0.553	0.647	0.193	**0.324**	0.727	0.672	NA	124	0.986	0.261	0.113	0.587	0.127
	7	0.426	0.566	0.494	0.607	0.304	0.353	0.659	0.602	NA	166	0.948	0.248	0.123	0.554	0.278
	2	0.357	0.854	0.709	0.730	0.163	0.911	0.054	**0.926**	0.726	0.223	0.667	0.701	0.402	0.756	0.144
	3	0.396	0.751	0.627	0.770	0.118	0.159	**0.893**	0.839	0.579	0.408	0.692	0.565	0.348	0.589	0.304
Indian Liver	2	0.053	0.985	0.971	**0.769**	0.155	**0.000**	**1.000**	**1.000**	0.768	0.158	0.688	0.694	0.388	0.786	**0.124**
Patient	3	0.468	0.654	0.481	0.619	0.139	0.264	0.820	0.730	NA	0.141	0.754	0.522	0.282	0.604	0.222
Iris	2	0.605	0.726	0.452	0.841	0.069	**0.106**	**0.953**	**0.906**	0.839	0.081	0.498	0.775	0.551	**0.859**	**0.068**
Normalized	3	0.594	0.587	0.380	0.741	0.198	0.228	0.838	0.757	NA	0.109	0.548	0.659	0.489	0.569	0.948
	4	0.589	0.508	0.344	0.422	0.476	0.242	0.782	0.710	0.520	324.639	0.610	0.554	0.406	0.340	1.186
Mental	3	0.351	0.773	0.660	0.576	0.244	**0.020**	**0.986**	**0.979**	0.408	0.602	0.790	0.495	0.243	0.586	0.199
Disorders	4	0.359	0.712	0.615	0.594	0.167	0.162	0.860	0.813	0.522	0.356	0.797	0.416	0.221	**0.596**	**0.133**
	5	0.355	0.774	0.661	0.577	0.243	0.023	0.984	0.976	0.444	0.824	0.788	0.497	0.246	0.588	0.199
Mental	3	0.530	0.633	0.449	0.496	0.177	**0.137**	**0.910**	**0.865**	0.442	0.472	0.818	0.473	0.209	0.521	0.204
Disorders	4	0.493	0.603	0.471	**0.527**	**0.165**	0.203	0.812	0.750	0.298	1.234	0.842	0.377	0.169	0.419	0.477
Augmented	5	0.481	0.562	0.452	0.499	0.226	0.108	0.888	0.860	0.353	0.798	0.859	0.314	0.142	0.492	0.272
Pima Indians	2	0.642	0.708	0.416	**0.508**	0.175	**0.186**	**0.915**	**0.829**	0.456	0.554	0.843	0.600	0.199	0.505	0.268
Diabetes	3	0.690	0.527	0.291	0.431	**0.152**	0.246	0.821	0.731	0.342	0.662	0.878	0.426	0.139	0.410	0.313
	2	0.636	0.711	0.421	**0.794**	**0.075**	**0.178**	**0.918**	**0.835**	0.766	0.115	0.576	0.736	0.473	**0.794**	0.083
Seeds	3	0.621	0.581	0.372	0.690	0.173	0.186	0.880	0.820	NA	0.211	0.642	0.595	0.393	0.683	0.246
	4	0.583	0.523	0.364	0.521	0.424	0.260	0.792	0.722	NA	112.158	0.677	0.501	0.334	0.368	1.793
Vehicle	3	0.633	0.574	0.361	0.584	**0.194**	**0.173**	**0.889**	**0.833**	0.464	0.643	0.774	0.503	0.255	**0.588**	0.250
silhouettes	4	0.602	0.511	0.348	0.495	0.198	0.267	0.781	0.708	NA	1.009	0.821	0.394	0.192	0.563	0.204
	5	0.571	0.474	0.342	0.437	0.287	0.339	0.674	0.593	0.323	7.560	0.836	0.324	0.155	0.482	4.413
Wine	2	0.557	0.748	0.497	0.545	0.188	**0.089**	**0.959**	**0.919**	0.467	0.479	0.784	0.635	0.269	0.561	0.202
Recognition	3	0.511	0.655	0.483	0.552	0.166	0.184	0.875	0.813	0.541	0.418	0.816	0.476	0.214	**0.583**	0.152
	4	0.516	0.575	0.433	0.539	**0.180**	0.196	0.843	0.790	0.411	0.541	0.856	0.366	0.155	0.573	0.189
Count		0/12	0/12	0/12	5/12	8/12	12/12	12/12	12/12	0/12	0/12	0/12	0/12	0/12	8/12	4/12

4   Cluster Structure Recovery Study

In this experiment we analyse the FS-AA algorithm to recover archetypes on synthetic data generated according to the typological assumptions of the FCPM model (5): (i) there exists a cluster structure underlying the model of data generation; (ii) in that structure each original prototype, say, o_i, is an 'ideal' point/'model' such that each data point, y_k shares a proportion of it. These data were generated from a proper data generator [16], the FCPM-DG, whose main characteristics are: (i) the data space dimensionality (p), the number of clusters c_0 and the number of data points generated within each cluster are randomly generated in pre-specified intervals chosen by the user; (ii) c_0 cluster directions are generated within pre-specified hyper-cubes in distinct orthants of the data space; (iii) the data points are randomly generated within two p-dimensional sampling boxes over the c_0 cluster direction vectors; (iv) all generated data including the original prototypes are normalized by centering to the mean of the data and scaling by the features ranges; (v) all the randomly generated items come from a Uniform distribution in the interval $[0, 1]$.

In this study we used 82 data sets from the whole collection of the simulation study presented in [15]. According to that study the data sets are categorized in three data clustering dimensionalities wrt the ratio $\frac{p}{c_0}$: low ($\frac{p}{c_0} \leq 5$); intermediate ($5 < \frac{p}{c_0} < 25$); and high ($\frac{p}{c_0} \geq 25$). From the 82 data sets 19 are of low, 52 of intermediate, and 12 of high cluster dimensionality.

To measure the quality of clustering recovery we apply the dissimilarity coefficient introduced in [15,16] to compare the retrieved prototypes ($c\prime$) of found partitions with the reference ones, c_0 (e.g. the original prototypes from which data were generated). The dissimilarity coefficient, D, between the sets $V' = \{\mathbf{v}_j\}_{j=1}^{c\prime}$ and $V = \{\mathbf{v}_i\}_{i=1}^{c}$, is defined as the squared relative quadratic mean error between the found prototypes and the original ones:

$$D(V', V) = \frac{\sum_{i=1}^{c_0} \sum_{h=1}^{p} (v'_{ih} - v_{ih})^2}{\sum_{i=1}^{c_0} \sum_{h=1}^{p} v_{ih}^2 + \sum_{i=1}^{c_0} \sum_{h=1}^{p} v'^2_{ih}}. \tag{11}$$

Coefficient $D(V', V)$ is always positive, and it equals 0 when $v_{ih} = v'_{ih}$, for all $i = 1, ..., c; h = 1, ..., p$. In a typical situation, when the components of each v_i and v'_i are in the same orthants, then D is small or equal to 1. The matching between found prototypes and the reference ones is calculated using the K-NN distances with $K = min(c\prime, c_0)$. In a scenario of tie in matching prototypes, one of the retrieved prototypes is matched to its next closest reference prototype.

To analyse the data recovery ability of FS-AA algorithm, it was run with the DG synthetic data. The dissimilarity index D, between the original DG prototypes and the retrieved archetypes, was calculated and compared with the prototypes obtained by FCPM [16].

Table 2 shows the average dissimilarity squared mean errors of FS-AA archetypes and the FCPM-DG originals (first column) as well as for the FCPM-2 and FCPM-0 found prototypes (following columns) for those selected FCPM-DG data sets. The results show how close FS-AA archetypes are to the original prototypes and, consequently, to the ones found by the FCPM-2 algorithm. This is concordant with the property of FCPM-2 always finding extreme prototypes, matching the FCPM-DG original ones [16]. For the FCPM-2 it is possible to see that, as the dimensionality increases, the prototypes became closer to the originals. In fact, in the high dimensional space, the prototypes found by the FCPM-2 are closer to the original ones, than the ones found by the AA. As for the FCPM-0, the dissimilarity values are higher than the ones of AA and FCPM-2. The reasons are that the FCPM-0 prototypes are typically central points and additionally removes one of its prototypes out of the data space in high dimensional spaces.

Table 2. Average dissimilarity (D) values of FS-AA to FCPM-DG V_{Org} and to FCPM-2, FCPM-0 prototypes.

Dim.	FS-AA vs V_{Org}	FS-AA vs FCPM-2	FS-AA vs FCPM-0	FCPM-2 vs V_{Org}	FCPM-0 vs V_{Org}
Low	0.008	0.006	0.109	0.021	0.156
Medium	0.007	0.001	0.167	0.011	0.200
High	0.006	0.000	0.217	0.005	0.228

5   Outlier Analysis

The goal of this experiment is to analyse how the presence of data points that are further away from the center of the data set influence the FS-AA, FCPM-2 and FCPM-0 clustering solutions. The motivation is two-fold: firstly, analyse how robust FS-AA and FCPM-2 algorithms are in the presence of outliers; secondly, analyse the behavior of FCPM-0 as an indicator of the number of clusters [16].

This study was conducted with data sets of the FCPM-DG simulated data described in the previous section, where each data set was augmented with one or two outliers generated by the Interquartile Range (IQR) method. Any point located $1.5 \times (IQR)$ or more below the 25th or above the 75th percentile (lower or upper fence, respectively), is considered an outlier. To add outliers, it is only necessary to create points above/below the upper/lower fence. The first outlier is calculated from the mean of each data feature vector added to five times the corresponding standard deviation, to guarantee it is above the upper fence. The second outlier is simply the symmetric of the first one with some random noise added (a factor of 0.2), to be below the lower fence. Figure 1 shows the box-plot of the features values for a data set of low dimensionality ($n = 37, p = 5, c = 3$) (Fig. 1a) and the corresponding plot of the data set augmented with two outliers: one outlier generated above the upper fence and the other one generated bellow the lower fence (Fig. 1b).

This experiment was conducted with 5 distinct parameter settings. The algorithms had to search for the same number of prototypes from which the data was generated, $k = c_0$, then for $k = c_0 + 1$, and finally for $k = c_0 + 2$. In the first two cases, the data set was augmented, first with one outlier ($out = 1$), and then, with two outliers ($out = 2$). In the third case, the data was augmented with two

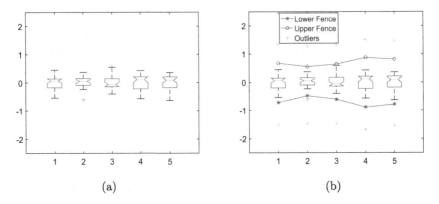

Fig. 1. Boxplot for a FCPM-DG data set of low dimensionality ($n = 37, p = 5, c = 3$): original data (1a); augmented data (1b) with the first outlier generated above the upper fence and second one generated bellow the lower fence.

outliers ($out = 2$). The FCPM-0 and FCPM-2 algorithms were initialized with the Futherst-Sum algorithm to compare with the FS-AA. The results shown in Table 3 are the average of Dissimilarity index D (11) for all the data sets and are summarized as follows:

Table 3. Average Dissimilarity D values for the data sets augmented with outliers

Dim.	$k=c_0$						$k=c_0 + 1$						$k=c_0 + 2$		
	Out=1			Out=2			Out=1			Out=2			Out=2		
	AA	FCPM-0	FCPM-2	AA	FCPM-0	FCPM-2	AA	FCPM-0	FCPM-2	AA	FCPM-0	FCPM-2	AA	FCPM-0	FCPM-2
Low	0.471	0.648	0.319	0.589	0.637	0.423	0.013	0.385	0.158	0.313	0.621	0.545	0.007	0.478	0.009
Medium	0.538	0.637	0.691	0.142	0.471	0.754	0.006	0.138	0.005	0.092	0.518	0.615	0.006	0.108	0.002
High	0.570	0.396	0.478	0.129	0.660	0.686	0.006	0.161	0.002	0.041	0.447	0.477	0.006	0.253	0.003

- For $k = c_0, out = 1$, the D index values are high for FS-AA/FCPM-2 algorithms because they put one archetype/prototype near the outlier and another between two clusters. The FCPM-0 removes one prototype out of data space, putting another prototype in the middle of two clusters which explains the higher values of dissimilarity D. Figure 2 illustrates these results for a FCPM-DG data set projected on the space of the three principal components.
- For $k = c_0, out = 2$, FS-AA results for low dimensionality data show an increase in D, as the algorithm puts one archetype in each outlier. But, in the intermediate and high dimensional data, the algorithm does not put archetypes in the outliers in most cases. More interesting is FCPM-2, since it always puts one prototype near each outlier, resulting in higher D values than in the first setting. The FCPM-0 exhibits the same behavior of shifting one prototype out of the data space.

- For $k = c_0 + 1, out = 1$, both FF-AA/FCPM-2 put one archetype/prototype in the outlier, but the others near the original prototypes. In the low dimensional space, the FCPM-2 puts two of the prototypes near the same original one. The FCPM-0, continues shifting one prototype out of the data space, but puts the others near the original prototypes. For a few data sets in the low dimensional space, the FCPM-0 shifts two prototypes resulting in a larger mean value for D.
- In case $k = c_0 + 1, out = 2$, the FS-AA puts one archetype in one outlier, and the other in the middle of the other outlier and an original prototype that doesn't have an archetype nearby. The other ones are near the original prototypes. As the dimensionality increases, the distance to the original prototype decreases. The FCPM-2 always puts one prototype in each outlier. The FCPM-0 shifts two prototypes out of the data space.
- In the last setting, $k = c_0 + 2, out = 2$, the results are similar to the third setting. Both the FS-AA and the FCPM-2 put one archetype/prototype in each outlier, and the rest near the original prototypes. As for the FCPM-0 in the low dimensional space, it typically shifts 3 prototypes to out of the data space, resulting in a larger D value. For the intermediate and high dimensional spaces, FCPM-0 removes the 2 extras prototypes out of the data space.

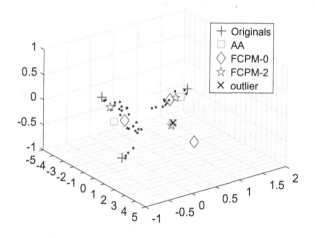

Fig. 2. 3D projection of a FCPM-DG low dimensional data set ($n = 37, p = 5, c = 3$), for the first setting ($k = c_0, Out = 1$) of outliers.

In summary, FS-AA and FCPM-2 algorithms are robust to outliers. As to the FCPM-0 it typically maintains the behavior of shifting as many prototypes out of the data space as the number of clusters with outliers; consequently, it accommodates the remaining prototypes between the clusters (settings 1, 2, and 4). In the remaining settings it shifts the extra prototypes out of the data space finding the true number of clusters.

6 Conclusions

This is a first study on comparing Archetypal analysis and Fuzzy Clustering with Proportional Membership. The study is a primer for the use of five popular unsupervised validation indices to access the quality of the fuzzy partitions of FS-AA and FCPM algorithms.

Taking advantage of simulated data from the FCPM data generator covering low, intermediate, and high dimensionality, the AA algorithm is able to retrieve cluster structure from data with archetypes very similar to the FCPM-2 ideal types. On the study of the robustness to outliers, we analyse that both FS-AA and FCPM-2 perform well with their presence. On the contrary, the FCPM-0 acts as an indicator of the right number of clusters by shifting one or more prototypes out of the data space. This is not surprising since the FCPM model is less restrictive than the FS-AA which imposes its prototypes to lye on the convex-hull of the data.

Work in progress suggests that other types of initialization, based on FCPM-0 properties can contribute to modeling the number of clusters to retrieve from data. Future work also include to improve the study on the influence of outliers.

Acknowledgments. This research is supported by FCT/MCTES, NOVA LINCS (UID/CEC/04516/2013). The authors are thankful to the anonymous reviewers for their insightful and constructive comments that allowed to improve the paper.

References

1. Bertsekas, D.: Nonlinear Programming. Athena Scientific, Belmont (1995)
2. Bezdek, J.: Pattern Recognition with Fuzzy Objective Function Algorithms, 1st edn. Springer, Heidelberg (1981). https://doi.org/10.1007/978-1-4757-0450-1
3. Campello, R., Hruschka, E.: A fuzzy extension of the silhouette width criterion for cluster analysis. Fuzzy Sets Syst. **157**(21), 2858–2875 (2006)
4. Chen, Y., Mairal, J., Harchaoui, Z.: Fast and robust archetypal analysis for representation learning. In: 2014 IEEE Conference on Computer Vision and Pattern Recognition, pp. 1478–1485 (2014)
5. Cutler, A., Breiman, L.: Archetypal analysis. Technometrics **36**(4), 338–347 (1994)
6. Eisenack, K., Lüdeke, M., Kropp, J.: Construction of archetypes as a formal method to analyze socialecological systems. In: Proceedings of the Institutional Dimensions of Global Environmental Change Synthesis Conference, Bali, Indonesia, 6–9 December 2006
7. Eugster, M., Leisch, F.: From spider-man to hero - archetypal analysis in R. J. Stat. Softw. **30**(8), 1–23 (2009)
8. Eugster, M.: Performance profiles based on archetypal athletes. Int. J. Perform. Anal. Sport **12**(1), 166–187 (2012)
9. Ferraro, M., Giordani, P.: A toolbox for fuzzy clustering using the R programming language. Fuzzy Sets Syst. **279**(8), 1–16 (2015)
10. Geng, L., Hamilton, H.: Interestingness measures for data mining: a survey. ACM Comput. Surv. **38**(3), 9 (2006)

11. Liu, Y., Li, Z., Xiong, H., Gao, X., Wu, J., Wu, S.: Understanding and enhancement of internal clustering validation measures. IEEE Trans. Cybern. **43**(3), 982–994 (2013)
12. Mirkin, B.: Mathematical Classification and Clustering. Nonconvex Optimization and Its Applications. Kluwer Academic Publishers, Dordrecht (1996)
13. Mirkin, B.: Clustering: A Data Recovery Approach, 2nd edn. Chapman & Hall/CRC Press, London/Boca Raton (2012)
14. Mørup, M., Hansen, L.: Archetypal analysis for machine learning and data mining. Neurocomputing **80**, 54–63 (2012)
15. Nascimento, S., Mirkin, B., Moura-Pires, F.: Modeling proportional membership in fuzzy clustering. IEEE Trans. Fuzzy Syst. **11**(2), 173–186 (2003)
16. Nascimento, S.: Fuzzy Clustering via Proportional Membership Model. IOS Press, Amsterdam (2005). 178 p
17. Nascimento, S., Franco, P.: Segmentation of upwelling regions in sea surface temperature images via unsupervised fuzzy clustering. In: Corchado, E., Yin, H. (eds.) IDEAL 2009. LNCS, vol. 5788, pp. 543–553. Springer, Heidelberg (2009). https://doi.org/10.1007/978-3-642-04394-9_66
18. Nascimento, S.: Applying the gradient projection method to a model of proportional membership for fuzzy cluster analysis. In: Goldengorin, B. (ed.) Optimization and Its Applications in Control and Data Sciences. SOIA, vol. 115, pp. 353–380. Springer, Cham (2016). https://doi.org/10.1007/978-3-319-42056-1_13
19. Porzio, C., Ragozini, G., Vistocco, D.: On the use of archetypes as benchmarks. Appl. Stoch. Models Bus. Ind. **54**(5), 419–437 (2008)
20. Tallón-Ballesteros, A.J., Correia, L., Xue, B.: Featuring the Attributes in Supervised Machine Learning. In: de Cos Juez, F., et al. (eds.) HAIS 2018. LNCS, vol. 10870. Springer, Cham (2018). https://doi.org/10.1007/978-3-319-92639-1_29
21. UCI Machine Learning Repository. http://archive.ics.uci.edu/ml. Accessed 27 July 2018

Bare Bones Fireworks Algorithm
for Medical Image Compression

Eva Tuba[1], Raka Jovanovic[2], Marko Beko[3,4], Antonio J. Tallón-Ballesteros[5], and Milan Tuba[1(✉)]

[1] Faculty of Informatics and Computing, Singidunum University,
Belgrade, Serbia
tuba@ieee.org
[2] Qatar Environment and Energy Research Institute,
Hamad bin Khalifa University, Doha, Qatar
[3] COPELABS, Universidade Lusófona de Humanidades e Tecnologias,
1749-024 Lisbon, Portugal
[4] CTS/UNINOVA, Campus da FCT/UNL, 2829-516 Caparica, Portugal
[5] Department of Languages and Computer Systems, University of Seville,
Seville, Spain

Abstract. Digital images are of a great importance in medicine. Efficient and compact storing of the medical digital images represents a major issue that needs to be solved. JPEG lossy compression algorithm is most widely used where better compression to quality ratio can be obtained by selecting appropriate quantization tables. Finding the optimal quantization tables is a hard combinatorial optimization problem and stochastic metaheuristics have been proven to be very efficient for solving such problems. In this paper we propose adjusted bare bones fireworks algorithm for quantization table selection. The proposed method was tested on different medical digital images. The results were compared to the standard JPEG algorithm. Various image similarity metrics were used and it has been shown that the proposed method was more successful.

Keywords: Medical image processing · Compression · JPEG
Quantization tables · Optimization · Bare bones fireworks algorithm

1 Introduction

Usage of digital images in medicine became almost necessary for diagnosis and treatment determination. Many computer aided diagnostic systems include machine learning algorithms that, based on different image features, detect various anomalies. Medical image analysis includes numerous images in high resolution, hence it represents source of big data. One of the data preprocessing steps

M. Tuba—This research is supported by the Ministry of Education, Science and Technological Development of Republic of Serbia, Grant No. III-44006.

© Springer Nature Switzerland AG 2018
H. Yin et al. (Eds.): IDEAL 2018, LNCS 11315, pp. 262–270, 2018.
https://doi.org/10.1007/978-3-030-03496-2_29

for these data can be image compression. In order to reduce the size of digital images different lossless and lossy compression methods can be applied. Special attention should be paid to medical images since they are different from standard everyday images obtained by cameras or mobile phones. Medical images are obtained from various sources such as magnetic resonance imaging (MRI), computed topography (CT), ultrasound, etc. Different sources produce images of very different characteristics and it is important to achieve high compression while preserving medically significant data, rather then preserving data for obtaining visually acceptable images.

Even though there are numerous compression algorithms, one of the most widely used lossy compression algorithms is JPEG which can reduce the size of image information 20 or even 50 times [1]. This compression algorithm was adjusted for different medical image compression such as dermatological images [2], 3D medical images [3], endoscopic images [4], etc. The main image size reduction is achieved by the quantization step that discards less important data by using a quantization table. JPEG standard contains recommended quantization tables, however medical digital images are rather specific, hence better results can be achieved by adjusting quantization tables. Finding the optimal elements of the quantization table is a combinatorial problem and it represents a hard optimization problem that cannot be solved by deterministic methods. For these kinds of problems various heuristics and metaheuristics are needed such as nature inspired algorithms that have been successfully used for finding acceptable solutions in a reasonable amount of time. In this paper we propose adjustment of the bare bones fireworks algorithm, the latest version of the fireworks algorithm, for selecting quantization table for medical image compression.

2   JPEG Compression Algorithm

JPEG algorithm conducts several steps. The first step in JPEG algorithm reduces the size of an image by using the fact that human eye is less sensitive to color then to light intensity changes, thus some color information is discarded. The second step is transformation of an image into the frequency domain which is achieved by applying the two dimensional discrete cosine transform (DCT). The DCT is performed on the light intensity component on non-overlapping blocks of the size 8×8. The result of the DCT transformation is 8×8 block, i.e. 64 frequency coefficients. The very first coefficient (DC component) represents the average light intensity of the block. Lower frequencies that keep the most information about the image are closer to the DC component, while higher frequencies that represents edges and noise are closer to the lower right corner of the block. Quantization step is performed on the DCT coefficients where the main compression is achieved.

Compression level, as well as the quality of the image compressed by the JPEG algorithm, are manly obtained by quantization step where quantization table is used. Compression is achieved by discarding some less important DCT coefficients and by reducing the precision of the remaining elements. In the quantization step, DCT coefficients are divided by the corresponding elements from the quantization table. Depending on the elements of the quantization table, different compression levels can be achieved which also results in different quality of the decompressed image.

There are empirically obtained recommended quantization tables named Q_x where x in $\{1, 2, \ldots, 100\}$ represents the scale for the quality of the decompressed image. The quantization table Q_{50} is used as the base table and all other quantization tables are calculated based on it.

Standard quantization tables Q_x are obtained based on the human perception which is not an objective measure for numerous applications where the compressed images have to be further analyzed and processed with the specific goal. In these cases, that specific goal is the quality metrics [5]. Quantization table selection problem was also considered for the medical image compression. In [6], three different quantization tables for ultrasound image compression were compared. Quantization tables were designed using Haddamard transform, generic psychovisual threshold and one was empirically determined. In [7], quantization tables for medical image compression based on coefficients obtained by framelet transformation were proposed. In this paper we search for quantization table elements that maximize the similarity between the original and decompressed images.

Finding the optimal elements for the quantization table is a combinatorial problem since 64 elements can take any value from some specific range which is, in practice, $[1, 255]$. Nature inspired algorithms, especially swarm intelligence algorithms, have been successfully used for finding good solutions for such problems. These algorithms were applied for quantization table selections, e.g. genetic algorithm [8], firefly algorithm [9], bacterial foraging optimization algorithm [10], etc. A brief review of the swarm intelligence algorithms proposed for JPEG algorithm adjustment is given in [11].

3   Bare Bones Fireworks Algorithm for JPEG Algorithm Optimization

Bare bones fireworks algorithm (BBFWA) represents recent swarm intelligence algorithm and the latest update of the fireworks algorithm (FWA) that was originally proposed by Tan and Zhu in 2010 [12]. BBFWA was developed by Li and Tan in 2018 [13]. FWA was widely used and applied to numerous problems, including image processing problems [14], wireless sensor networks localization and coverage problem [15, 16], etc.

BBFWA uses only one firework that produces fixed number of sparks around itself. Exploration and exploitation were implemented by increasing and decreasing the space around the best solution where the random solutions (sparks) would be generated. In each iteration of the BBFWA only the best solution is saved for the next iteration. Size of the search space around that solution is regulated by two parameters, C_a and C_r. The first solution is randomly generated and the initial size of the search space around it is set to cover the whole search space.

If Lb and Ub represent the lower and upper bounds of the search space, respectively, then random solutions are generated in the hyper-rectangle whose center is the best solution and the size of side is $2 * A$ where initially $A = Ub - Lb$. In further iterations, size of the hyper-rectangle is changed for factor $C_a > 1$ (increased) if the new best solution was found because the search space is not explored enough, thus wider search is needed. On the other hand, if the best solution remains the same in two iterations, then the size of hyper-rectangle is reduced for factor $C_r < 1$ which enables the fine search of the promising area. Pseudo code of the BBFWA is presented in Algorithm 1.

Algorithm 1. Bare bones fireworks algorithm [13]

Sample $x \sim U(Lb, Ub)$
Evaluate $f(x)$
$A = Ub - Lb$
repeat
    for $i = 1$ to N **do**
        Sample $s_i \sim U(x - A, x + A)$
        Apply mapping operator to s_i
        Evaluate $f(s_i)$
    end for
    if $\min_{i=1,2,\ldots,N}(f(s_i)) < f(x)$ **then**
        $x = \operatorname{argmin}(f(s_i))$
        $A = C_a A$
    else
        $A = C_r A$
    end if
until maximal iteration number is reached
return x

In this paper, we adjusted the BBFWA for quantization table selection. The goal is to achieve the same compression level as the one obtained by the recommended quantization tables but with better quality, according to the selected metrics which is very important for medical images. One solution is 64-dimensional vector that represents elements of the quantization table. Theoretical bounds for DCT coefficients are 1 and 1023, but in the practice they are rarely larger then 255 thus, range for the elements is set to be [1, 255]. Elements of the quantization table are integers while BBFWA uses the real numbers, thus we rounded the obtained solutions to the nearest integer.

Compression level that need to be obtained can be defined differently and in this paper it was considered that the same level of the compression is achieved with the quantization tables whose sum of elements is the same. By introducing this condition, considered problem becomes constrained optimization problem. Equality constraints such as this one are the most difficult, since finding the feasible solution is almost impossible. For the considered problem, equality constraint can be easily converted into inequality one: if the sum of elements of quantization table is greater or equal to some constant, then higher or the same level of the compression will be obtained. Since higher compression means lower quality, the BBFWA algorithm will be reaching equality constraints by optimizing the fitness function which represents the quality of the compressed image.

The BBFWA algorithm will not always generate feasible solutions even for the relaxed constraint. The sum of the elements can be larger then the given constant thus it is necessary to define comparison between two solutions and mechanism for handling infeasible solutions. For comparing of two solutions, we used Deb's rules, i.e. feasible solution is always better then infeasible one, infeasible solution that violate constraint less is better when comparing two infeasible solutions and between two feasible solutions, better solution has the better value of the fitness function. For the considered problem, solution with sum of elements closer to the given constant violets constraint less. Calculating the fitness function is rather expensive operation because it includes compression and decompression, solutions that has sum of elements less then the given number were not considered but new solution was generated.

Quality of the compressed image was measured by metrics commonly used for medical images [6,7]: mean square error and normalized cross correlation. Mean square error (MSE) and peak signal to noise ratio (PSNR) are defined as:

$$MSE = \frac{1}{NM} \sum_{i=1}^{N} \sum_{j=1}^{M} (x_{i,j} - x'_{i,j})^2 \qquad PSNR = 10 \log \frac{255^2}{MSE + \epsilon} \qquad (1)$$

where $x_{i,j}$ and $x'_{i,j}$ are the intensity values of the pixel (i, j) in the original image and compressed image, respectively, while M and N represent the dimension of image. Parameter ϵ represents a small constant that prevents division by zero. MSE value is equal or larger than zero where two identical images hava $MSE = 0$ thus the goal is to minimize MSE. Normalized cross correlation (NK) is the second fitness function that is used in this paper to determine the quality of the compressed image. It is defined as:

$$NK = \frac{\sum_{i=1}^{N} \sum_{j=1}^{M} x_{i,j} x'_{i,j}}{\sum_{i=1}^{N} \sum_{j=1}^{M} x_{i,j}^2} \qquad (2)$$

For the two identical images, NK is equal to 1 while less similar images have smaller NK (minimal possible value is -1). The goal is to minimize -NK.

The proposed method can be easily adjusted for specific medical images by replacing the used metrics by appropriate ones. Quality metric of the decompressed image defines fitness function and the optimal quantization tables will be searched to satisfy that condition.

4   Simulation Results

The proposed BBFWA for the JPEG quantization table selection for medical image compression was implemented in Matlab version R2016b and simulations were performed on Intel ® Core^TM i7-3770K CPU @ 4 GHz, 8 GB RAM computer with Windows 10 Professional OS. Population size for the BBFWA was 200, maximal number of the fitness function evaluation was 10,000. Parameters C_a and C_r were set to 1.2 and 0.7, respectively.

We tested our proposed BBFWA algorithm on brain MRI slice available on the web-based medical image depository [17], phantom ultrasound image

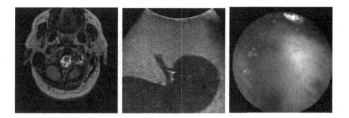

Fig. 1. Original test images

publicly available at www.yezitronix.com/kidney_phantom.html and endoscopic image from free dataset [18]. Images used in this paper are shown in Fig. 1.

First, we tested the proposed algorithm for the higher compression level. We used the recommended quantization table Q_{10}. The proposed BBFWA was used to find the optimal elements of the quantization table Q_{10_opt}, while keeping the same compression level. We used MSE as fitness function and the quantization tables are presented in Table 1.

Table 1. Quantization table Q_{10} (left) and Q_{10_opt} obtained by the BBFWA (right)

80	55	50	80	120	200	255	255		7	5	15	39	238	239	153	255
60	60	70	95	130	255	255	255		8	13	17	56	136	230	255	255
70	65	80	120	200	255	255	255		10	13	15	196	214	255	255	255
70	85	110	145	255	255	255	255		34	36	245	226	255	255	255	255
90	110	185	255	255	255	255	255		159	247	236	255	255	255	255	255
120	175	255	255	255	255	255	255		194	223	255	255	255	255	255	255
245	255	255	255	255	255	255	255		241	255	255	255	255	255	255	255
255	255	255	255	255	255	255	255		255	255	255	255	255	255	255	255

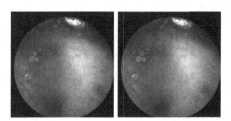

Fig. 2. Decompressed image by Q_{10} (left) and quantization table Q_{10_opt} (right) obtained by BBFWA with MSE as fitness function

Recommended quantization table Q_{10} and the quantization table obtained by our proposed algorithm are very different above the anti-diagonal. All elements under the anti-diagonal are 255. Our proposed BBFWA found that is better to save DCT coefficients in the upper left corner (lower frequencies). Sum of the elements in the Q_{10_opt} is 12,625 which is by 13 larger the sum of elements in the Q_{10} (12,612), hence even with larger compression, visually better image was obtained which can be seen in Fig. 2. Moreover, MSE for the MRI image when the Q_{10} table is used was 83.0964 while the MSE for the image compressed by the quantization table obtained by the proposed method is 52.5501 which is significantly better. Similar situation is for the ultrasound and endoscopic images. For the ultrasound image and Q_{10}, MSE was 59.0486 while with the Q_{10_opt} MSE was reduced to 23.7515. MSE for the endoscopic image and Q_{10} was 19.6122 while with the optimized Q_{10_opt} it was only 5.7582.

In the second experiment, we tested with the compression level obtained by the Q_{20} quantization table. In this experiment, again quantization tables that perform the same compression level are significantly different. Sum of the elements in the Q_{20} is 9326 which is again a little less then the sum of the elements in the Q_{20_opt} obtained by the proposed BBFWA, 9378. MSE for the MRI image compressed by the Q_{20} is 45.0917 while by using the proposed Q_{20_opt} quantization table MSE is reduced to 22.3347 which is more then double reduction. For the ultrasound image, MSE was reduced from 30.1042 to 13.0285 if the optimized quantization table is used, while in case of the endoscopic image, MSE with standard quantization table was 9.0847 and with the optimized one was 3.0428. All the results are summarized in the Table 2. Our proposed method outperformed the JPEG algorithm for all test images and for all used metrics.

Table 2. Comparison of the similarity metrics by using recommended JPEG standard and quantization tables obtained by the proposed BBFWA

Image	MSE		PSNR		NK	
	Q_{10}	Q_{BBFWA}	Q_{10}	Q_{BBFWA}	Q_{10}	Q_{BBFWA}
Slice 022	83.0964	52.5501	30.3999	30.9251	0.9832	0.9894
Ultrasound	59.0486	23.7515	30.4187	34.3739	0.9880	0.9951
Endoscopic	19.6122	5.7582	35.2055	40.5280	0.9978	0.9994
	Q_{20}	Q_{BBFWA}	Q_{20}	Q_{BBFWA}	Q_{20}	Q_{BBFWA}
Slice 022	45.0917	22.3347	31.5898	34.6410	0.9907	0.9955
Ultrasound	30.1042	13.0285	33.3445	36.9819	0.9939	0.9973
Endoscopic	9.0847	3.0428	38.5477	43.2981	0.9990	0.9997

5   Conclusion

Compression is a common preprocessing step for medical image analyses. The bare bones fireworks algorithm, was adjusted for finding optimal quantization tables of JPEG algorithm for medical image compression. Two different fitness function were considered. Quantization table was searched so that certain compression level was obtained and at the same time the quality of the decompressed image was maximized in terms of two similarity metrics, the mean square error and normalized cross correlation. Our proposed bare bones fireworks based method significantly improved the quality of the compression compared to the standard JPEG algorithm. In the further work, compression level can be measured by different metric such as bits needed for storing the compressed image and also other similarity metrics can be used.

References

1. Gupta, M., Garg, A.K.: Analysis of image compression algorithm using DCT. Int. J. Eng. Res. Appl. (IJERA) **2**, 515–521 (2012)
2. Amri, H., Khalfallah, A., Lapayre, J.C., Bouhlel, M.S.: REPro. JPEG: a new image compression approach based on reduction/expansion image and JPEG compression for dermatological medical images. Imaging Sci. J. **65**, 98–107 (2017)
3. Dubey, V.G., Singh, J.: 3D medical image compression using Huffman encoding technique. Int. J. Sci. Res. Publ. **2** (2012)
4. Cheng, C., Liu, Z., Hu, C., Meng, M.Q.H.: A novel wireless capsule endoscope with JPEG compression engine. In: IEEE International Conference on Automation and Logistics (ICAL), pp. 553–558 (2010)
5. Thai, T.H., Cogranne, R., Retraint, F.: JPEG quantization step estimation and its applications to digital image forensics. IEEE Trans. Inf. Forensics Secur. **12**, 123–133 (2017)
6. Zimbico, A., Schneider, F., Maia, J.: Comparative study of the performance of the JPEG algorithm using optimized quantization matrices for ultrasound image compression. In: 5th ISSNIP-IEEE Biosignals and Biorobotics Conference: Biosignals and Robotics for Better and Safer Living (BRC), pp. 1–6 (2014)
7. Sabri, A.A., Sultan, N.H.: Quantization matrix for medical image compression using framelet transform. Kufa J. Eng. **5**, 77–92 (2014)
8. Lazzerini, B., Marcelloni, F., Vecchio, M.: A multi-objective evolutionary approach to image quality/compression trade-off in JPEG baseline algorithm. Appl. Soft Comput. **10**, 548–561 (2010)
9. Tuba, M., Bacanin, N.: JPEG quantization tables selection by the firefly algorithm. In: International Conference on Multimedia Computing and Systems (ICMCS), pp. 153–158. IEEE (2014)
10. Dua, R.L., Gupta, N.: Fast color image quantization based on bacterial foraging optimization. In: Fourth International Conference on Advances in Recent Technologies in Communication and Computing (ARTCom), pp. 100–102 (2012)
11. Viswajaa, S., Kumar, V., Karpagam, G.R.: A survey on nature inspired metaheuristics algorithm in optimizing the quantization table for JPEG baseline algorithm. Int. Adv. Res. J. Sci. Eng. Technol. **2**, 114–123 (2015)
12. Tan, Y., Zhu, Y.: Fireworks algorithm for optimization. In: Tan, Y., Shi, Y., Tan, K.C. (eds.) ICSI 2010. LNCS, vol. 6145, pp. 355–364. Springer, Heidelberg (2010). https://doi.org/10.1007/978-3-642-13495-1_44
13. Li, J., Tan, Y.: The bare bones fireworks algorithm: a minimalist global optimizer. Appl. Soft Comput. **62**, 454–462 (2018)
14. Tuba, E., Tuba, M., Dolicanin, E.: Adjusted fireworks algorithm applied to retinal image registration. Stud. Inf. Control **26**, 33–42 (2017)
15. Tuba, E., Tuba, M., Simian, D.: Wireless sensor network coverage problem using modified fireworks algorithm. In: International Wireless Communications and Mobile Computing Conference (IWCMC), pp. 696–701. IEEE (2016)

16. Tuba, E., Tuba, M., Beko, M.: Node localization in ad hoc wireless sensor networks using fireworks algorithm. In: 5th International Conference on Multimedia Computing and Systems (ICMCS), pp. 223–229. IEEE (2016)
17. Johnson, K.A., Becker, J.A.: The whole brain atlas, June 1999. http://www.med.harvard.edu/AANLIB/
18. Deeba, F.: Bleeding images and corresponding ground truth of CE images (2016). https://sites.google.com/site/farahdeeba073/Research/resources

EMnGA: Entropy Measure and Genetic Algorithms Based Method for Heterogeneous Ensembles Selection

Souad Taleb Zouggar[1] and Abdelkader Adla[2(✉)]

[1] Department of Economics, University of Oran 2, Bir El Djir, Algeria
souad.taleb@gmail.com
[2] Department of Computer Science, University of Oran 1, Oran, Algeria
adla.abdelkader@univ-oran.dz

Abstract. Generating ensembles of classifiers increase the performances in classification and prediction but on the other hand it increases the storage space and the prediction time. Selection or simplification methods have been proposed to reduce space and time while maintaining or improving the performance of initial ensemble. In this paper we propose a method called EMnGA that uses a diversity-based entropy measure and a genetic algorithm-based search strategy to simplify a heterogeneous ensemble of classifiers. The proposed method is evaluated against its prediction performance and is compared to the initial ensemble as well as to the selection methods of heterogeneous ensembles in the literature using a sequential way.

Keywords: Classification · Diversity · Homogeneous ensemble
Heterogeneous ensemble · Ensemble selection · Genetic algorithms
Fitness function

1 Introduction

The purpose of classification is to learn a target function that links a set of data to a set of predefined categories (classes). The constructed target functions are also called classification models.

To address the problem of instability of some classifiers [1, 2], it is asserted that the use of an ensemble classifiers generally gives better results than the use of a singular classification model [3–6]. These ensembles can be either homogeneous or heterogeneous depending on the nature of the models used. The homogeneous ensembles are obtained by performing different executions of the same learning algorithm; for example these ensemble can be obtained by varying the parameters of the learning algorithm or by manipulating the data (input or output) [2, 7–9]. As for heterogeneous models, they are obtained by executing different learning algorithms on the same set of data. These models have different visions of the data which allows obtaining diverse sets.

Generating a large number of predictors allows exploring the solution space largely, and by aggregating all the predictions, we recover a predictor that accounts for all this

© Springer Nature Switzerland AG 2018
H. Yin et al. (Eds.): IDEAL 2018, LNCS 11315, pp. 271–279, 2018.
https://doi.org/10.1007/978-3-030-03496-2_30

exploration. The increase in the number of models can be done without risk of over-learning [10]. However, a very large number of models require a large storage space and an important time for the prediction due to the interrogation of all the predictors.

The simplification of ensembles of classifiers, called ensemble selection, reduces the size of ensembles prior to their integration or combination [11–16]. In this paper, we use a diversity-based entropy measure combined to a genetic algorithm-based search strategy to simplify an ensemble of heterogeneous classifiers.

The rest of the paper is organized as follows: In Sect. 2, we present the recent works in the ensemble selection field and especially those related to heterogeneous ensembles. Section 3 details the proposed method describing the entropy measure and search strategy used. In Sect. 4, we give the experimentation results and a comparative study with heterogeneous ensemble selection method in literature. Finally in Sect. 5, we conclude and give some future work.

2   Literature Review

Ensemble methods have attracted a growing interest since their appearance. These methods have been applied in several areas namely, statistics, pattern recognition, and machine learning. These methods consist of two main phases: a model production phase and an aggregation or combination phase of these models. In the first phase, variable performance models are added arbitrarily. The reduced performance models negatively affect the ensemble performance. An ensemble may contain several similar models which reduces its diversity.

Several recent works on ensemble selection [3, 17–19] have been developed in the literature. Dai [20] proposes an improvement of the ensemble selection method of the same authors. This method uses backtracking in depth, which is perfectly adapted to systematically seek solutions to combinatorial problems of great magnitude. This improvement concerns the response time of this method, which has been considerably improved in this study.

Bhatnagar et al. [21] perform ensemble selection using a performance-based and diversity-based function that considers the individual performance of classifiers as well as the diversity between pairs of classifiers. A bottom-up search is performed to generate the sub ensembles by adding various pairs of classifiers with high performance.

Simplifying a set of classifiers usually involves reducing the number of trees while maximizing performance. Based on the approximate ensembles, Guo et al. [22] propose a new framework for ensemble selection. In this context, the relationship between attributes in an approximate space is considered a priori as well as their degree of maximum dependence. This effectively reduces the search space and increases the diversity of selected sub-ensembles. Finally, to choose the appropriate sub-ensemble, an evaluation function that balances diversity and precision is used. The proposed method allows repetitively changing the search space of the relevant sub-ensembles and selecting the next ensemble from a new search space.

The selection methods were applied for either homogeneous or heterogeneous cases [16, 23]. The heterogeneous ensembles are developed for sensitive application domains. Haque et al. [24] propose a search based on genetic algorithms to find the optimal combination to form a heterogeneous ensemble from an ensemble of classifiers. For this, they develop an algorithm that uses decimal cross-validation on the learning ensemble to evaluate the quality of each candidate ensemble. The proposed method uses a random resampling approach to balance the class distribution and is particularly used for class imbalance cases.

Pölsterl et al. [25] use a heterogeneous ensemble including survival models with the purpose of predicting the survival of patients with stage 3 prostate cancer. The results demonstrate that the constructed ensemble can predict the date of death of patients with this disease.

Partalas et al. [26] propose a diversity measure and a forward selection search strategy that considers all the possible cases that may exist when adding a certain model h_t to an ensemble. The measure called FES (Focused Ensemble Selection) considers 4 cases when adding.

$$FES_{Eval}(h_k, SUB) = \sum_{i=1}^{|Eval|} (\alpha * I(y_i = h_k(x_i)ET y_i \neq Sub(x_i)) - \beta * I(y_i \neq h_k(x_i)ET y_i = Sub(x_i)) + \beta * I(y_i = h_k(x_i)ET y_i = Sub(x_i)) - \alpha * I(y_i \neq h_k(x_i)ET y_i \neq Sub(x_i))).$$

Eval: the sample of evaluation or selection;

$I(y_i = h_k(x_i)ET y_i \neq Sub(x_i)) = 1$ if the instance x_i is well classified by the model h_k and is not properly classified by the current ensemble Sub and 0 otherwise.

The factors α, β represent respectively the number of models in the *Sub* ensemble correctly classifying the instance (x_i, y_i) and the number of models not properly classifying the same instance.

3   The EMnGA Method

3.1   Entropy Measure

The key idea in this approach is to generate only trees that have maximum diversity (they are less correlated with each other). This is based on the principle that the generalization error of the random forest will be on the wane while diversity among the trees increases.

Let Ω_V be a sample of individuals with their labels (classes), $|\Omega_V| = n$, $\Omega_V = \{v_1, \ldots, v_n\}$, and $\Omega_V = \{v_1, \ldots, v_n\}$. Each individual v_j is described by m variables denoted $x_{1j}, \ldots, x_{mj}$. Let Ci be a classifier of the classifiers ensemble $\{C_1, \ldots, C_i, \ldots, C_T\}$ represented by a n-dimensional binary vector $y_i = (y_{1i}, \ldots, y_{ni})^T$ such that $y_{ji}=1$ if the classifier C_i classifies the individual v_j and 0 otherwise. The entropy function f_E measures the diversity within an ensemble [27]. Given an individual $x_j \in \Omega_V$, if half of the classifiers T/2 don't misclassify x_j then the other half T-T/2 misclassifies it necessarily and vice versa. We speak in this case of maximum diversity.

We note $nc(x_j)$ the number of classifiers of T which correctly classify x_j, $nc(x_j) = \sum_{i=1}^{T} y_{ij}$. The entropy measure f_E is written as:

$$f_E = \frac{1}{n} \sum_{j=1}^{n} \frac{1}{T - \frac{T}{2}} \min\{nc(x_j), T - nc(x_j)\}$$

$f_E \in [0, 1]$ where 1 indicates a very large diversity and 0 a lack of diversity. Thus, the goal is to maximize the f_E function.

3.2 Genetic Algorithms

Genetic algorithms are a preferred technique for selection because they are inspired by natural selection. They generate individuals that optimize an evaluation function also called fitness function.

A genetic algorithm is defined by [28]:

- Individual also called chromosome or sequence represents a potential solution of the problem. In our case, a solution of the problem corresponds to a binary string of size T (corresponds to the number of trees composing the forest). A chromosome is noted $ch = (val_1 val_2 \ldots val_T)$ where $val_i = 1$ if the tree is present in the selected chromosome and 0 otherwise;
- Population corresponds to all the chromosomes representing all possibilities of 1 and 0 in a binary chain of size T;
- Environment represents the search space $|ER| = 2^T$.
- Fitness function corresponds to the function $ff_E = f_E$ (f_E is the diversity function defined above). The goal is to maximize the value of ff_E.

Calculate the ff_E fitness function for chromosome ch_1 is equivalent to calculate the function f_E:

$$f_E = \frac{1}{n} \sum_{j=1}^{n} \frac{1}{T - \frac{T}{2}} \min\{nc(x_j), T - nc(x_j)\}$$

For instance, if $n = 2 = |\Omega_v|$, T = 2 (the classifiers to which correspond the value 1 T_1 and T_3), x_1 and x_2 are the individuals of Ω_v classified respectively $(1\ 0)^t$ and $(0\ 1)^t$ by T_1 et T_3.

$nc(x_1) = 1$ (the number of trees that correctly classify instances x_1).
$nc(x_2) = 1$ (the number of trees that correctly classify instances x_2).
$f_E = \frac{1}{2} \left(\frac{1}{2 - \frac{2}{2}} * \min(1, 2, -1) + \frac{1}{2 - \frac{2}{2}} * \min(1, 2, -1) \right) = 1$ and $ff_E = f_E = 1$, ff_E takes its minimum equal to 0 when the trees are diverse and its maximum 1 when they disagree; hence $ff_E \in [0\ 1]$.

We give hereafter the EMnGA algorithm that uses a genetic algorithm-based search strategy and an entropy-based fitness function. It is described by the following pseudo code:

Algorithme AG_{fE} ;

Input :
$C=\{C_1,...,C_i,...,C_T\}$: a heterogeneous ensemble of classifiers ;
$y_i=(y_{1i}...,y_{ni})^T$: a classification vector associated with C_i on Ω_V;
Ω_V: a validation or selection sample;
ch_i: chromosome i of the search space;
$ff_E(ch,\Omega_V)$: fitness function ;
Output :
Ch_sol : solution chromosome;
Begin
Generate a population of bits of size T;
Evolve the population where the fitness of a chromosome ch_i is calculate by $ff_E(ch_i,\Omega_V)$;
$ch\_sol :=argmax_{chi}(ff_E(ch_i,\Omega_V))$
End.

4   Experiments and Results

In this section, we describe information about the datasets used to carry out our experiments. We experienced 8 benchmarks datasets downloaded from the UCI Repository [29] as depicted in Table 1.

Table 1. Description of datasets

Dataset	Size	Variables	Classes
cmc	1473	9	3
kr-vs-kp	3196	36	2
credit-g	1000	21	2
tic-tac-toe	958	9	2
vehicle	946	18	4
vowel	990	13	11
Hypothyroïde	3772	30	4
segment	2310	20	7

All data sets contain enough data to split them into three samples: a learning sample Ω_L (40% of the initial sample size), a validation or selection sample (20% of the size of the initial sample) and the remaining 20% are used for the test.

We adopt a similar approach to that proposed in [26]. We generate, for each dataset, a heterogeneous ensemble of 200 models (classifiers) containing:

- 60 Kppv: with 22 values for K ranging from 1 to the size of the training sample. Three weighting methods are applied: no weighting, inverse weighting and similarity weighting;
- 110 SVM: we use 10 values for the complexity parameter $\{10^{-7}, 10^{-6}, 10^{-5}, 10^{-4}, 10^{-3}, 10^{-2}, 0.1, 1, 10, 100\}$, and 10 different kernels (2 polynomials of degree 2 and 3 and, 8 radial with gamma $\in \{0.001, 0.005, 0.01, 0.05, 0.1, 0.5, 1, 2\}$;
- 2 naive bayes: one model is built with the default parameters and another with a kernel estimate;
- 4 decision trees: two values are used for the confidence factor $\{0.25, 0.5\}$;
- 24 MLPs with a single hidden layer containing 6 neurons.

For each dataset, the generation of the ensemble (composed of 200 models) is repeated 10 times and the majority vote is used for the combination of the different models.

In Table 2, we give the performance results or success rates obtained (calculated on an average of 10 iterations for each dataset) on the 8 datasets. Comparisons are made between the Initial ensemble EI, the method EMnGA, and FES [26].

Table 2. Success rates obtained respectively by EI, EMnGA and FES for the 8 datasets

	EI	EMnGA	FES
cmc	47.6%	**53.2%**	52.7%
kr-vs-kp	95.6%	98.55%	**99%**
credit-g	70.8%	**85.32%**	74.4%
tic-tac-toe	63.9%	97.43%	**98.7%**
vehicle	75.3%	**83.02%**	81.1%
vowel	90.7%	**98.5%**	90.3%
Hypothyroïde	91.9%	98.6%	**99.3%**
segment	97.8%	**98.66%**	96.9%
Average	74%	86%	83%

The EMnGA method gives better performance compared to FES on 5 of the 8 datasets, with improvements ranging from 0.5% to 10.92%. FES is doing better than the proposed method on 3 dataset with improvements of 0.45% for kr-vs-kp and segment, and 1.27% for tick-to-toe. On average, across all datasets, EMnGA improves the performance of FES by 3% and the initial ensemble EI by 12% (Table 3).

We note that FES method generate ensembles with reduced sizes in 6 cases among the 8 with reductions ranging from 0.2 to 2.3. EMnGA reduces the size of the credit-g and tick-to-toe datasets by 0.1 and 1.8, respectively. On average, on the 8 datasets, FES allows a reduction of 0.5 compared to EMnGA.

Table 3. Sizes of ensembles obtained by EMnGA and FES on the 8 datasets

	EMnGA	FES
cmc	20.6	**18.3**
kr-vs-kp	14	**13.4**
credit-g	**15.6**	15.7
tic-tac-toe	**26.1**	27.9
vehicle	10.3	**8.6**
vowel	8.7	**8.5**
hypothyroïde	12.5	**11.7**
segment	21.3	**20.5**
Average	15.9	15.4

5   Conclusion and Future Work

In this paper we proposed an ensemble selection method using diversity based entropy measurement combined to a genetic algorithm based strategy search. The proposed method was compared with the FES method [27] which uses a forward selection course for heterogeneous ensembles. Based on the comparison made against the criteria of performance and size of the ensemble generated, we have observed that the EMnGA method generate larger ensembles in most cases compared to FES due mainly to the strategy search method used to explores more possibilities but gives better performance.

In future work, we will compare the measure with other methods in the literature using other search strategies and will use more datasets in order to formally validate our results with appropriate statistical tests. We will also experience a measure based on diversity and performance in order to see the impact of jointly using both criteria. Finally, we intend to apply the EMnGA method in random forest ensemble simplification.

References

1. Breiman, L., Friedman, J.H., Olshen, R.A., Stone, C.J.: Classification and Regression Trees. Chapman & Hall, New York (1984)
2. Geurts, P.: Contributions to decision tree induction: bias/variance tradeoff and time series classification. Ph.D. thesis, Department of Electronical Engineering and Computer Science, University of Liège, May 2002
3. Margineantu, D., Dietterich, T.: Pruning adaptive boosting. In: Proceedings of the 14th International Conference on Machine Learning, pp. 211–218 (1997)
4. Dietterich, T.G.: Ensemble methods in machine learning. In: Kittler, J., Roli, F. (eds.) MCS 2000. LNCS, vol. 1857, pp. 1–15. Springer, Heidelberg (2000). https://doi.org/10.1007/3-540-45014-9_1
5. Banfield, R.E., Hall, L.O., Bowyer, K.W., Kegelmeyer, W.P.: A comparison of decision tree ensemble creation techniques. IEEE Trans. Pat. Recog. Mach. Int. **29**, 173–180 (2007)

6. Bian, S., Wang, W.: On diversity and accuracy of homogeneous and heterogeneous ensembles. Intl. J. Hybrid Intell. Syst. **4**, 103–128 (2007)

7. Breiman, L.: Bagging predictors. Mach. Learn. **26**(2), 123–140 (1996)

8. Schapire, R.E.: The strength of weak learnability. Mach. Learn. **5**(2), 197–227 (1990)

9. Tang, E.K., Suganthan, P.N., Yao, X.: An analysis of diversity measures. Mach. Learn. **65** (1), 247–271 (2006)

10. Breiman, L.: Random forests. Mach. Learn. **45**, 5–32 (2001)

11. Banfield, R.E., Hall, L.O., Bowyer, K.W., Kegelmeyer, W.P.: Ensemble diversity measures and their application to thinning. Inf. Fusion **6**(1), 49–62 (2005)

12. Martinez-Munoz, G., Suarez, A.: Aggregation ordering in bagging. In: International Conference on Artificial Intelligence and Applications (IASTED), pp. 258–263. Acta Press (2004)

13. Martinez-Munoz, G., Suarez, A.: Pruning in ordered bagging ensembles. In: 23rd International Conference in Machine Learning (ICML-2006), pp. 609–616. ACM Press (2006)

14. Caruana, R., Niculescu-Mizil, A., Crew, G., Ksikes, A.: Ensemble selection from libraries of models. In: Proceedings of the 21st International Conference on Machine Learning (2004)

15. Partalas, I., Tsoumakas, G., Katakis, I., Vlahavas, I.: Ensemble pruning using reinforcement learning. In: Antoniou, G., Potamias, G., Spyropoulos, C., Plexousakis, D. (eds.) SETN 2006. LNCS (LNAI), vol. 3955, pp. 301–310. Springer, Heidelberg (2006). https://doi.org/10.1007/11752912_31

16. Partalas, I., Tsoumakas, G., Vlahavas, I.: An ensemble uncertainty aware measure for directed hill climbing ensemble pruning. Mach. Learn. **81**, 257–282 (2010)

17. Fan, W., Chu, F., Wang, H., Yu, P.S.: Pruning and dynamic scheduling of cost- sensitive ensembles. In: Eighteenth National Conference on Artificial Intelligence, American Association for Artificial Intelligence, pp. 146–151 (2002)

18. Zhang, Y., Burer, S., Street, W.N.: Ensemblepruning via semi-definite programming. J. Mach. Learn. Res. **7**, 1315–1338 (2006)

19. Hernández-Lobato, D., Martínez-Muñoz, G.: A statistical instance-based pruning in ensembles of independant classifiers. IEEE Trans. Pattern Anal. Mach. Intell. **31**(2), 364–369 (2009)

20. Dai, Q.: An efficient ensemble pruning algorithm using one-path and two-trips searching approach. Knowl.-Based Syst. **51**, 85–92 (2013). ISSN 0950-7051

21. Bhatnagar, V., Bhardwaj, M., Sharma, S., Haroon, S.: Accuracy-diversity based pruning of classifier ensembles. Prog. Artif. Intell. **2**(2–3), 97–111 (2014)

22. Guo, Y., et al.: A novel dynamic rough subspace based selective ensemble. Pattern Recognit. **48**(5), 1638–1652 (2015). ISSN 0031-3203

23. Partalas, I., Tsoumakas, G., Vlahavas, I.: A study on greedy algorithms for ensemble pruning. Technical report TR-LPIS-360-12, Department of Informatics Aristotle University of Thessaloniki, Greece (2012)

24. Haque, M.N., Noman, N., Berretta, R., Moscato, P.: Heterogeneous ensemble combination search using genetic algorithm for class imbalanced dataclassification. PLoS ONE **11**(1), e0146116 (2016). https://doi.org/10.1371/journal.pone.0146116

25. Pölsterl, S., et al.: Heterogeneous ensembles for predicting survival of metastatic, castrate-resistant prostate cancer patients. F1000Research **5**(2676) (2016). *PMC*.Web 13 October 2017

26. Partalas, I., Tsoumakas, G., Vlahavas, I.: Focused ensemble selection: a diversity- based method for greedy ensemble selection. In: Ghallab, M., Spyropoulos, C.D., Fakotakis, N., Avouris, N.M. (eds.) ECAI 2008 – Proceedings of the 18th European Conference on Artificial Intelligence. Frontiers in Artificial Intelligence and Applications, Patras, Greece, vol. 178, pp. 117–121 (2008)

27. Kuncheva, L.I., Whitaker, C.J.: Measures of diversity in classifier ensembles and their relationship with the ensemble accuracy. Mach. Learn. **51**, 181–207 (2003)

28. Lerman, I., Ngouenet, F.: Algorithmes génétiques séquentiels et parallèles pour une représentation affine des proximités, Rapport de Recherche de l'INRIA Rennes - Projet REPCO 2570, INRIA (1995)

29. Asuncion, A., Newman, D.: UCI machine learning repository (2007). http://www.ics.uci.edu/»mlearn/MLRepository.html

Feature Selection and Interpretable Feature Transformation: A Preliminary Study on Feature Engineering for Classification Algorithms

Antonio J. Tallón-Ballesteros[1](✉) (iD), Milan Tuba[2] (iD), Bing Xue[3] (iD), and Takako Hashimoto[4]

[1] University of Seville, Seville, Spain
atallon@us.es
[2] Singidunum University, Belgrade, Serbia
[3] Victoria University of Wellington, Wellington, New Zealand
[4] Chiba University of Commerce, Konodai Ichikawa City, Chiba, Japan

Abstract. This paper explores the limitation of consistency-based measures in the context of feature selection. These kinds of filters are not very widespread in large-dimensionality problems. Typically, the number of selected of attributes is very small and the ability to do right predictions is a drawback. The principal contribution of this work is the introduction of a new approach within feature engineering to create new attributes after the feature selection stage. The experimentation on multi-class problems with a feature space in the order of tens of thousands shed light on that some improvements took place with the new proposal. As a final insight, some new relationships were discovered due to the combined application of feature selection and feature transformation. Additionally, a new measure for classification problems which relates the number of features and the number of classes or labels is also proposed.

Keywords: Classification · Feature engineering · Feature selection
Data mining · Feature discovery · Feature transformation

1 Introduction

Classification studies problems where every object is categorised with a label represented in a discrete domain and therefore the number of values is limited [1]. The goal of classification is to predict the output variable value of unseen data given instances with values for the input and output variable(s). We have chosen two problems from Bioinformatics where the number of features is higher than twelve thousands and the number of classes is greater or equal than seven. As we are proposing a refinement of a previous approach, we have picked up two data sets with an error rate higher than a twenty percent and in the worst scenario in the sense that we are only considering the problems in a not very fruitful initial situation. According the reported results in a previous paper [2], the feature selection based on a stochastic search procedure such as scatter search considering a consistency measure may be a fast approach although at the

© Springer Nature Switzerland AG 2018
H. Yin et al. (Eds.): IDEAL 2018, LNCS 11315, pp. 280–287, 2018.
https://doi.org/10.1007/978-3-030-03496-2_31

same time the number of selected characteristics is very low and the potential results are worse than with the application of classical correlation measures. This paper utilises as a baseline feature subset the resulting one with a scatter-search meta-heuristic with a consistency-based measure which is only applied in the training set. Data preparation is conducted by means of feature selection in order to decrease the number of input feature space. Then, the features are augmented

Our goal is to improve the averaged test accuracy and the Cohen's kappa. The rest of this paper is organized as follows: Sect. 2 remembers some concepts about feature engineering and data mining; Sect. 3 introduces the proposal; Sect. 4 describes the experimental design; Sect. 5 depicts the results; finally, Sect. 6 states the conclusions.

2 Feature Engineering and Data Mining

CRoss Industry Standard Process for Data Mining (CRISP-DM) proposes a framework to conduct data mining projects in an independent way of both the industry sector and the technology used [3]. This paper makes use of two visual tools to cover tasks from Data Analytics (DA) within Data Engineering (DE) [4] such as Waikato Environment for Knowledge Analysis (WEKA) [5] and RapidMiner [6]. Feature engineering (FE) plays an outstanding role in DA; it encompasses many fields such as feature transformation, feature generation, feature selection, feature analysis and many others. Machine learning algorithms cannot operate without data. As stated by G. Dong and H. Liu in the newly book published in 2018, "little can be achieved if there are few features to represent the underlying data objects, and the quality of the results of those algorithms largely depends on the quality of the available features" [7]. This idea has many connections with an earlier study from 2017 which put on the table the real situation that as times the feature selection may get a very small number of attributes which may not be enough to achieve reliable predictions [8]. Nowadays the data generation is going faster and faster and a data preparation task is an important step [9]. This paper puts emphasis on the feature perspective. Feature transformation (FT) is a process through which a new set of features is created [10]. Feature selection (FS) is a crucial task to conduct the training of classifiers with a reduced number of inputs and, at the same time, the outcome predictions could be more reliable. Nonetheless, a final characteristic-space with very small number of features constitutes an undesirable situation. There are some works accounting for the limited quality of the predictions and even the indiscernibility of the classes in scenarios with a very simple feature subset [8].

The bibliographical review and our previous experience motivate us to start the current research. A simple approach may be to create new features from the full feature space and then applying feature selection. Alternatively, we have typically applied feature selection and then we have applied supervised machine learning algorithms from the classification scope. Some papers aimed at increasing the final feature space combining solutions from different types of feature selection approaches [11] or even iterating in a particular feature selection method more than once [12].

Thinking abstractly about all the aforementioned ideas and considering that an interpretable feature space is convenient we opt to deepen FT which is a way to augment the feature space combining groups or subgroups of features. FT could be conducted from the starting point with the whole feature space. On the first contribution in this field, FT was thought as an initial step and the application of FS may retrieve only the important properties of a study [10]. Our particular view is to establish a trade-off between FS and FT to be applied in contexts with continuous features. It is very well-known that for most of the problems in the nature, correlation-based relationships [13] are more common that consistency-based ones [14–16].

Basically, classifiers could be grouped in decision trees, neural networks, ruled based and classifiers based on the k nearest neighbours (kNN). We have chosen two reference algorithms such as kNN and Support Vector Machine (SVM) due to its good performance and their high consideration within data mining community.

3 Proposal

The proposal is based on the application of two kinds of FE procedures. The former carries out feature selection. The latter takes as input the selected features to conduct a feature transformation combining every different pair of selected attributes and applying to both operands a binary arithmetical operator (e.g. sum) to get one new attribute for every possible pair. Then, the selected features and the transformed features are merged to establish a new final characteristic space. As the proposal framework applies an operator which keeps the interpretative perspective on the data, we have called it Feature Selection and Interpretable Feature Transformation (FS-IFT). Figure 1 depicts the approach proposed. There are no requirements for the current proposal although we would like to remark some important facts: (i) any kind of feature selection approach may be conducted in the step 1 which is related with the initial feature selection. It means that feature ranking or feature subset selection are perfectly applicable, (ii) the number of selected features must not be very high especially whether afterwards is going to be conducted in the feature transformation a pair combination of selected features; as a recommendation, no more than 25 attributes. For instance, a number of features around twenty is going to generate a number of new attributes which is close to two hundred.

This paper also introduces a new measure for classification to relate the number of features and the number of classes to characterise better classification problems. The Feature-to-class ratio (FtoC ratio) is a normalised ratio which represents how many features are approximately for every class label. The aforementioned ratio is of especial interest for feature engineering since the number of attributes after the data pre-processing is going to be altered in any way. It aids to analyse in a more detailed way the final outcome of the data preparation procedure. FtoC ratio is defined as follows:

$$Feature-to-class-ratio = No.features/No.classes \qquad (1)$$

FtoC gives a measurement between the dimensionality and the complexity in terms of number of different labels for a problem at hand. The number of patterns is also another interesting value to analyse; nonetheless we are working with Bioinformatics problems and their number is from almost one hundred to typically no more than five hundreds.

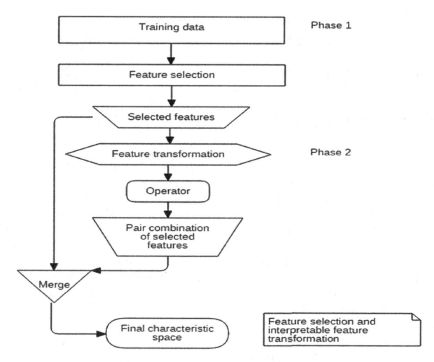

Fig. 1. Proposal framework within feature engineering.

4 Experimentation

Two challenging multiple class classification problems from Bioinformatics have been used to assess the proposal. Global Cancer Map (GCM) is a very difficult problem from the context of Bioinformatics which started to be studied in the beginning of the current century although its interest, even now, is very wide especially to the prevalence of very common diseases. GCM is populated with 190 tumor objects that have one out of the 14 possible labels such as Breast, Lung, Bladder to cite some of them in a very large dimensional space. Subtypes of Acute Lymphoblastic Leukemia (SALL) deals with leukemia and contains 327 samples distributed in the six known leukemia subtypes (T-cell, E2APBX1, TEL-AML1, MLL, BCR-ABL and hyper-diploid) and an extra category for those instances which do not belong to none of the featured subtypes. The order of magnitude of the number of attributes is exactly the same as SALL which is the order of tens of thousands. Additionally, the Feature-to-class ratio is similar and its values are in-between one and two thousands of attributes with a distance to the cut-off point of an exact thousand around two hundred. Table 1 illustrates the most representative properties of the test bed. The experimental design follows a stratified hold-out cross validation procedure with three and one quarters for the training and testing sets, respectively.

Table 1. Summary of problems.

Problem	Patterns	Features	Classes	Feature-to-class ratio	Distribution of classes
GCM	190	16063	14	1147.4	10(3), 11(8), 20(1), 22(1), 30(1)
SALL	327	12558	7	1794.0	15(1), 20(1), 27(1), 43(1), 64(1), 79(2)
Average	258.50	14310.50	10.50	1470.68	

Table 2 details the setting of the FE methods within the FS-IFT approach. Specifically, the proposal has been configured as follows: the first phase carries out feature subset selection based on a consistency-based measure with a scatter search to guide the exploration; the second phase takes as input the reduced training subset according to the previous step and creates new attributes with the arithmetical combination of the feature values that are merged to the selected attributes in the first step to create a new feature space to be used as the training set to train the classifier. As the baseline approach for feature selection, Scatter Search under a CoNsistency-based feature Selection (SS-CNS) [2] has been considered and the parameter that we have changed is the operator included in the second stage, hereinafter we refer to a concrete configuration of the proposal as FS-IFT (SS-CNS, op) where op could take as values the sum (+) or the multiplication (*). These operators are appropriate since the test bed consists of problems with continuous features. The feature engineering methods have been conducted for five different seeds to smooth the results and get reliable results due

Table 2. Feature engineering methods within feature selection and interpretable feature transformation (FS-IFT).

Phase and name	Descriptive property	Value/s
Phase 1: Feature selection	Input file	Training set
	Input	Full feature space
	Output	Subset of features
	Attribute evaluation measure	Consistency
	Type of search	Scatter search
	Population size	250
Phase 2: Feature transformation	Input file	Reduced training set
	Input	Reduced feature space
	Output	Augmented reduced feature space
	Number of attributes to act as operands	2
	Operator	Sum(+), Multiplication(*)
	Way to obtain new attributes	All possible combinations excluding self-combinations

to the stochastic nature of the first procedure and the indirect stochasticity that is inherited into the second phase from first one. The combination of a stochastic step and a non-stochastic one further than all that is incorporated in the first stage has been applied within feature selection but not in the context of FE merging to subtypes of methods [17]. Table 3 reports the average number of attributes included in the feature space in the starting point, after the first phase of FS-IFT and at the end of FS-IFT which is also related to the number of classes to get an overview about what is convenient for a concrete problem.

Table 3. Average number of attributes and feature-to-class ratio in every phase of FS-IFT.

Problem	Features	Classes	Initial feature-to-class ratio	Selected features				Feature-to-class ratio			
				FS-IFT (Phase 1)		FS-IFT (Both phases)		FS-IFT (Phase 1)		FS-IFT (Both phases)	
				Avg.	SD	Avg.	SD	Avg.	SD	Avg.	SD
GCM	16063	14	1147.4	12.6	3.65	91.0	49.76	0.9	0.26	**6.5**	3.55
SALL	12558	7	1794.0	8.8	0.84	43.4	7.89	1.3	0.12	**6.2**	1.13

5 Results

We have compared the baseline results (SS-CNS) to the new proposal in two scenarios such as FS-IFT (SS-CNS, +) and FS-IFT (SS-CNS, *). kNN (with $k = 1$) and SVM classifiers are applied to two different databases in the contexts aforementioned in Sect. 4. Both classifiers have been assessed with two performance measures that are averaged with five seeds: accuracy and Cohen's kappa (CK). Tables 4 and 5 show the test results for kNN and SVM including the mean and the Standard Deviation (SD). The sum operator helps to overcome the baseline results for GCM with kNN concerning the mean accuracy and/or the CK. There is one tie in the mean accuracy, although the SD is a bit slower what indicates more homogeneous solutions. For SALL, the improvements only happen with the multiplication operator which may reveal that the interaction of measures is more profitable to study Leukemia cases. Clearly, the new proposal is very convenient for SVM as there are gains in both scenarios for both problems. It is especially remarkable that with sum operator the improvements are very outstanding. As a final breakthrough, a Feature-to-class ratio in the environment of 1 (average values between 0.9 and 1.3) is not very promising to get accurate predictions according to the results. The experimentation has shown that with average values close to 6 the prediction ability of the classifier is considerably higher. Additionally, as one reviewer suggested, we have tried an extra feature selection procedure after FS-IFT. We have applied CoNsistency-based feature selection [15] and the number of selected attributes is even lower than the initial values after Phase 1.

Table 4. *K*NN classifier: test results.

Problem	Feature engineering procedures											
	SS-CNS				FS-IFT(SS-CNS, +)				FS-IFT(SS-CNS, *)			
	Accuracy		CK		Accuracy		CK		Accuracy		CK	
	Mean	SD	Mean	SD	Mean	SD	Mean	SD	Mean	SD	Mean	SD
GCM	41.30	3.64	0.3630	0.0390	**43.91**	3.19	**0.3910**	0.0340	41.30	3.37	0.3640	0.0350
SALL	80.98	3.24	0.7666	0.0038	80.73	3.03	0.7630	0.0360	**81.91**	3.40	**0.7770**	0.0410
Average	61.14		0.5648		62.32		0.5770		61.61		0.5705	

Table 5. SVM classifier: test results.

Problem	Feature engineering procedures											
	SS-CNS				FS-IFT(SS-CNS, +)				FS-IFT(SS-CNS, *)			
	Accuracy		CK		Accuracy		CK		Accuracy		CK	
	Mean	SD	Mean	SD	Mean	SD	Mean	SD	Mean	SD	Mean	SD
GCM	30.43	5.67	0.2200	0.0750	**33.04**	5.40	**0.2690**	0.0570	31.30	6.24	0.2350	0.0770
SALL	81.46	3.96	0.7670	0.0490	**83.90**	4.11	**0.7999**	0.0520	82.93	4.29	0.7870	0.0550
Average	55.95		0.4935		58.47		0.5345		57.12		0.5110	55.95

6   Conclusions

This paper presented a framework within feature engineering including two kinds of methods: the first step reduces the dimensionality via feature selection and the second one expands the feature space by means of feature transformation. The proposed reported very competitive results with remarkable improvements in one or two assessment measures. Additive and multiplicative relations are very noticeable. In kNN there are some cases where the sum is better, whereas in others the multiplication is better, while in SVM the sum is substantially better in all cases. Concretely, for GCM the classifier kNN obtained an improvement of 2.61 in accuracy with the + operator and in SALL it is obtained an improvement of 0.93 with * operator. With the classifier SVM, in GCM it is obtained an improvement of 2.61 as well as in SALL a gain of 2.44 with is also accompanied with very relevant overcoming in CK evaluation measure. Additionally, the feature-to-class ratio must be similar to the number of classes as happens in the second most difficult case as SALL is or close to the half as in the most challenging problem (GCM). Though the current proposal has been focused on two complex problems from Bioinformatics, we strongly believe that the proposed approach could be extensively used for other classification problems involving an important number of attributes and multiple classes and even for other classifiers. Moreover, this contribution may represent an important push for promoting the analysis of low dimensionality spaces that may be achieved after feature subset selection especially with consistency measures or sometimes with correlation metrics.

Acknowledgment. This work has been partially subsidised by TIN2014-55894-C2-R, TIN2017-88209-C2-R (Spanish Inter-Ministerial Commission of Science and Technology

(MICYT)), P11-TIC-7528 projects ("Junta de Andalucía" (Spain)) and FEDER funds. It has also been supported by the Ministry of Education, Science and Technological Development of Republic of Serbia, Grant no. III-44006.

References

1. Alpaydin, E.: Introduction to Machine Learning. MIT press, Cambridge (2014)
2. Tallón-Ballesteros, A.J., Ibiza-Granados, A.: Simplifying pattern recognition problems via a scatter search algorithm. Int. J. Comput. Methods Eng. Sci. Mech. **17**(5–6), 315–321 (2016)
3. Wirth, R., Hipp, J.: CRISP-DM: towards a standard process model for data mining. In: Proceedings of the 4th International Conference on the Practical Applications of Knowledge Discovery and Data Mining, pp. 29–39 (2000)
4. Cho, S.-B., Tallón-Ballesteros, Antonio J.: Visual tools to lecture data analytics and engineering. In: Ferrández Vicente, J.M., Álvarez-Sánchez, J.R., de la Paz López, F., Toledo Moreo, J., Adeli, H. (eds.) IWINAC 2017. LNCS, vol. 10338, pp. 551–558. Springer, Cham (2017). https://doi.org/10.1007/978-3-319-59773-7_56
5. Frank, E., Hall, M., Holmes, G., Kirkby, R., Pfahringer, B., Witten, I.H., Trigg, L.: Weka-a machine learning workbench for data mining. In: Maimon, O., Rokach, L. (eds.) Data Mining and Knowledge Discovery Handbook, pp. 1269–1277. Springer, Boston (2009). https://doi.org/10.1007/978-0-387-09823-4_66
6. Akthar, F., Hahne, C.: Rapidminer 5 operator reference. Rapid-I GmbH **50**, 65 (2012)
7. Dong, G., Liu, H.: Feature Engineering for Machine Learning and Data Analytics. CRC Press, Boca Raton (2018)
8. Tallón-Ballesteros, A.J., Riquelme, J.C.: Low dimensionality or same subsets as a result of feature selection: an in-depth roadmap. In: FV, J.M., Álvarez-Sánchez, J.R., de la Paz López, F., Toledo Moreo, J., Adeli, H. (eds.) IWINAC 2017. LNCS, vol. 10338, pp. 531–539. Springer, Cham (2017). https://doi.org/10.1007/978-3-319-59773-7_54
9. Tallón-Ballesteros, A.J., Li, K. (eds.): Fuzzy Systems and Data Mining III: Proceedings of FSDM 2017, vol. 299. IOS Press, Amsterdam (2017)
10. Liu, H., Motoda, H.: Feature transformation and subset selection. IEEE Intell. Syst. **2**, 26–28 (1998)
11. Tallón-Ballesteros, A.J., Riquelme, J.C., Ruiz, R.: Merging subsets of attributes to improve a hybrid consistency-based filter: a case of study in product unit neural networks. Connect. Sci. **28**(3), 242–257 (2016)
12. Tallón-Ballesteros, A.J., Correia, L., Xue, B.: Featuring the attributes in supervised machine learning. In: de Cos Juez, F., et al. (eds.) HAIS 2018. Lecture Notes in Computer Science, vol. 10870, pp. 350–362. Springer, Cham (2018). https://doi.org/10.1007/978-3-319-92639-1_29
13. Hall, M.A.: Correlation-based feature selection for machine learning (1999)
14. Shin, K., Kuboyama, T., Hashimoto, T., Shepard, D.: sCwc/sLcc: highly scalable feature selection algorithms. Information **8**(4), 159 (2017)
15. Shin, K., Xu, X.M.: Consistency-based feature selection. In: Velásquez, J.D., Ríos, S.A., Howlett, R.J., Jain, L.C. (eds.) KES 2009. LNCS (LNAI), vol. 5711, pp. 342–350. Springer, Heidelberg (2009). https://doi.org/10.1007/978-3-642-04595-0_42
16. Arauzo-Azofra, A., Benitez, J.M., Castro, J.L.: Consistency measures for feature selection. J. Intell. Inf. Syst. **30**(3), 273–292 (2008)
17. Tallón-Ballesteros, Antonio J., Correia, L., Cho, S.-B.: Stochastic and non-stochastic feature selection. In: Yin, H., et al. (eds.) IDEAL 2017. LNCS, vol. 10585, pp. 592–598. Springer, Cham (2017). https://doi.org/10.1007/978-3-319-68935-7_64

Data Pre-processing to Apply Multiple Imputation Techniques: A Case Study on Real-World Census Data

Zoila Ruiz-Chavez[1]([✉]), Jaime Salvador-Meneses[1], Jose Garcia-Rodriguez[2], and Antonio J. Tallón-Ballesteros[3]

[1] Universidad Central del Ecuador, Ciudadela Universitaria, Quito, Ecuador
{zruiz,jsalvador}@uce.edu.ec
[2] Universidad de Alicante, Ap. 99., 03080 Alicante, Spain
jgarcia@dtic.ua.es
[3] University of Seville, Seville, Spain
atallon@us.es

Abstract. Improving accuracy or reducing computational cost are the main approaches of machine learning techniques, but it depends heavily on the test data used. Even more so when it comes to from real-world data such as censuses, surveys or tokens that contain a high level of missing values. The data absence or presence of outliers are problems that must be treated carefully prior to any process related to data analysis. The following work presents an overview of data pre-processing and aims at presenting the steps to follow prior to process large volumes of high-dimensionality data with categorical variables. As part of the dimensionality reduction process, when there is a high level of missing values present in one or more variables, we use the Pairwise and Listwise Deletion methods. Thus, the generation of m-clusters using the Kohonen Self-Organizing Maps (SOM) algorithm with H2O over R is also considered as a division of data into similar groups, which are used as cluster to apply Multiple Imputation algorithms, creating different m-values to impute a missing value.

Keywords: Machine learning · Data preparation · Data mining
Imputation methods

1 Introduction

Any study involving data analysis undoubtedly requires a data pre-processing stage. In order to obtain relevant information and avoid biases, we must take into account some characteristics of the data, such as the type of variable to be used (also called attribute), the data source, the data frequency, among others. The preparation of data prior to the application of statistics techniques or machine learning techniques, requires extreme care and attention to ensure the quality of the data to obtain adequate results. Another reason to perform a data pre-processing step is that usually when dealing with real-world data, these present

© Springer Nature Switzerland AG 2018
H. Yin et al. (Eds.): IDEAL 2018, LNCS 11315, pp. 288–295, 2018.
https://doi.org/10.1007/978-3-030-03496-2_32

problems that must be solved before analysis. In our work we use census data as a source of information that allows us to make decisions based on projections [2].

Census information works with what is called "variable", and its main characteristic is to have a high dimensionality (great number of variables generally of categorical type). As it is information from censuses or surveys, it is necessary to take into account a very common problem related to this type of information: the presence of *missing values* or outliers [1], which can introduce bias into the estimation or significantly reduce the sample size. Dealing with these problems is part of any study, imputing these values by replacing them with machine learning techniques requires a training stage of the algorithm to be used. Prior to train the algorithm, we need a reliable data set to calculate the algorithm's accuracy. This leads us to review the steps we must take at the data pre-processing stage to achieve this objective.

This paper is organized as follows: Sect. 2 summarizes the work done in the area, Sect. 3 a brief description of the techniques and data set used is presented, Sect. 4 describes the steps to be followed in the data analysis prior to the extraction of information, classification or data imputation and it describes the pre-processing carried out to obtain a complete data set, Sect. 5 explains the procedure to generate complete *m sets* using SOM to apply multiple imputation algorithms and, finally, some conclusions and future work are stated in Sect. 6.

2   Related Work

There are studies oriented to the analysis of data pre-processing techniques such as [4], which describes 14 techniques used for this task. One of the problems faced at this stage is the lack of data, very common in real data. There are three different types: Completely Random Absence (MCAR), Random Absence (MAR) and Non-Random Absence (NMAR) [1]. The amount of missing data present in the attributes can lead to considerable bias or unacceptably reduce the sample size. To address the problem of missing values, we will review two widely discussed and used methods: Listwise Deletion and Pairwise Deletion [3]. From the work written by Rubin (1976, 1987) [11], it is clear that it is necessary to deal with data imputation methods instead of eliminating all observations that contain missing data. The considerable sample decrease when using this alternative can produce biased or poor quality results, in the work of Cheema-Jehanzeb in 2014, a revision of some methods used to impute data is carried out, as an alternative to the Listwise Deletion method [3]. Similarly, we can find many studies that focus on finding the best alternatives when imputing data –different from the traditional ones such as replacing the missing value by the median or mode depending on the type of variable (numerical or categorical)–, as the Fuzzy K-means Clustering method for data imputation proposed by Li and Deougun [7]. Most studies, that use different machine learning algorithms to process the data, previously analyze in the pre-processing stage problems with the data that can be solved by eliminating the sensitive variables, changing the sample to be used or selecting the representative variables allowing the dimensionality reduction [6]. Any alternative chosen is intended to ensure the quality of the information to be processed. The amount of missing data present in the attributes can lead to considerable bias or unacceptably reduce the sample size. To address the problem of missing values, we will review two widely discussed and used methods: Listwise Deletion and Pairwise Deletion [3].

3   Concepts of the Techniques and Data Sets Used

Before analyzing the data, it is advisable to review the basic concepts of the techniques that we will use in the following sections.

- **Pairwise Deletion:** This method comprises using all the available data in the analysis of each variable, which is why it is known as the Available Data Method. It is possible to calculate covariances between each pair of variables using all available cases. If one of the variables contains too much missing data, the variable is deleted [8].
- **Listwise Deletion:** This method is based on discarding rows that contain at least one missing value. This method is also called the Full Case Method (McKnight et al. 2007), as it considers only the complete subset of data. We must bear in mind that if the data come from a probability sample, this method is not the most appropriate. However, the biggest drawback occurs when the data contains a large number of variables hence the complete subset that is obtained may be insufficient or may obtain biased results [10].
- **Self-Organizing Maps (SOM):** It is a neural network model formed by a set of nodes organized in a matrix. SOM uses competitive learning, which defines a spatial neighborhood for each output unit while preserving its topology; nearby input patterns produce nearby output units on the map [5].

3.1   Data Set

In order to carry out studies that involve data analysis, either by statistical methods or machine learning techniques, there are fundamental steps that must be taken to obtain adequate results. These steps can be executed as long as we can assume that the data are concrete, complete and correct; in most cases, the data require a prior process to be considered acceptable for analysis. Therefore, data pre-processing is essential to determine the validity of the results obtained from the prediction or whether were importantly altered because not having conducted an adequate treatment of the raw data.

For this study, demographic information corresponding to the 2010 population and housing census of Ecuador is used. In this particular case, one province was selected for each region, with a total population of 1048575, according to the 2010 census. This information has 154 variables, mostly categorical, grouped into:

- Housing: Information associated with the dwelling (28 attributes)

- Home: Household information (26 attributes)
- Individual: Information associated with people (95 attributes)
- Geographic Information: DPA Administrative Politics Division Information (5 attributes).

Since we have data from censuses with high dimensionality, to solve two very common problems in this type of data, some steps should be applied in the pre-processing stage: dimensionality reduction and sample preparation for training.

4   Pre-processing to Generate a Complete Data Set

Four phases can be defined within the data pre-processing: Cleanliness, Integration, Reduction and Transformation. Each one of them is oriented to treat the data in such a way that, at the end of the procedure, the data we obtain are reliable and useful for the study proposed: as a study case, the categorical variable *"graesc"* for the prediction has been selected. This variable has 25 different categories (1 to 25), this represents a greater difficulty in predicting absent values. In the pre-processing stage, the information was subjected to the following phases:

Integration: Initially all available variables were taken from the four groups, described in Sect. 3. DPA is made up of 4 regions which in turn are made up of provinces (24 in total). The subset of data corresponding to the Insular Region and three provinces has been selected, one for each region: Imbabura, Napo and Los Ríos. As a result of this stage we have a data set with 154 attributes and 1048575 observations. We then analyze the data obtained.

Cleanliness: In order for the data to be consistent, it has been detected if there are lost, inconsistent or out of range values to correct, impute or, if necessary, delete. Let us take as an example the variable p23 described in Table 2. This variable does not contain lost values, however, a simple frequency shows us values outliers (1795 cases), these cases correspond to people who do not respond, so they are considered as lost values. In the same way, each of the variables is analyzed, the valid ranges are defined for each one of them and when cases are detected whose values are out of range, they are replaced with NA (not applicable).

Reduction: An important part of data analysis is to have a complete data set that allows us to determine the accuracy of each algorithm prior to the classification or imputation of data. Choosing the representative variables allows to reduce the complexity and computational cost within any model. Before executing a Feature Selection algorithm as part of the pre-processing, we perform an analysis of the number of missing values present in each attribute.

Then, using the Pairwise Deletion method, the variables for which there are a high number of NA values were eliminated (the proportion of data absence is so large that the variable should not be considered). Table 1 shows some features with a high percentage of missing values. It has been chosen 76% as cut-off point, hence variables with a higher percentage of missing values were discarded (63 variables); in consequently we obtain a data set with 91 variables.

Table 1. Percentage of missing values per attribute

Name	Description	Percentage NA
p143p	Father speaks indigenous language	99.40
p144m	Mother speaks Spanish	99.77
p151	Speak indigenous language	92.74
p153	Speak foreign language	98.61
p154	Does not speak any language	96.96
p17	Nationality or indigenous people to which they belong	89.97

As a next step, we apply the Listwise Deletion method, which eliminates rows where at least one column contains NA values. With this we obtain a complete data set with 182323 rows and 91 variables. The available information is used to consider the variables that may influence the calculation of the *graesc* variable. We perform selection of significant variables by applying Linear Regression through backward elimination (P.Values > 0.05), recursively eliminating the less significant variables within the model. This procedure reduces the dimension to 18. We must keep in mind that the generated data set is used in the training stage of the algorithms, then the prediction is made on the original data containing missing values. Table 2 depicts the list of selected variables that contain missing values in the original data set:

Table 2. Selected attributes after Listwise Deletion

Variable	Description	Initial NA (%)
p08	Has a permanent disability for more than one year?	5496 (3.01)
p23	What is the highest level of education you are attending?	1795 (0.9)
p24	What is the highest grade, course or year you passed?	12263 (6.7)
p36	How many live-born children have you had in your lifetime?	1292 (0.7)
p37	Total number of children currently alive	95050 (52.1)
p38	At what age did you have your first child?	95050 (52.1)
graesc	Level of education	20131 (11.0)

Transformation: Once a complete and concrete data set has been obtained, we carry out the necessary transformations depending on the algorithm to be used, so that the processing can be carried out in an adequate way. In the case of SOM all variables are transformed to binary.

Finally, as a result of the pre-processing stage we obtain a complete data set with 182323 observations and 18 dimensions. This data set allows us to train the different machine learning algorithms and calculate the accuracy of each one of them.

5   Pre-processing to Apply Multiple Imputation

One of the ways of assigning missing values is the Multiple Imputation (MI) which consists of assigning several values to each missing data ($m > 1$); to do that it generates m complete data sets, in each set the value of the imputed data usually varies, which results in a procedure with a certain level of complexity. At the end, an ordinary statistical analysis of the m complete data sets is performed and the results are combined using the Little and Rubin combination rules (1987). A variant of this method is Multiple Imputation regression based, each of the imputations being based on regression parameters from the observed data [9]. These regression imputation parameters would be assumed to be the true population parameters.

The proposal is based on the generation of m data sets (clusters) through Auto-Organizing Networks (SOM), achieving an optimal number of clusters by mapping the original data to a smaller set of values, close to the original values (Codebook).

SOM: Next we present the process used to generate clusters using SOM, which only operates with binary variables, so dummy variables are created for each categorical variable. The execution of SOM was parameterized with the following values: Topology: rectangular, Grid size: 5×5, Number of iterations: 100, Learning rate: 0.05–0.01.

The approximate time in the execution of the model is 89 seconds. We use SOM in Clustering mode, which consists of using the assignment of each observation to the representative vector as a cluster. Each cell corresponds to a cluster and each observation is assigned to the nearest vector inside the grid, 25 clusters are generated.

Figure 1 shows the relationship between the count map and the distribution by cell, which allows a better analysis of the distribution of the variables according to the generated clusters.

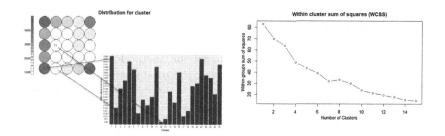

Fig. 1. Count map-clusters and WCSS

After analyzing the error distribution graph according to the number of clusters (WCSS), we find a break point in the value corresponding to the number of clusters equal to 7. This value is used to generate the codebooks that best represent groups with similar characteristics and that are taken as complete data sets to apply Multiple Imputation algorithms. With this process we generate 7 clusters that allow us to generate m different values for the variable to be imputed. Figure 2 represents the data distribution in these clusters.

Fig. 2. Clusters distribution (Codebook)

5.1   Results

The purpose of this work is to describe the steps in the pre-processing stage, however, we will quickly mention the result of the Random Forest execution to impute the data tha was generated at this stage. Once the data is prepared, Random Forest (RF) is applied for the imputation of the *graesc* variable. The algorithm is trained in H2O over R, with the following parameters: ratio = 0.1, number of trees = 50, two sets are generated, the training set (boostrap sample) and the validation set (out-of-bag OOB). The division into subsets of data is done with the clusters generated by SOM. In each cluster the elements are similar, each set generated contributes to reduce the correlation of the classifiers but maintains their precision and therefore reduces the error of the imputation. RF is executed in two steps, the first one obtains the ranking of importance of each of the variables and the second one carries out the imputation of the variable according to the other variables. The result is an estimated error corresponding to 17.07%, the error that was obtained prior to the generation of clusters, execution with original dataset was 23.66%.

6   Conclusions and Future Work

Although there is no established sequence or methodology for data pre-processing, any study should take due care at the pre-processing stage to obtain accurate, concrete and reliable data on which to apply any data analysis technique.

The definition of the steps to follow before data analysis, is more a strategy result of the experience that is had in the data analysis field.

The lack of adequate approaches when preparing the data prior to the classification, clustering or imputation of data causes problems when applying the algorithms or the results obtained are not the most reliable.

As future work, we plan to apply different machine learning techniques for data imputation, allowing us to compare the results obtained with the original data and those resulting from the proposed pre-processing.

References

1. Bar, H.: Missing data–mechanisms and possible solutions. Cultura y Educación **29**(3), 492–525 (2017)
2. Chackiel, J.: Métodos de estimaciones demográficas de pueblos indígenas a partir de censos de población: La Fecundidad y la Mortalidad. Pueblos indigenas y afrodescendientes de América Latina y el Caribe: relevancia y pertinencia de la informacion sociodemografica para politicas y programas, p. 30 (2005)
3. Cheema, J.R.: A review of missing data handling methods in education research. Rev. Educ. Res. **84**(4), 487–508 (2014)
4. Famili, A., Shen, W.-M., Weber, R., Simoudis, E.: Data preprocessing and intelligent data analysis. Intell. Data Anal. **1**(1), 3–23 (1997)
5. Fessant, F., Midenet, S.: Self-organising map for data imputation and correction in surveys. Neural Comput. Appl. **10**(4), 300–310 (2002)
6. Kamiran, F., Calders, T.: Data preprocessing techniques for classification without discrimination. Knowl. Inf. Syst. **33**(1), 1–33 (2012)
7. Li, D., Deogun, J., Spaulding, W., Shuart, B.: Towards missing data imputation: a study of fuzzy k-means clustering method. In: Tsumoto, S., Słowiński, R., Komorowski, J., Grzymała-Busse, J.W. (eds.) RSCTC 2004. LNCS (LNAI), vol. 3066, pp. 573–579. Springer, Heidelberg (2004). https://doi.org/10.1007/978-3-540-25929-9_70
8. Myers, T.A.: Goodbye, listwise deletion: presenting hot deck imputation as an easy and effective tool for handling missing data. Commun. Methods Meas. **5**(4), 297–310 (2011)
9. Newman, D.A.: Longitudinal modeling with randomly and systematically missing data: a simulation of ad hoc, maximum likelihood, and multiple imputation techniques. Organ. Res. Methods **6**(3), 328–362 (2003)
10. Nishanth, K.J., Ravi, V.: Probabilistic neural network based categorical data imputation. Neurocomputing **218**, 17–25 (2016)
11. Rubin, D.B.: Multiple Imputation for Nonresponse in Surveys, vol. 81. Wiley, New York (2004)

Imbalanced Data Classification Based on Feature Selection Techniques

Paweł Ksieniewicz[(✉)] and Michał Woźniak

Department of Systems and Computer Networks,
Wrocław University of Science and Technology, Wrocław, Poland
{pawel.ksieniewicz,michal.wozniak}@pwr.edu.pl

Abstract. The difficulty of the many classification tasks lies in the analyzed data nature, as disproportionate number of examples from different class in a learning set. Ignoring this characteristics causes that canonical classifiers display strongly biased performance on imbalanced datasets. In this work a novel classifier ensemble forming technique for imbalanced datasets is presented. On the one hand it takes into consideration selected features used for training individual classifiers, on the other hand it ensures an appropriate diversity of a classifier ensemble. The proposed method was tested on the basis of the computer experiments carried out on the several benchmark datasets. Their results seem to confirm the usefulness of the proposed concept.

Keywords: Machine learning · Classification
Imbalanced data · Feature selection · Random search

1 Introduction

Most of classifier training methods assume that the numbers of objects from each classes are roughly equal in a learning set. However, in many real-life decision tasks this assumption is not fulfilled. We may deal with examples from classes being abundant and easy to collect and with the classes where number of examples is small and hard to access [1,11]. Therefore, there is a need of constructing effective predictive systems which can take into consideration imbalanced data distributions [3]. Let us present shortly the main groups of algorithms in imbalanced data classification.

Data Preprocessing. Such methods modify the learning set, before a classifier is being trained [12]. They should manipulate learning examples to obtain a balanced dataset. One may achieve this by either removing samples from the majority classes (*undersampling*), or adding new object from the minority ones (*oversampling*). One have also mention techniques of dimensionality reduction as feature selection which may be also applied to this task [4].

Algorithm-Level Methods. They modify the classifier learning procedure to take into consideration imbalanced data distributions. Usually, they use non-symmetric loss-function [7] to assign higher cost to the error committed on

© Springer Nature Switzerland AG 2018
H. Yin et al. (Eds.): IDEAL 2018, LNCS 11315, pp. 296–303, 2018.
https://doi.org/10.1007/978-3-030-03496-2_33

minority class objects. Another approaches employ one-class classifier learning techniques, where a given class is learned only and the objects which do not belong to it are treated as outliers.

Hybrid Solutions. They are trying to exploit the strengths of the previously discussed methods and to combine them with other techniques. Usually ensemble learning is used [13], which is able to train set of diverse individual predictors, which may take into consideration the data imbalance and propose such combination rule which can make a high quality decision on complex data.

In this paper, we introduce a novel hybrid technique that employs feature selection techniques to train a pool of individual classifiers used by a classifier ensemble. To avoid the overfitting a regularization techniques are used which on the one hand ensures that the pool of classifier is diverse, i.e., subsets of features should be different for each classifier and on the other hand the number of features used by all individuals should be as small as possible. To train the pool we use simple techniques based on random search, but experimental study carried out on a number of benchmarks prove that the proposed method is able to return satisfactory performance.

2   Proposed Algorithm

Selecting appropriate set of the feature is an important data preprocessing step in classifier learning [8] and Chawla et al. [4] underlined its crucial role when classification model is train on the basis of imbalanced data. Traditional feature selection techniques usually use criterion based on the accuracy with a factor responsible for regularization, i.e., discourages learning too complex model to avoid the overfitting. While the methods dedicated for imbalanced data [9,14] usually employs metrics related with binary problem, as *g-mean* [6].

To present the proposed solution, firstly let us formulate the classification problem.

2.1   Problem Formulation

Classifier Ψ makes a decision by assigning an observed object into one of predefined classes derived from the set of possible labels $\mathcal{M} = \{1, 2, ..., M\}$ [7]. Each object is described by the set of attributes (features) gathered in the feature vector x belonging to d dimensional feature space $\mathcal{X}$

$$x = [x^1, x^2, \ldots, x^d]^T \in \mathcal{X} \subseteq \mathcal{R}^d, \tag{1}$$

The aim of a feature selection algorithm is to choose k valuable features only to avoid so-called *course of dimensionality* [5].

$$x = \begin{bmatrix} x^{(1)} \\ x^{(2)} \\ ... \\ x^{(d)} \end{bmatrix} \rightarrow \overline{x} = \begin{bmatrix} \overline{x}^{(1)} \\ \overline{x}^{(2)} \\ ... \\ \overline{x}^{(k)} \end{bmatrix}, \; k < d \tag{2}$$

Let us also define a pool of individual classifiers

$$\Pi = \{\Psi_1, \Psi_2, \ldots, \Psi_K\} \tag{3}$$

where Ψ_k denotes the k-th elementary classifier. To ensure diversity of the pool we will train the individuals on the basis of the different set of features. Let's propose the following representation of the classifier pool Π as word of bits:

$$\Pi = \left[\left[b_1^1, b_1^2, \ldots, b_1^d\right] \left[b_2^1, b_2^2, \ldots, b_2^d\right] \ldots \left[b_K^1, b_K^2, \ldots, b_K^d\right]\right] \tag{4}$$

where b_i^j denotes if the jth feature is used by the ith classifier.

2.2 Criterion

In our algorithm we use the following optimization criterion based on *Balanced Accuracy*

$$Q(\Pi) = BAC(\Pi) - \alpha * \frac{no - features(\Pi)}{d} + \beta * \frac{av - Hamming(\Pi)}{d} \tag{5}$$

where $BAC(\Pi)$ denotes balanced accuracy of the classifier ensemble based on the ensemble represented by Π. The first regularization factor $no - features(\Pi)$ is responsible for the number of the selected features, i.e., number of features used by all individual in the ensemble, while the second regularization factor $av - Hamming(\Pi)$ is the average Hamming distance between the words represented individuals in Π. It is a kind of a diversity measure [13], which encourages to select different features by different individuals. α and β are the parameters of the algorithm, which should be set experimentally.

2.3 Algorithm Description

Firstly, the algorithm randomly generates the population of ensembles

$$Population = \{\Pi_1, \Pi_2, \ldots, \Pi_S\} \tag{6}$$

A size of the *Population* is an input parameter. Its value S is set arbitrary, but we have to take into consideration, that on the one hand the larger S guarantees the more comprehensive optimization, but on the other hand, the larger S requires the higher computational effort.

Individuals in the population are evaluated by criterion Eq. 5 calculated on the basis of Algorithm 1 using samples stored in the training set. We decide to select the best evaluated ensemble only.

The *combination rule* of chosen ensemble is carried out by *averaging* the *support vectors* received from the members of a pool. It is important, that for such approach, it is necessary to use a *probabilistic classification model*. Three combination rules are proposed for further analysis:

1. **R**—basic accumulation of support without weighing the committee members.

Algorithm 1. Criterion count

1: **Input:** pool of individual classifiers Π, training set $\mathcal{TS}$
2: **Parameters:** α, *beta*
3: **Output:** value of criterion 5 for Π
4:
5: $counter \leftarrow 0$
6: $nobits \leftarrow 0$
7: $word \leftarrow [00..0]$
8: **for** $i \leftarrow 1$ **to** $K - 1$ **do**
9:     **for** $j \leftarrow i$ **to** K **do**
10:         $counter \leftarrow counter + 1$
11:         $nobits \leftarrow nobits +$ number of bits of $\left[b_i^1, b_i^2, ..., b_i^d\right] XOR \left[b_j^1, b_j^2, ..., b_j^d\right]$
12:     **end for**
13:     $word \leftarrow wordOR \left[b_i^1, b_i^2, ..., b_i^d\right]$
14: **end for**
15: $word \leftarrow wordOR \left[b_K^1, b_K^2, ..., b_K^d\right]$
16: $no - features \leftarrow$ number of bits in $word * \frac{1}{d}$
17: $av - Hamming \leftarrow \frac{nobits}{counter*d}$
18: $BAC \leftarrow$ balanced accuracy of Π calculated on $\mathcal{TS}$
19: $criterion \leftarrow BAC - \alpha * no - used - features + \beta * av - Hamming - dist$
20: **return** $criterion$

2. **W**— weighted aggregation, where weights are proportional to balanced accuracy values achieved by individual classifiers.
3. **N**—weighted aggregation, where weights are proportional to balanced accuracy values achieved by individual classifiers and additionally weights are subjected to *MinMax* scaling.

3   Experimental Study

Experimental investigations, backed up with statistical analysis of the results, were conducted to evaluate the practical usefulness of the proposed strategy. In the remainder of this section we describe set-up of the study, present obtained results and discuss achieved outcomes.

3.1   Set-Up

For the experimental evaluation of the proposed method, a series of benchmark datasets available on the KEEL repository [2] were used. Selection was made to ensure wide scope of 35 binary problems with *Imbalance Ratio* IR varying from 1 to around 40. The overview of chosen datasets, informing about their IR and number of features, was included in Table 1.

To allow a reliable comparison of literature methods, datasets from KEEL repository are pre-divided into folds. It led to employ *k-fold cross-validation* with 5 folds in the experimental procedure. Due to strong bias of regular classification metrics towards majority class, to ensure reliable results, all scores are presented as *balanced accuracy*, according to its implementation from the development version (0.20.dev0) of the *scikit-learn* library [10].

Implementation of the experimental procedure, as well as the implementation of the method itself, has been prepared according to the *scikit-learn* library API, using *Gaussian Naive Bayes* as a base classifier. Besides the variations of a method, to provide a comparative result, each problem was also evaluated on a full-featured representation of a dataset. To analyze a paired dependency between the classifiers outputs, the signed-rank *Wilcoxon* test was employed.

The implementation of the method proposed in following paper, as well as the script allowing to reconstruct conducted research can be found in repository[1].

3.2   Results

First step of experimental evaluation was optimization procedure to obtain the best α and β values in the context of *balanced accuracy*. It has been conducted with a *Grid Search* approach, analyzing 7 values evenly dividing the range from 0 to 1. Example visualizations of results for three datasets are presented on Fig. 1. Presentation for all datasets is available at website[2].

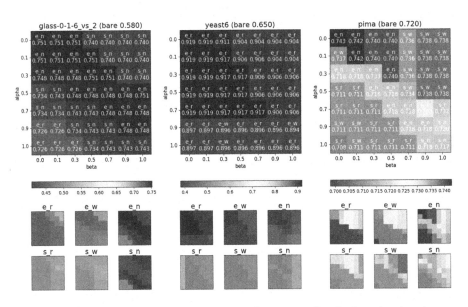

Fig. 1. Examples of α and β influence on classification quality for best (top) and every (bottom) approach. Blue indicates result worse than full-featured classification, red – a better result. (Color figure online)

[1] https://github.com/w4k2/ideal2018.
[2] http://w4k2.github.io/ideal2018.

Table 1. Balanced accuracy scores obtained with the optimized hyperparameters α and β on datasets selected to experimental evaluation. **Full** stands for the results of the classifier using all features, **Ensemble** stands for results of classifier ensemble using different combination rules described in Sect. 2.3, and **Best in ensemble** stands for balance accuracy of the best individual in the ensemble.

Dataset	IR	F.	α	β	Full	Ensemble			Best in ensemble		
						E_R	E_W	E_N	S_R	S_W	S_N
australian	1	14	.0	.0	0.777	0.852	0.855	0.877	0.878	0.876	0.861
heart	1	13	.0	.0	0.838	0.870	0.878	0.874	0.847	0.847	0.868
glass0	2	9	.3	.5	0.700	0.746	0.746	0.749	0.750	0.750	0.763
glass1	2	9	.1	.0	0.671	0.721	0.719	0.710	0.738	0.723	0.732
pima	2	8	.0	.0	0.720	0.736	0.737	0.743	0.731	0.732	0.736
wisconsin	2	9	.0	.0	0.969	0.976	0.976	0.976	0.967	0.967	0.974
yeast1	2	8	.0	.0	0.519	0.695	0.699	0.680	0.654	0.659	0.660
glass0123vs456	3	9	.0	.0	0.869	0.891	0.898	0.910	0.900	0.910	0.910
hepatitis	5	19	.0	.0	0.687	0.880	0.903	0.877	0.872	0.872	0.881
glass6	6	9	.0	.0	0.891	0.939	0.942	0.945	0.945	0.959	0.959
yeast3	8	8	.0	.0	0.605	0.904	0.895	0.915	0.841	0.840	0.813
glass015vs2	9	9	.0	.0	0.519	0.711	0.728	0.765	0.691	0.696	0.710
glass04vs5	9	9	.0	.0	0.994	0.994	0.994	0.994	0.994	0.994	0.994
yeast0256vs3789	9	8	.0	.3	0.670	0.689	0.689	0.737	0.753	0.748	0.771
yeast02579vs368	9	8	.0	.0	0.577	0.912	0.911	0.900	0.878	0.878	0.894
yeast0359vs78	9	8	.0	.0	0.557	0.668	0.662	0.621	0.607	0.600	0.607
yeast05679vs4	9	8	.0	.0	0.504	0.780	0.763	0.720	0.706	0.702	0.710
yeast2vs4	9	8	.3	.0	0.561	0.897	0.887	0.892	0.838	0.904	0.885
glass016vs2	10	9	.0	.0	0.580	0.726	0.726	0.751	0.705	0.700	0.731
vowel0	10	13	.5	.3	0.917	0.898	0.914	0.911	0.924	0.929	0.933
glass0146vs2	11	9	.1	.0	0.577	0.747	0.761	0.739	0.724	0.746	0.773
glass06vs5	11	9	.0	.0	0.945	0.995	0.995	0.995	0.960	0.960	0.995
glass2	12	9	.0	.0	0.591	0.767	0.775	0.747	0.718	0.721	0.721
shuttlec0vsc4	14	9	.0	.0	0.991	1.000	1.000	1.000	1.000	1.000	1.000
glass4	15	9	.1	.0	0.587	0.609	0.609	0.718	0.768	0.766	0.753
pageblocks13vs4	16	10	.0	.0	0.763	0.786	0.866	0.949	0.867	0.879	0.928
glass016vs5	19	9	.0	.0	0.941	0.991	0.991	0.989	0.989	0.989	0.989
shuttlec2vsc4	20	9	.0	.0	0.996	1.000	1.000	1.000	0.996	0.996	1.000
yeast1458vs7	22	8	.0	.0	0.547	0.592	0.588	0.574	0.556	0.556	0.569
glass5	23	9	.0	.0	0.938	0.988	0.988	0.988	0.988	0.988	0.988
yeast2vs8	23	8	.0	.0	0.657	0.799	0.810	0.774	0.774	0.774	0.774
yeast4	28	8	.0	.0	0.551	0.817	0.783	0.797	0.679	0.670	0.651
yeast1289vs7	31	8	.0	.0	0.544	0.683	0.701	0.706	0.629	0.628	0.606
yeast5	33	8	.0	.0	0.831	0.963	0.964	0.973	0.954	0.947	0.945
yeast6	41	8	.0	.0	0.650	0.919	0.905	0.903	0.821	0.857	0.858

The results of the evaluation after optimization procedure are presented in Table 1, which has been divided to present a *balanced accuracy* obtained on different variations of the method, using whole ensemble or just its best member, according to the different fusers (R – regular, W – weighted and N – normalized weights). Such division led to the number of 6 analyzed approaches.

Scores for the method were supplemented by the quality of a single model trained on a whole possible feature space. The green color in table indicates the statistical dependency to the best result and underline – the highest *balanced accuracy* obtained on a given dataset.

The presented results clearly showed that feature selection plays important role for imbalanced data classification. Our proposition usually outperforms the results obtained by the classifier using the whole set of features. It also behaves better (22 out of 35 datasets) than the best classifier in the pool. It has been probably caused by very naive optimization method (random search) used in this work.

4   Conclusions and Future Directions

The novel hybrid classification method for imbalanced data classification was presented. It employs ensemble learning to increase performance (*balanced accuracy*) of the combined classifier. To ensure the appropriate level of diversity, each individual is trained on the basis of selected features. Nevertheless, in contrast with well-known methods based on randomly chosen features (as *Random Subspaces*), the choice of the features is the results of the optimization procedure. The optimization criterion takes into consideration not only the performance of the classifier, but to protect against *overfitting* it encourages to build the ensemble of diverse individuals which do not use too many features. Additionally, we observed that the proposed method can significantly outperform the classifier based on whole set of features.

As the future works we are going to use more sophisticated optimization procedure based on genetic approach, but we also realize that it will negatively impact computational complexity, therefore we will focus on the method which can be run in distributed computing systems as GPU or SPARK. Additionally, we plan to definitely extend the scope of experiments to compare our methods with other methods as *Random Subspaces* or *Decision Forrest*.

Acknowledgments. This work was supported by the Polish National Science Center under the grant no. UMO-2015/19/B/ST6/01597 as well as Statutory Found of the Faculty of Electronics, Wroclaw University of Science and Technology.

References

1. Ahmed, F., Samorani, M., Bellinger, C., Zaïane, O.R.: Advantage of integration in big data: feature generation in multi-relational databases for imbalanced learning. In: 2016 IEEE International Conference on Big Data, BigData 2016, Washington DC, USA, 5–8 December 2016, pp. 532–539 (2016)
2. Alcalá-Fdez, J., Fernández, A., Luengo, J., Derrac, J., García, S., Sánchez, L., Herrera, F.: Keel data-mining software tool: data set repository, integration of algorithms and experimental analysis framework. J. Multiple-Valued Logic Soft Comput. **17** (2011)

3. Branco, P., Torgo, L., Ribeiro, R.P.: A survey of predictive modeling on imbalanced domains. ACM Comput. Surv. **49**(2), 1–50 (2016)
4. Chawla, N.V., Japkowicz, N., Kotcz, A.: Editorial: special issue on learning from imbalanced data sets. SIGKDD Explor. Newsl. **6**(1), 1–6 (2004)
5. Domingos, P.: A few useful things to know about machine learning. Commun. ACM **55**(10), 78–87 (2012)
6. Du, L.M., Xu, Y., Zhu, H.: Feature selection for multi-class imbalanced data sets based on genetic algorithm. Ann. Data Sci. **2**(3), 293–300 (2015)
7. Duda, R.O., Hart, P.E., Stork, D.G.: Pattern Classification, 2nd edn. Wiley, New York (2001)
8. Guyon, I., Elisseeff, A.: An introduction to variable and feature selection. J. Mach. Learn. Res. **3**, 1157–1182 (2003)
9. Maldonado, S., Weber, R., Famili, F.: Feature selection for high-dimensional class-imbalanced data sets using support vector machines. Inf. Sci. **286**, 228–246 (2014)
10. Pedregosa, F., Varoquaux, G., Gramfort, A., Michel, V., Thirion, B., Grisel, O., Blondel, M., Prettenhofer, P., Weiss, R., Dubourg, V., Vanderplas, J., Passos, A., Cournapeau, D., Brucher, M., Perrot, M., Duchesnay, E.: Scikit-learn: machine learning in Python. J. Mach. Learn. Res. **12**, 2825–2830 (2011)
11. Porwik, P., Doroz, R., Orczyk, T.: Signatures verification based on PNN classifier optimised by PSO algorithm. Pattern Recogn. **60**, 998–1014 (2016)
12. Triguero, I., Galar, M., Merino, D., Maillo, J., Bustince, H., Herrera, F.: Evolutionary undersampling for extremely imbalanced big data classification under apache spark. In: IEEE Congress on Evolutionary Computation, CEC 2016, Vancouver, BC, Canada, 24–29 July 2016, pp. 640–647 (2016)
13. Wozniak, M., Graña, M., Corchado, E.: A survey of multiple classifier systems as hybrid systems. Inf. Fusion **16**, 3–17 (2014)
14. Yin, L., Ge, Y., Xiao, K., Wang, X., Quan, X.: Feature selection for high-dimensional imbalanced data. Neurocomputing **105**, 3–11 (2013)

Special Session on New Models of Bio-inspired Computation for Massive Complex Environments

Design of Japanese Tree Frog Algorithm for Community Finding Problems

Antonio Gonzalez-Pardo$^{(\boxtimes)}$ and David Camacho

Computer Science Department, Universidad Autonoma de Madrid, Madrid, Spain
{antonio.gonzalez,david.camacho}@uam.es

Abstract. Community Finding Problems (CFPs) have become very popular in the last years, due to the high number of users that connect everyday to Social Networks (SNs). The goal of these problems is to group the users that compose the SN in several communities, or circles, in such a way similar users belong to the same community, whereas different users are assigned to different communities. Due to the high complexity of this problem, it is common that researchers use heuristic algorithms to perform this task in a reasonable computational time. This paper is focused on the applicability of a novel bio-inspired algorithm to solve CFPs. The selected algorithm is based on the real behaviour of the Japanese Tree Frog, that has been successfully used to colour maps and extract the Maximal Independent Set of a graph.

Keywords: Community finding problems · Japanese Tree Frog
Swarm intelligence

1 Introduction

The amount of data generated in the most popular Social Networks (SNs) is increasing everyday, because there is a huge number of users that daily connects to these networks and interacts with other users. In spite of the source of the data is well-known (like relationships between users, or actions performed to content published in the SN) there exists valuable knowledge hidden in this data. That is the reason behind the high interest of the research community, and the industry, in the application of data mining techniques to discover this hidden knowledge.

The most popular examples of the hidden knowledge that can be extracted from the SN are the following: (1) It is possible to identify the key users within a network in such a way if these users are removed, the network is splitted in the maximum number of disconnected sub-networks [13]. (2) Other studies are focused on the information flows within the SN [3]. (3) Other works try to measure some specific characteristics of the users based on their interactions [7,11,12]. (4) The goal of other approaches is the identification of groups of users depending on several factors like their interactions in the SN [9,16].

This last application, that is called Community Finding Problem (CFP) is the application domain of this work. The goal of any CFP is to group the different users of a SN in different communities in such a way, similar individuals

© Springer Nature Switzerland AG 2018
H. Yin et al. (Eds.): IDEAL 2018, LNCS 11315, pp. 307–315, 2018.
https://doi.org/10.1007/978-3-030-03496-2_34

belong to the same comminities whereas different users are grouped in different communities.

In order to solve any CFP, the SN needs to be modelled into a graph $G = (V, E)$, where V is the set of vertex of the graph and contains the users that compose the network; and E represents the edges of the network in such a way two nodes $(u_1, u_2 \in V)$ will be connected $(\exists\, e(u_1, u_2) \in E)$ if their corresponding users have a relation in the SN.

The meaning of this 'relation' may differ depending on the SN taken into account and even on the dataset available. For example, working with data extracted from Facebook, two nodes can be connected if the corresponding users are 'friends' in the SN; whereas if the dataset is extracted from Twitter, two nodes can be connected in the graph if one user is follower of the other, or if one user 'retweets' the tweets published by the other... etc.

Although the goal of any CFP is to find the group of users with common characteristics, there is a wide range of applications around the CFPs such as discovering functionally related objects [8,19], studying the different interactions between the objects [2], or predicting unobserved connections [6], among others.

In any of the just mentioned works, any researcher that wants to solve any CFP must face two problems derived from the complexity of the task and the complexity of the dataset. One of this problems is that due to the huge amount of data available in the SNs the classical algorithms for solving clustering problems are not a valid approach because standalone approaches cannot handle all the data contained in the dataset. To solve this problems there are two approaches: the first one consist of applying new methodologies based on the well-known *Map-Reduce* approach. This type of solutions split the dataset in smaller subsets that are *mapped* to different computers that try to solve this small sub-problem. Once all the computers have finished, the different solutions are merged in the *Reduce* phase to build the final global solution. Another approach to reduce the amount of data is to work with '*Ego Networks*'. This work uses the Ego Networks to represent the different networks of the dataset. Although these networks are described in Sect. 2, it can be briefly described as the subgraph of the network that contains a center node (called *ego*), the different users connected to this ego (called *alters*), and the different relations between the alters.

Finally, the second problem that must be faced is related to the complexity of the task. Due to the complexity of grouping the different users of a SN into several communities, it is really common that researchers adapt approaches belonging to Computational Intelligence research field to perform this task. In this sense, [9] has modified the classical Ant Colony Optimization (ACO) algorithm to perform this tasks. But other works, like [15] had applied a Genetic Algorithm (GA), or a Particle Swarm Optimization (PSO) [5]. One of the last published paper [16], presented promising results applying an Iterated Greedy Algorithm.

This work analyse the adaptation of a new algorithm inspired by the calling behaviour of male Japanese tree frogs to solve CFPs. This algorithm has been successfully applied to find the large independent sets in graphs [4].

This paper is structured as follows, Sect. 2 describes the '*Ego Networks*' and how these networks are generated. Section 3 provides a description of the calling behaviour of male Japanese tree frogs and also how the algorithm has been adapt to solve CFPs. Section 4 describes the dataset used, and the experimental evaluation carried out in this paper. Finally, Sect. 5 shows the main conclusions and future lines of this work.

2   Social Network Representation by Using Ego Networks

As it has been said, the execution of any algorithm to solve any CFP is quite complex because the volume of the data can be extracted from any SN is really big. Thus, the data must be reduced in such a way the reduced data can be handled by any computer.

In this sense, there are two different approaches. The first one is the adaptation of the selected algorithm to the '*Map-Reduce*' paradigm. In this way, the whole dataset is split in smaller sub-problems that are solved independently in the Map phase for each computer that compose the cluster. When the map phase is finished, the '*Reduce*' phase starts recovering all the independent solution and merging then into the final global solution.

The other approach (that is the one used in this paper) is the generation of the different '*Ego Networks*' contained in the network. An Ego Network [18] is a social network composed by one user centering the graph (called 'Ego'), all the users connected to this Ego (called 'Alters') and all the relations between the alters. For this reason, in any given SN there are as many *Ego Network* as users are involved in the network.

The usage of Ego Networks provides two main advantages. The first one is that the dataset is reduced drastically, because the algorithm will handle only the specific Ego Network for each user instead of working with the whole network. The second advantage is that the usage of Ego Network makes sense for solving CFPs because in this way the algorithm is discovering the communities within a contact list of a specific user.

3   Japanese Tree Frog

This section describes the Japanese Tree Frog behaviour that has motivated the design of the algorithm, and then, the details about how this algorithm has been adapted to solve CFPs.

It is known that male Japanese tree frogs use their call to attract females [17], and females are able to determine the location of the male just by recognising the source of the call. The problem appears when two, or more, males are too close in space and they communicate at the same time, because the female is not able to distinguish the source of the call. To solve this problem, males have evolved in such a way that the are able to desynchronize their calls.

This behaviour have been formally modelled by Aihara et al. [1] that represented each frog with an oscillator that adapts its frequency. In the just mentioned work, Aihara et al. were able to connect a couple of oscillators. This work was later extended by Mutazono et al. [14] that defined an anti-phase synchronization for the purpose of collision-free transmission scheduling in sensor networks. This model defined by Mutazono et al. was applicable to larger topologies and the system reaches stable solutions more easily. For a more detailed description of this behaviour we refer the reader to [10]. This algorithm has been successfully applied to the graph coloring problem and the identification of the Maximum Independent Set [4].

The Japanese tree frog is a decentralized algorithm characterized because the different frogs build a local solution, and send it to its neighbours. In this sense, this algorithm needs the definition of a graph ($G = (V, E)$) and it will locate a frog in each node of the graph. Each frog contains a message queue ($\mathcal{Q}$) where each frog will store the received messages that are taken into account to build its local solution. The pseudo-code for this algorithm is shown in Algorithm 1.

Algorithm 1. Pseudo-code for a given frog i.

---
1     $\theta_i \leftarrow calculateNewTheta()$
2     $s_i \leftarrow buildLocalSolution()$
3     $sendMessageToNeighbours()$
4     $clearMessageQueue()$

A frog i in the network contains a θ_i value and the local solution s_i. The θ_i value represent the moment in which the frog is going to be executed (using a bio-inspired analogy, this value will determine exact the moment when the frog starts the call). θ_i is defined in the range $[0, 1)$ and the frog i will build its local solution at time $t + \theta_i$.

Frogs are able to communicate to their neighbours sending a message, m, that will be stored in the message queue, Q_i, of the receiver. Any message is defined as follows: $m = \langle \theta_m, s_m \rangle$, where θ_m is the theta value of the sender, and s_m is the local solution build by frog m. The local solution of the sender is needed because each frog will take into account the local solutions received to build its local solution, and thus, perform this self-organization.

The θ_m is also needed, because each θ is updated with the values stored in the message queue (Line 1). This adaptation is done as follows:

$$\theta_i = \theta_i - \alpha \sum_{m \in \mathcal{Q}_i} \frac{\sin(2\pi \cdot (\theta_m - \theta_i))}{2\pi} \tag{1}$$

where $\alpha \in [0, 1]$ is a parameter used to control the convergence of the system.

In the second step (Line 3), each frog will build its local solution taking into account the local solutions stored in the message queue. This step is dependent on the problem to be solved. In this work, the local solution of each frog will determine the community to which the corresponding user is attached to. For this reason, each frog will analyse the different communities received in the message and will analyse what community is better for the corresponding node.

In this work each the local solution contained in each message is defined as a community $(s_m = \{n : n \in V\})$ and frog i selects the community according to Eq. 2.

$$s_i = \arg \max_{s_m \in Q_i} s_m \cap N_i \tag{2}$$

where N_i is the set of nodes connected to node i. In other words, each frog selects the best community for the user represented by the frog. This selection will be done taking into account the overlapping between the members of the received community and the friends of the represented user in the SN.

Once, the frog i has selected the community, the frog sends to its neighbours the message composed by the θ_i, and the community selected. Note that the community selected in Eq. 2 needs to be updated with the current node where the frog i is located: $s_i = s_m \cup n_i$. Finally, when the message is sent, frog s remove all the messages from its message queue.

When all the frogs have been executed, the system checks whether with the updated theta values, the execution order of the frogs has changed. If a change is produced, those frogs whose position in the list have changed repeat the execution showed in Algorithm 1. This is repeated until the adaption of the theta does not produce any change in the execution order of the frogs.

4   Experimental Phase

This section provides a detailed description of the different experiments carried out in this work. The goal of these experiments is to validate the proposed Japanese Tree Frog algorithm applied to community finding tasks and also to analyze the performance using data extracted from Facebook.

In order to do that, we have selected 4 different networks extracted from the SNAP project[1]. The selected datasets varies in the complexity of the network, and its information is shown in Table 1.

Table 1. Description of the dataset used in this work. For each dataset, this table shows the number of nodes and edges that composed the network.

DataSet	Nodes	Edges
1–3980	60	292
2–698	67	540
3–414	160	3386
4–348	228	6384

---

[1] Stanford Network Analysis Project (SNAP): http://snap.stanford.edu/index.html.

For each dataset, we have executed several times the Japaneses tree frog algorithm varying the value of α. The goal of this parameter tunning is to understand the influence that the parameter exerts on both: the quality of the solutions and the execution time of the algorithm.

For initializing the θ of each frog, we have used the initialization proposed in [4]: $\theta_i = 1 - 1.1^{-d_i + \epsilon}$, where ϵ is a uniform random from $(0, 0.01]$, and d_i denotes the degree of the node i.

The evaluation of the global solution built by the frogs is carried out using the intercluster and intracluster densities of each community. In this sense, given a solution S composed by several communities ($S = \{c_i, \ i \geq 0\}$) the function that defines the quality of this solution is defined as:

$$f(S) = \frac{\sum_{c_i \in S} Q(c_i)}{|S|} \tag{3}$$

$$Q(c_i) = \mathbb{I}(c_i) - \mathbb{T}(c_i) \tag{4}$$

$$\mathbb{I}(c_i) = \frac{m_{c_i}}{n_{c_i}(n_{c_i} - 1)/2} \tag{5}$$

$$\mathbb{T}(c_i) = \frac{C_{c_i}}{n_{c_i}(n - n_{c_i})} \tag{6}$$

$$m_{c_i} = |\{(u, v) \ : \ u \in c_i, v \in c_i\}| \tag{7}$$

$$n_{c_i} = |c_i| \tag{8}$$

$$C_{c_i} = |\{(u, v) \ : \ u \in c_i, v \notin c_i\}| \tag{9}$$

Equation 3 computes the average quality for each community. This quality ($Q(\cdot)$) is measured as the difference between the intercluster density, $\mathbb{I}(\cdot)$ (Eq. 5), and the intracluster density, $\mathbb{T}(\cdot)$ (Eq. 6).

The intercluster density is computed as the number of internal edges of the community (m_{c_i}) divided by the maximum number of internal edges that could be generated in the community ($n_{c_i}(n_{c_i} - 1)/2$).

In the same way, the intracluster density (also known as Cut Ratio) is measured as the number of edges pointing out the community (C_{c_i}), divided by all the possible number of external edges in the community ($n_{c_i}(n - n_{c_i})$).

The results obtained by the frog algorithm are shown in Table 2. This table shows for each dataset, the number of repetitions, the value for the parameter α, the best value obtained by Eq. 3, and the mean and standard deviation for all the solutions generated by the algorithm. Note that a better solution is those that provides a higher value for Eq. 3, because it means that the communities found show high internal density and low connectivity with the rest of communities.

For the first dataset '1–3980', the best performance is obtained with $\alpha = 0.5$. Although the best value is equal to the best value obtained with $\alpha = 0.75$, the mean quality value is higher for $\alpha = 0.5$. Note that given two similar performances obtained by two different configurations, the best configuration will be those with lower α, because the lower α, the faster convergence.

Regarding the datasets '2–698' and '3–414', the best configuration ($\alpha = 0.25$ and $\alpha = 0.1$, respectively) are selected according to the best value obtained,

Table 2. Experimental results obtained by the Japanese Tree Frog with different values for α. This table shows for each dataset, the number of repetitions, the value of α, the best value obtained taking into account all the solutions built, and the mean value for the solutions with its corresponding standard deviation.

DataSet	Rep	α	Best value	Mean
3980	100	0.01	0.2647	0.2647 ± 0.00
		0.1	0.3189	0.2653 ± 0.0005
		0.25	0.3216	0.2608 ± 0.0131
		0.5	**0.3685**	**0.2765 ± 0.0248**
		0.75	0.3685	0.2708 ± 0.0188
		1	0.3216	0.2692 ± 0.015
698	100	0.01	0.1493	0.1493 ± 0.00
		0.1	0.2074	0.1509 ± 0.0086
		0.25	**0.4720**	**0.1836 ± 0.0615**
	50	0.5	0.4206	0.1876 ± 0.0636
		0.75	0.2731	0.1636 ± 0.0335
	20	1	0.3501	0.1779 ± 0.0672
414	100	0.01	0.2150	0.1883 ± 0.0033
		0.1	**0.2765**	**0.1994 ± 0.0210**
		0.25	0.2372	0.2013 ± 0.0205
	50	0.5	0.2365	0.1904 ± 0.0094
		0.75	0.2354	0.1890 ± 0.0069
	20	1	0.1894	0.1881 ± 0.0020
348	100	0.01	0.1475	0.1475 ± 0.0000
		0.1	**0.4095**	**0.1853 ± 0.0643**
		0.25	0.1475	0.1475 ± 0.0000
	50	0.5	0.3876	0.1981 ± 0.0681
		0.75	0.4106	0.1721 ± 0.0629
	20	1	0.3589	0.1841 ± 0.0809

though these configuration does not provide the best mean value. Nevertheless, in both cases, there are not many difference between the mean value for the selected configuration and the maximum mean value.

For the hardest dataset ('4–348') the best solution is built with $\alpha = 0.1$. In this case, $\alpha = 0.5$ also provided a good best solution and the mean is the higher. But the computational time for $\alpha = 0.1$ is really reduced compared to $\alpha = 0.5$.

5   Conclusions and Future Work

Japanese Tree Frog algorithm [10] is a decentralized algorithm inspired by the calling behaviour of this animal. This new algorithm has been successfully applied to the Coloring Graph Problem and the identification of the Maximum Independent Set [4].

This paper present an adaptation of this algorithm to solve Community Finding Problems (CFPs). This type of problems tries to group the different users of a Social Network (SN) in several communities, in such a way users belonging to the same community are similar to each others according to a specific metric.

For testing the performance of the proposed algorithm, we have selected 4 different dataset extracted from the Stanford Network Analysis Project (SNAP). Experimental results reveals that the proposed algorithm is able to detect communities that exhibit high internal density and few connections to the rest of communities.

The proposed algorithm only depends on one parameter, $\alpha \in [0, 1]$, that affects to the convergence of the algorithm. Experimental results have showed the algorithm generates better solutions with lower values of α if the complexity of the network is high.

As a general conclusion, the Japanese tree frog algorithm can be used to detect communities in SNs. Nevertheless, there are some aspects that need to be studied carefully in a future work.

The most critical aspect is related to the definition of α. As it has been demonstrated, this value affect to the quality of the solutions, the convergence of the algorithm and thus, to the computational time. Also, different datasets provides better results with different values of α. For this reason, it seems interesting the definition of a procedure that adapts the value of this parameter as long as the execution of the algorithm.

Finally, a comparison against the algorithm extracted from the State-of-the-Art is needed. In this way, we can compared the performance of the proposed algorithm against some of the algorithms that are used to solve CFPs.

Acknowledgements. This work has been co-funded by the following research projects: DeepBio (TIN2017-85727-C4-3-P) Spanish Ministry of Economy and Competitivity; CIBERDINE S2013/ICE-3095, under the European Regional Development Fund FEDER; and Justice Programme of the European Union (2014–2020) 723180 – RiskTrack – JUST-2015-JCOO-AG/JUST-2015-JCOO-AG-1. The contents of this publication are the sole responsibility of their authors and can in no way be taken to reflect the views of the European Commission.

References

1. Aihara, I.: Modeling synchronized calling behavior of Japanese tree frogs. Phys. Rev. E **80**, 011918 (2009)
2. Airoldi, E.M., Blei, D.M., Fienberg, S.E., Xing, E.P.: Mixed membership stochastic blockmodels. J. Mach. Learn. Res. **9**, 1981–2014 (2008)
3. Bakshy, E., Rosenn, I., Marlow, C., Adamic, L.: The role of social networks in information diffusion. In: Proceedings of the 21st International Conference on World Wide Web, pp. 519–528. ACM (2012)

4. Blum, C., Calvo, B., Blesa, M.J.: FrogCOL and FrogMIS: new decentralized algorithms for finding large independent sets in graphs. Swarm Intell. **9**(2), 205–227 (2015)
5. Cai, Q., Gong, M., Ma, L., Ruan, S., Yuan, F., Jiao, L.: Greedy discrete particle swarm optimization for large-scale social network clustering. Inf. Sci. **316**, 503–516 (2015)
6. Chang, J., Blei, D.M.: Hierarchical relational models for document networks. Ann. Appl. Stat. **4**(1), 124–150 (2010)
7. Fernandez, M., Asif, M., Alani, H.: Understanding the roots of radicalisation on Twitter. In: Proceedings of the 10th ACM Conference on Web Science, pp. 1–10 (2018)
8. Girvan, M., Newman, M.E.J.: Community structure in social and biological networks. Proc. Nat. Acad. Sci. **99**(12), 7821–7826 (2002)
9. Gonzalez-Pardo, A., Jung, J.J., Camacho, D.: ACO-based clustering for ego network analysis. Future Gener. Comput. Syst. **66**, 160–170 (2017)
10. Hernández, H., Blum, C.: Distributed graph coloring: an approach based on the calling behavior of Japanese tree frogs. Swarm Intell. **6**(2), 117–150 (2012)
11. Lara-Cabrera, R., Gonzalez-Pardo, A., Benouaret, K., Faci, N., Benslimane, D., Camacho, D.: Measuring the radicalisation risk in social networks. IEEE Access **5**, 10892–10900 (2017)
12. Lara-Cabrera, R., Gonzalez-Pardo, A., Camacho, D.: Statistical analysis of risk assessment factors and metrics to evaluate radicalisation in Twitter. Future Gener. Comput. Syst. (2017). https://doi.org/10.1016/j.future.2017.10.046. ISSN 0167-739X
13. Lozano, M., García-Martnez, C., Rodríguez, F.J., Trujillo, H.M.: Optimizing network attacks by artificial bee colony. Inf. Sci. **377**, 30–50 (2017)
14. Mutazono, A., Sugano, M., Murata, M.: Frog call-inspired self-organizing antiphase synchronization for wireless sensor networks. In: 2009 2nd International Workshop on Nonlinear Dynamics and Synchronization, pp. 81–88 (2009)
15. Pizzuti, C.: GA-net: a genetic algorithm for community detection in social networks. In: Rudolph, G., Jansen, T., Bcume, N., Lucas, S., Poloni, C. (eds.) PPSN 2008. LNCS, vol. 5199, pp. 1081–1090. Springer, Heidelberg (2008). https://doi.org/10.1007/978-3-540-87700-4_107
16. Sánchez-Oro, J., Duarte, A.: Iterated Greedy algorithm for performing community detection in social networks. Future Gener. Comput. Syst. **88**, 785–791 (2018). https://doi.org/10.1016/j.future.2018.06.010. ISSN 0167-739X
17. Wells, K.D.: The social behaviour of anuran amphibians. Anim. Behav. **25**, 666–693 (1977)
18. Xie, J., Kelley, S., Szymanski, B.K.: Overlapping community detection in networks: the state-of-the-art and comparative study. ACM Comput. Surv. **45**(4), 43:1–43:35 (2013)
19. Yang, J., Leskovec, J.: Overlapping community detection at scale: a nonnegative matrix factorization approach. In: Proceedings of the Sixth ACM International Conference on Web Search and Data Mining, WSDM 3, pp. 587–596. ACM (2013)

An Artificial Bee Colony Algorithm for Optimizing the Design of Sensor Networks

Ángel Panizo[1]([✉]), Gema Bello-Orgaz[1], Mercedes Carnero[2], José Hernández[2], Mabel Sánchez[3], and David Camacho[1]

[1] Computer Science Department, Universidad Autónoma de Madrid,
28049 Madrid, Spain
{angel.panizo,gema.bello,david.camacho}@uam.es
[2] Facultad de Ingenieria, UNRC, Campus Universitario, 5800 Río Cuarto, Argentina
{mcarnero,jlh}@ing.unrc.edu.ar
[3] Planta Piloto de Ingeniería Química, UNS - CONICET, Bahía Blanca, Argentina
msanchez@plapiqui.edu.ar

Abstract. The sensor network design problem (SNDP) consists of the selection of the type, number and location of the sensors to measure a set of variables, optimizing a specified criteria, and simultaneously satisfying the information requirements. This problem is multimodal and involves several binary variables, therefore it is a complex combinatorial optimization problem. This paper presents a new Artificial Bee Colony (ABC) algorithm designed to solve high scale designs of sensor networks. For this purpose, the proposed ABC algorithm has been designed to optimize binary structured problems and also to handle constraints to fulfil information requirements. The classical version of the ABC algorithm was proposed for solving unconstrained and continuous optimization problems. Several extensions have been proposed that allow the classical ABC algorithm to work on constrained or on binary optimization problems. Therefore the proposed approach is a new version of the ABC algorithm that combines the binary and constrained optimization extensions to solve the SNDP. Finally the new algorithm is tested using different systems of incremental size to evaluate its quality, robustness, and scalability.

Keywords: Artificial Bee Colony · Sensor network design
Combinatorial optimization

This work has been co-funded by the following research projects: EphemeCH (TIN2014-56494-C4-4-P), DeepBio (TIN2017-85727-C4-3-P) projects (Spanish Ministry of Economy and Competitivity, under the European Regional Development Fund FEDER) and in part by the Justice Programme of the European Union (2014-2020) 723180, RiskTrack, under Grant JUST-2015-JCOO-AG and Grant JUST-2015-JCOO-AG-1.

H. Yin et al. (Eds.): IDEAL 2018, LNCS 11315, pp. 316–324, 2018.
https://doi.org/10.1007/978-3-030-03496-2_35

1   Introduction

The optimally selection of the type, number and location of the sensors needed to measure a set of variables is known as the sensor network design problem (SNDP). The main objective of this type of problem is to design an optimal sensor network for monitoring processes (measuring certain variables of interest). In addition, the sensors have a cost and give certain precision. Also, the measured variables have associated a series of precision restrictions. Therefore, this problem can be formulated in terms of binary variables, where each of them indicates the presence or absence of a sensor, giving rise to a problem of combinatorial optimization for minimizing a specific criterion (the cost) and simultaneously satisfying the information constraints of the measured variables.

This type of problems can be solved using tree search algorithms, like Branch and Bound-type strategies, without relaxation in the nodes. The algorithms use intelligent forms of exploration by defining appropriate levels and stop criteria. Nguyen and Bagajewicz (2011) solved a small size problem, which consisted of only 24 variables, for different designs with degrees of low, moderate and high specification. An exact method based on a hybrid technique that combines depth- width searches was proposed. The authors considered that the SNDPs for the problems with a high degree of specification were the most complex to solve. It was not analysed how the resolution methodology performs in larger scale problems.

Other classic approaches have formulated the problem as a mixed integer nonlinear programming problem (MINLP). Kelly and Zyngier (2008) proposed a MILP formulation that allows to consider three aspects related to an optimal design of a sensor network: observability, redundancy and precision. In the objective function, both the instrumentation cost associated with the primary measurement of the i-th flow is considered, as well as the cost of measuring this variable with more than one sensor. The examples shown are small. The first corresponds to a process with 8 streams and for this case the full MINLP is solved. The second, is an upgrade instrumentation problem for a 63 streams process, of which 29 already have sensors assigned. These classic methods guarantee to find the optimal solution to the problem, but their computational time is too high for large size problems, being inefficient in these cases. Meta-heuristics algorithms have been traditionally used for solving combinatorial optimization problems in a large number of different domains such as Data Mining [3], Mission Planning [11], or Constraint Satisfaction Problems (CSP) [7]. Especially, meta-heuristic algorithms inspired by natural systems such as evolutionary algorithms, ant colony optimization and particle swarm optimization, have become a very active and relevant topic of research in recent years. These algorithms do not guarantee that the global optimally solution is found, but provide a sufficiently good solution with much less computational effort than traditional optimization algorithms.

Different approaches based on bio-inspired algorithms have been presented in the literature to solve SNDP. Sen et al. [13] presented one of the first methods based on Genetic Algorithms (GAs) using concepts from graph theory to select flow meters for non-redundant Sensor Networks (SNs). This approach optimized only a single criteria such as cost, reliability or estimation accuracy. Afterwards, Carnero et al. [5] proposed a GA whose operators were modified based on linear algebra concepts, and also allowed to optimize single or multiple criteria for the same type of SNs.

The GAs are population based methodologies where the solutions to the optimization problem are the individuals in a population, and the algorithm tries to improve them using a fitness function that determines their quality. Currently, there are other population-based methods that have demonstrated good performance in solving complex combinatorial problems, such as the Estimation of Distribution Algorithms (EDAs) or the Artificial Bee Colony (ABC) Algorithms which belongs to a family of algorithms called Swarm Intelligence. A strategy that combines univariate EDAs, such as Pbil [2] and the advantages of Tabu Search was presented by Carnero et al. [4] which is able to run in parallel, solving high scale designs of SNs. It has also been studied in Carnero et al. [6] the applicability of more complex EDAs, which include structural learning, to solve design problems in process plants that can be represented by linear and nonlinear models. One of the most recently population based methods is the ABC algorithm, however few research works have focused on its application to solve the SNDP. The classical version of the ABC algorithm was proposed for solving unconstrained and continuous optimization problems. Therefore, it cannot be applied directly to solve the SNDP, because it requires the optimization of structured binary problems, as well as the handling of constraints to fulfil the information requirements of the sensors.

Karaboga and Akay [9] proposed a modification of the classical ABC for continuous constrained optimization problems. This new algorithm uses three simple heuristic rules and a probabilistic selection scheme for feasible solutions based on their fitness values and infeasible solutions based on the number of constraints violated. On the other hand, a new version of ABC, called DisABC [10], was designed for binary optimization by Husseinzadeh et al. Therefore, this paper presents a new ABC algorithm, that combines the two strategies mentioned above to work on constrained binary optimization problems, that solves the SNDP.

The rest of the paper has been structured as follows: Sect. 2 describes the problem formulation to design SNs. Section 3 presents the proposed ABC algorithm for solving SNDPs, and the fitness functions used. Section 4 provides a description of the test cases used, the experimental setup of the proposed algorithm, and the experimental results obtained. In Sect. 5 some final remarks and future research lines are presented.

2 Problem Formulation

The minimum cost SNDP that satisfies precision and estimability constraints for a set of key variables is stated by Eqs. 1–2, where $\mathbf{q}$ is an n dimensional vector of binary variables such that: $q_i = 1$ if variable i is measured, and $q_i = 0$

otherwise, $\mathbf{c}^T$ is the cost vector; $\hat{\sigma}_k$ is the estimate standard deviation of the k-th variable contained in S_σ after a data reconciliation procedure [12] is applied, and E_l stands for the degree of estimability of the l-th variable included in S_E [1]. For this formulation E_l is set equal to one, consequently only a variable classification procedure run is needed to check its feasibility. Furthermore S_σ and S_E are the set of key process variables with requirements in precision and estimability, respectively.

$$
\min \mathbf{c}^T \mathbf{q}
$$
$$
\text{s.t.} \hat{\sigma}_k(\mathbf{q}) \leq \sigma_k^*(\mathbf{q}) \forall k \in S_\sigma \qquad (1)
$$

$$
E_l \geq 1 \forall l \in S_E
$$
$$
\mathbf{q} \in \{0,1\}^n \qquad (2)
$$

It is assumed that a linearized algebraic model represents a plant operation, measurements are subject to non-correlated random errors, there is only one potential measuring device for each variable and, there are no restrictions for the localization of instruments.

3   Algorithm Description

Artificial Bee Colony (ABC) is an optimization algorithm inspired by the foraging behaviour of honey bees [8]. In a real bee colony, bees try to maximize the nectar amount unloaded to the food stores in the hive by searching for food sources with a lot of nectar. A 'food source' in our simulated hive is a possible solution of the problem, in our case a set of sensors codify as a binary vector, and the amount of 'nectar' that a 'food source' has corresponds to the quality (cost and variable precision) of this solution. In a simulated hive there are three different types of bees: *employed bees*, *onlooker bees* and *scout bees*. Pseudo-code for the ABC algorithm can be found on Algorithm 1.

Employed bees have assigned a specific food source and are responsible for exploiting the area around it to find better solutions (line 7 to 14 of Algorithm 1). To decide how far to look for the new food source, the food source assigned to an other *employed bee* (randomly selected) of the hive is used. Once a new food source has been found and evaluated (calculated the cost and constraints met), it is compared against the previous food source been exploited by that bee. If the new food source is better than the old one, the latter is forgotten. Finally the *employed bees* go back to the hive to tell its findings.

Onlooker bees don't have assigned a specific food source, they wait in the hive for the *employed bees* to return with their findings. Depending on the information share by the *employed bees*, the *onlooker bees* select one of the food sources exploited by them (line 15 to 19 of Algorithm 1). This selection process takes into account the quality of food sources, and those sources with higher quality will have a higher chance of being selected to be explored (line 16). Once a food source has been selected, the *onlooker bee* explores the area around it in the same way as the *employed bee* does. If a *onlooker bee* finds a better solution in that area she tells the hive so the *employed bees* can update their knowledge. Finally if a solution being exploited stops to improve, the *employed bee* exploiting it becomes a *scout bee* and will search the environment to find a new food source to exploit (line 20 to 22 of Algorithm 1).

Algorithm 1. ABC algorithm for binary constraint optimization

---

1: **function**   ABC(SENSOR_COST,   V_CONSTRAINTS,   NBEES_EMPLOYED, NBEES_ONLOOKER, MAXITRS, MAXTRIALS)
2:      $best \leftarrow null$
3:      $hive \leftarrow \emptyset$
4:      **for all** $i \in [0, NBEES\_EMPLOYED]$ **do**   ▷ populates the hive with bees that exploit random solutions
5:          $hive \leftarrow hive \cup makeBeeRandomSolution()$
6:      **for all** $i \in [0, MAXITRS]$ **do**
7:          **for all** $i \in [0, NBEES\_EMPLOYED]$ **do**   ▷ send employed bees
8:              $otherBee \leftarrow selectRandomBee(hive)$
9:              $newBee \leftarrow generateNewSolution(hive[i], otherBee)$
10:             $evaluate(newBee, SENSOR\_COST, V\_CONSTRAINTS)$
11:             **if** $isBetter(newBee, bee)$ **then**
12:                 $hive[i] = newBee$
13:             **else**
14:                 $hive[i].trials+ = 1$
15:             **for all** $j \in [0, NBEES\_EMPLOYED]$ **do**
16:                 $hive[i].probability \leftarrow calculateProbability(hive[i])$
17:             **for all** $i \in [0, NBEES\_ONLOOKER]$ **do**   ▷ send onlooker bees
18:                 $i \leftarrow rouletteSelection(hive)$ ▷ selects a bee using roulette wheel selection and bee's probability values
19:                 repeat lines 8-14...
20:             **for all** $i \in [0, NBEES\_EMPLOYED]$ **do**
21:                 **if** $hive[i].trials \geq MAXTRIALS$ **then**   ▷ check if it is necessary to send scout bees
22:                     $hive[i] \leftarrow makeBeeRandomSolution()$
23:                 $best \leftarrow getBetter(best, findBest(hive))$ ▷ stores the best solution found so far
24:     **return** $best$

ABC original algorithm was firstly proposed for unconstrained continuous optimizations. As the SNDP is a constrained binary optimization some modifications need to be added. First we introduce the modifications proposed by Karaboga and Akay in [9] for handling continuous constrained optimization problems. This new version of the ABC algorithm applies Deb's rule to compare solutions (functions $isBetter(*), getBetter(*), findBest(*)$ of Algorithm 1). This rule follows the next criteria: A feasible solution, one that met all the constraints, always dominates an infeasible one; Between two infeasible solutions, the one that violates less constraints dominates the other; Between two feasible solutions, the one with higher fitness dominates the other.

Also, since infeasible solutions are allowed to populate the colony, an other modification is included to assign probability values to both feasible and infeasible solutions (function $calculateProbability(*)$ of Algorithm 1 in line 16). Feasible solutions will have probabilities between 0.5 and 1.0, depending on their cost. Infeasible solutions will have probabilities between 0.0 and 0.5, depending on the number of constraints violated. Equations 3 and 4 indicates how to calculate the probability of a solution where: *sol* is a solution of the problem, SN is the number of solutions being exploited in the hive, $cost(*)$ is the cost of a solution and

$violations(*)$ is the number of constraints that a solution violates. Introducing this modifications implies that the algorithm could deliver an infeasible solution if no feasible one have been found.

$$P(sol) = \begin{cases} 0.5 + \left(\dfrac{fitness(sol)}{\sum_{j=0}^{SN} fitness(sol_j)} \right) * 0.5 & \text{if } sol \text{ is feasible} \\[4ex] \left(1 - \dfrac{violation(sol)}{\sum_{j=0}^{SN} violation(sol_j)} \right) * 0.5 & \text{if } sol \text{ is infeasible} \end{cases} \qquad (3)$$

$$fitness(sol) = 1/(1 + cost(sol)) \qquad (4)$$

Second, we introduce the modifications proposed by Kashan et al. in [10] for binary optimization. These modifications change the way of finding new food sources (function $generateNewSolution(*)$ of Algorithm 1 in line 9). Once a solution, X, is selected to be exploited, an other solution from the hive, Y, is selected too. This solution Y will determine how far from X should the new solution be. Equation 6 indicates this relation, where: X is the solution to be exploited, Y is an other solution selected from the hive, Z is the new solution to exploit next and φ is a positive random scaling factor. In order to measure the dissimilarity between two solutions, the Jaccard's coefficient is used. This coefficient is defined as Eq. 5, where X and Y are two solutions, M_{11} is the number of bits where both X and Y have a value of 1, M_{01} is the number of bits where X have a value of 0 and Y have a value of 1 and M_{10} is the number of bits where X have a value of 1 and Y have a value of 0.

$$Dissimilarity(X,Y) = 1 - Jaccard(X,Y) = 1 - \frac{M_{11}}{M_{01} + M_{10} + M_{11}} \qquad (5)$$

$$Dissimilarity(X,Z) \approx \varphi * Dissimilarity(X,Y) \qquad (6)$$

To produce the new solution Z we must determine: the number of bits where X and Z are both 1 (M_{11}) and the number of bits with value 1 in Z and 0 in X (M_{10}). The rest of bits in Z must be 0. In order to find the values that better fit Eq. 6 the following integer programming model is used, defined in Eqs. 7–10. Where n_1 and n_0 are the total number of 1 and 0 in X respectively.

$$min \left| 1 - \frac{M_{11}}{M_{01} + M_{10} + M_{11}} - \varphi * Dissimilarity(X,Y) \right| \qquad (7)$$

$$M_{11} + M_{01} = n_1 \quad (8) \qquad M_{10} \leq n_0 \quad (9) \qquad \begin{array}{c} M_{11}, M_{10}, M_{01} \geq 0 \\ (10) \end{array}$$

Once the most suitable M_{11} and M_{10} are calculated, initialize solution Z as a vector of the same size of X fill with zeros. Then, select M_{11} bits whose value is 1 in X and set those bits to 1 in Z. Finally, select M_{10} bits whose value is 0 in X and set those bits to 1 in Z.

Finally, to generate a new random solution (function *makeBeeRandom Solution*() of Algorithm 1), a vector following a Bernoulli distribution with probability equals to 0.5 is generated. We don't use any kind of local search in this version of the algorithm.

4   Experimental Results

The algorithm has been evaluated using three test cases of incremental size. Each test case represents one process flow diagram[1] with the following features: 11 units - 28 sensors - 8 constraints, 19 units - 52 sensors - 17 constraints, and 47 units - 82 sensors - 17 constraints. It is assumed that variables are related only by mass balance equations. The precision of the sensors are 2.5%, 2% and 2% of the corresponding true flowrates, respectively. The analysis of the results takes into account a solution's quality and variability. The quality of the solutions obtained by the proposed ABC algorithm are compared against the ones obtained by two algorithms proposed by Carnero et al. [4] called pPBIL and pPBIL-SOTS. Both algorithms are EDAs with the difference that pPBIL-SOTS also applies a local search to solve the SNDP.

The parameters of the ABC algorithm used for each case can be seen on Table 1. These parameters have been tune up experimentally. As [10] recommends we have used a dynamic φ that decreases linearly from 0.9 to 0.5 following Eq. 11, where t is the actual iteration number and φ^t is the φ of the tth iteration.

$$\varphi^t = 0.9 - \left(\frac{0.4}{MAXITRS}\right) * t \tag{11}$$

Table 1. ABC parameters for SNDP.

Case	Parameter	Description	Value
1,2,3	$NBEES\_EMPLOYED$	Number of employed bees	50
1,2,3	$NBEES\_ONLOOKER$	Number of onlooker bees	50
1	$MAXTRIALS$	Number trials before abandoning a food source	2000
2,3	$MAXTRIALS$	Number trials before abandoning a food source	2500
1	$MAXITRS$	Number of iterations the algorithm run	200
2,3	$MAXITRS$	Number of iterations the algorithm run	5000

We have run each algorithm 30 times for each test case. Table 2 presents the results obtained for each algorithm (ABC, pPBIL and pPBIL-SOTS), showing the minimum cost found, the mean cost of the 30 executions, and the standard deviation. As shown in Table 2, *pPBIL-SOTS* achieves the best results for all the

---

[1] The interested reader can get access to the files containing information about the case studies from https://drive.google.com/file/d/1FvPwDxW06xhcrEcX7RgUhMV0Eh 4lrY1p/view?usp=sharing.

three cases tested. This highlights the benefits of pairing a meta-heuristics with a local search method. Comparing pPBIL with the ABC algorithm, we can see that pPBIL obtains a lower costs but with a higher standard deviation than the ABC algorithm. It means that the solutions found by pPBIL vary more between different executions of the algorithm. Taking into account that the average costs for pBIL and ABC algorithms are closer, and the standard deviation is lower for ABC, it can be concluded that according to the stability of the solutions found, the ABC algorithm obtains better results to solve the problem.

Table 2. Results obtained by ABC, pPBIL and pPBIL-SOTS.

Case	pPBIL-SOTS		pPBIL		ABC	
	Best	Mean	Best	Mean	Best	Mean
1	**1297.39**	**1297.39 ± 0.0**	1297.39	1424.24 ± 113.62	1297.39	1323.53 ± 32.33
2	**1154.34**	**1154.34 ± 0.0**	1154.34	1408.28 ± 381.17	1164.59	1475.62 ± 161.52
3	**107377.13**	**107377.32 ± 0.006**	109905.35	128824.61 ± 18919.26	159151.02	190542.97 ± 14042.01

5   Conclusions

In this article we propose an ABC algorithm for solving the SNDP, which can be solve as a binary constrained optimization problem. The original ABC algorithm works for continuous unconstrained optimization problems, so the new algorithm combines two extensions of the standard one in order to solve this kind of problems. The first extension allows ABC to work with constraints optimizations, and the second one to work with binary optimizations. To ascertain the performance of the proposed algorithm we have tested it against three process flowsheets of incremental size, and the results obtained have been compared against two other algorithms from the state of the art, called pPBIL and pPBIL-SOTS. The results show that pPBIL-SOTS reaches the best results for all the three cases. This highlights the benefits of combining a meta-heuristic with a local search. If we compare ABC and pPBIL, we can conclude that pPBIL achieves slightly better results than ABC at the expense of having a higher variability between runs. Future research will focus on implementing local search methods to work with the ABC algorithm in order to test the true capabilities of this meta-heuristic to solve SNDP.

References

1. Bagajewicz, M., Sánchez, M.: Cost optimal design and upgrade of non-redundant and redundant linear sensor networks. AIChE J. **45**(9), 1927–1938 (1999)
2. Baluja, S.: Population-based incremental learning: a method for integrating genetic search based function optimization and competitive learning. Technical report-CMU-CS-94163, Carnegie Mellon University, Pittsburgh, PA (1994)

3. Bello-Orgaz, G., Salcedo-Sanz, S., Camacho, D.: A multi-objective genetic algorithm for overlapping community detection based on edge encoding. Inf. Sci. **462**, 290–314 (2018)
4. Carnero, M., Hernández, J., Sánchez, M.: A new metaheuristic based approach for the design of sensor networks. Comput. Chem. Eng. **55**, 83–96 (2013)
5. Carnero, M., Hernández, J., Sánchez, M., Bandoni, A.: An evolutionary approach for the design of nonredundant sensor networks. Indus. Eng. Chem. Res. **40**(23), 5578–5584 (2001)
6. Carnero, M., Hernández, J.L., Sánchez, M.: Optimal sensor location in chemical plants using the estimation of distribution algorithms. Indus. Eng. Chem. Res. (2018). https://doi.org/10.1021/acs.iecr.8b01680
7. Gonzalez-Pardo, A., Ser, J.D., Camacho, D.: Comparative study of pheromone control heuristics in ACO algorithms for solving RCPSP problems. Appl. Soft Comput. **60**, 241–255 (2017)
8. Karaboga, D.: An idea based on honey bee swarm for numerical optimization. Technical report-tr06, Engineering Faculty, Computer Engineering Department, Erciyes University (2005)
9. Karaboga, D., Akay, B.: A modified artificial bee colony (ABC) algorithm for constrained optimization problems. Appl. Soft Comput. **11**(3), 3021–3031 (2011)
10. Kashan, M.H., Nahavandi, N., Kashan, A.H.: DisABC: a new artificial bee colony algorithm for binary optimization. Appl. Soft Comput. **12**(1), 342–352 (2012)
11. Ramirez-Atencia, C., Bello-Orgaz, G., R-Moreno, M.D., Camacho, D.: Solving complex multi-UAV mission planning problems using multi-objective genetic algorithms. Soft Comput. **21**(17), 4883–4900 (2017)
12. Romagnoli, J., Sánchez, M.: Data Processing and Reconciliation for Chemical Process Operations. Academic Press, Cambridge (2000)
13. Sen, S., Narasimhan, S., Deb, K.: Sensor network design of linear processes using genetic algorithms. Comput. Chem. Eng. **22**(3), 385–390 (1998)

Community Detection in Weighted Directed Networks Using Nature-Inspired Heuristics

Eneko Osaba[1]([✉]), Javier Del Ser[1,2,3], David Camacho[4], Akemi Galvez[5,6], Andres Iglesias[5,6], Iztok Fister Jr.[7], and Iztok Fister[7]

[1] TECNALIA, 48160 Derio, Spain
eneko.osaba@tecnalia.com
[2] University of the Basque Country (UPV/EHU), 48013 Bilbao, Spain
[3] Basque Center for Applied Mathematics (BCAM), 48009 Bilbao, Spain
[4] Universidad Autonoma de Madrid, 28049 Madrid, Spain
[5] Universidad de Cantabria, 39005 Santander, Spain
[6] Toho University, Funabashi, Japan
[7] University of Maribor, Maribor, Slovenia

Abstract. Finding groups from a set of interconnected nodes is a recurrent paradigm in a variety of practical problems that can be modeled as a graph, as those emerging from Social Networks. However, finding an optimal partition of a graph is a computationally complex task, calling for the development of approximative heuristics. In this regard, the work presented in this paper tackles the optimal partitioning of graph instances whose connections among nodes are directed and weighted, a scenario significantly less addressed in the literature than their unweighted, undirected counterparts. To efficiently solve this problem, we design several heuristic solvers inspired by different processes and phenomena observed in Nature (namely, Water Cycle Algorithm, Firefly Algorithm, an Evolutionary Simulated Annealing and a Population based Variable Neighborhood Search), all resorting to a reformulated expression for the well-known modularity function to account for the direction and weight of edges within the graph. Extensive simulations are run over a set of synthetically generated graph instances, aimed at elucidating the comparative performance of the aforementioned solvers under different graph sizes and levels of intra- and inter-connectivity among node groups. We statistically verify that the approach relying on the Water Cycle Algorithm outperforms the rest of heuristic methods in terms of Normalized Mutual Information with respect to the true partition of the graph.

Keywords: Bio-inspired computation · Community detection

1 Introduction

With the advent of Social Networks, the spectrum of tools and techniques capable of achieving insights from the interrelations between their users has increased

© Springer Nature Switzerland AG 2018
H. Yin et al. (Eds.): IDEAL 2018, LNCS 11315, pp. 325–335, 2018.
https://doi.org/10.1007/978-3-030-03496-2_36

considerably in the last decade [2]. The acquired knowledge by virtue of such methods range from the quantification of the level of influence of a node within the network (*centrality*) to the discovery of shortest paths between a given pair of nodes or the derivation of enriched ways to visualize a network given the weight distribution of its edges. Many of the functionalities that Social Network users enjoy nowadays build upon this algorithmic portfolio, with a late prominence noted around other practical goals (e.g. child abuse [6,28,30] or the detection of radicalization risk [17]).

In this context, inferring communities within the nodes of a given network is arguably one of the most addressed tasks in the related literature. In this regard, a *community* refers to a set of nodes that comply with the general principles of strong intra-connectivity (namely, high degree/strength of the links among nodes belonging to the community) and weak inter-connectivity (correspondingly, low degree/strength of edges between nodes belonging to different communities). These measured parameters can be redefined as per the characteristics of the network at hand (directed, weighted, multiple edges, self loops), so that they ultimately quantify the cohesiveness of any proposed partition of the network. This evaluation is rather done based on diverse metrics proposed in related studies, each composing differently how connectivity is analyzed to yield a single quality value for the partition under different assumptions, e.g. Newman and Girvan's Modularity [21], Permanence [5], Surprise [1] and others alike [4].

Algorithmically speaking, many contributions have hitherto gravitated on the development of different heuristic approaches to find communities towards – implicitly or explicitly – optimizing one of the aforementioned metrics. This is the case, for instance, of iterative greedy methods capable of inferring a hierarchy of communities in a constructive fashion, similarly to agglomerative hierarchical clustering techniques [3]. Interestingly for the scope of this work, a growing strand of literature is currently devoted to the use of heuristic optimization algorithms directly adopting one modularity metric as their objective function. Examples abound, each focusing on assorted combinations of network instances, metric functions and algorithmic approximations. Genetic Algorithms are arguably among those more recurrently explored to date for discovering communities in networks of different characteristics [10,23,27]. However, many other solvers within the Evolutionary Computation and Swarm Intelligence fields have been also employed for this same purpose: to cite a few, Differential Evolution [16], Particle Swarm Optimization [25] or Ant Colony Optimization [14][24]. More recently, the research attention has steered towards the use of modern nature-inspired solvers for community detection in graphs, such as the Firefly Algorithm [7], Bat Algorithm [13] or Artificial Bee Colony [11], among others.

The work presented in this manuscript takes a step further over the state of the art exposed above by elaborating on several new research directions: (1) we address the problem of detecting communities in weighted directed networks, far less studied than other graph instances; (2) the adoption of the Hamming distance as a measure to compute the similarity between different partitions, which can be exploited during the search process of the overall heuristic; and (3) the assessment of these algorithmic ingredients with a diversity of nature-inspired solvers: Water Cycle Algorithm (WCA, [8]), Firefly Algorithm (FA, [31]), Evolutionary Simulated Annealing (ESA, [32]) and Population based Variable Neighborhood Search (PVNS, [29]). Thorough details on how each of these methods has been tailored to benefit from the developed operators are given, along with a justification of their expected benefits in terms of convergence. In order to assess their comparative performance, results obtained over 24 synthetically generated datasets are presented and discussed on the basis of their capability to discover their ground-of-truth partition. The significance of the performance gaps found in this benchmark is verified by two statistical tests (Friedman's and Wilcoxon), from which we conclude that WCA outperforms the rest of heuristic approaches in most of the networks.

The rest of the paper is structured as follows: in Sect. 2 the problem of finding communities in weighted directed networks is mathematically formulated, whereas the heuristic solvers are described in Sect. 3. The experimentation is introduced in Sect. 4. Finally, Sect. 5 concludes the paper.

2   Problem Statement

We begin by modeling a weighted network as a graph $\mathcal{G} \doteq \{\mathcal{V}, \mathcal{E}, f_{\mathcal{W}}\}$, where $\mathcal{V}$ denotes the set of $|\mathcal{V}| = V$ nodes or vertices of the network, $\mathcal{E}$ correspond to the set of links or edges connecting every pair of nodes, and $f_{\mathcal{W}} : \mathcal{V} \times \mathcal{V} \mapsto \mathbb{R}^+$ is a function assigning a non-negative weight to the edge connecting every pair of nodes. We assume that $f_{\mathcal{W}}(v, v) = 0$ (i.e. no self loops), and that $f_{\mathcal{W}}(v, v') = 0$ if nodes v and v' are not connected. For notational convenience we define $f_{\mathcal{W}}(v, v') \doteq w_{v,v'}$, yielding a $V \times V$ adjacency matrix $\mathbf{W}$ given by $\mathbf{W} \doteq \{w_{v,v'} : v, v' \in \mathcal{V}\}$ and fulfilling $\mathrm{Tr}(\mathbf{W}) = 0$. The directed nature of the network is ensured by overriding any assumption on the symmetry of $\mathcal{W}$, e.g. $w_{v,v'}$ is not necessarily equal to $w_{v',v}$.

With the above notation in mind, the general problem of detecting communities in a graph $\mathcal{G}$ can be conceived as the partition of the vertex set $\mathcal{V}$ into a number of disjoint, non-empty groups, each with an arbitrary size. Let M denote the number of groups of partition $\widetilde{\mathcal{V}} \doteq \{\mathcal{V}_1, \dots, \mathcal{V}_M\}$, such that $\cup_{m=1}^{M} \mathcal{V}_m = \mathcal{V}$ and $\mathcal{V}_m \cap \mathcal{V}_{m'} = \emptyset \; \forall m' \neq m$ (i.e. no overlapping communities). By extending this notation, the community to which node v belongs can be denoted as $\mathcal{V}^v \in \widetilde{\mathcal{V}}$.

The weighted directed nature of the network imposes a redefinition of the conventional in-degree and out-degree to the input and output *strength* of a given node of the network, which are correspondingly given by

$$s_v^{in} = \sum_{v' \in \mathcal{V}} w_{v',v}, \qquad s_v^{out} = \sum_{v' \in \mathcal{V}} w_{v,v'}, \tag{1}$$

namely, as the sum of the weights of the incident (outgoing) edges to (from) node v. It is important to note that these quantities reflect both the directivity and the weighted nature of adjacency matrix $\mathbf{W}$. Thereby, they should play an important role in the definition of the communities in a similar fashion to the in- and out-degree when clustering undirected, unweighted networks. Following this rationale, a measure of the quality of a given partition $\widetilde{\mathcal{V}}$ can be formulated departing from the definition of *modularity* for undirected graphs introduced in [18,20]. By defining a binary function $\delta : \mathcal{V} \times \mathcal{V} \mapsto \{0,1\}$, such that $\delta(v,v') = 1$ if $\mathcal{V}^v = \mathcal{V}^{v'}$ as per the partition set by $\widetilde{\mathcal{V}}$ (and 0 otherwise), the measure of modularity in weighted directed graphs can be computed by:

$$Q(\widetilde{\mathcal{V}}) \doteq \frac{1}{|\sum_{\mathbf{W}}|} \sum_{v \in \mathcal{V}} \sum_{v' \in \mathcal{V}} \left[w_{v,v'} \frac{s_v^{in} s_{v'}^{out}}{|\sum_{\mathbf{W}}|} \right] \delta(v,v'), \tag{2}$$

where $|\sum_{\mathbf{W}}|$ denotes the sum weight of all edges of the network. Finding a *good* partition $\widetilde{\mathcal{V}}^*$ of a weighted directed network $\mathcal{G}$ can be then casted as:

$$\widetilde{\mathcal{V}}^* = \arg \max_{\widetilde{\mathcal{V}} \in \mathcal{B}_V} Q(\widetilde{\mathcal{V}}), \tag{3}$$

where $\mathcal{B}_V$ denotes the set of possible partitions of V elements into nonempty subsets (i.e. the solution space of the above combinatorial problem). The cardinality of this set, given by the V-th Bell number [12] is huge, thus calling for the adoption of heuristics for its efficient exploration. For example, a network instance with $V = 20$ nodes can be partitioned in approximately $517.24 \cdot 10^{12}$ different ways. Provided that the above metric could be computed in 1 microsecond on average, we would need more than one and a half years to exhaustively check all possible partitions.

3   Proposed Nature-Inspired Solvers

To efficiently solve the problem posed above, several discrete solvers are proposed. Before a deep detail of each method, common design aspects such as the encoding strategy, the solution repair method and the distance to compare different solutions are first described in what follows.

The first issue to be tackled is the encoding of solutions or individuals. In this work we adopt a label-based representation [15]: each solution is represented as a permutation $\mathbf{x} = [c_1, c_2, \ldots, c_V]$ of V integers from the range $[1, \ldots, V]$, where we recall that V represents the number of nodes in the network. The value of c_v denotes the cluster label to which node v belongs. For example, assuming a $V = 10$ network, one feasible solution could be $\mathbf{x} = [1,2,1,1,2,2,2,3,3,3]$, meaning that the partition represented by this vector is $\widetilde{\mathcal{V}} = \{\mathcal{V}_1, \mathcal{V}_2, \mathcal{V}_3\}$, where $\mathcal{V}_1 = \{1,3,4\}$, $\mathcal{V}_2 = \{2,5,6,7\}$ and $\mathcal{V}_3 = \{8,9,10\}$. In order to avoid ambiguities in the representation, a repairing procedure has been developed partly inspired from the one proposed in [9]. By applying this process to every newly produced solution, ambiguities such as those between $\mathbf{x} = [4,2,4,4,2,2,2,3,3,3]$ and $\mathbf{x} = [7,1,7,7,1,1,1,4,4,4]$ (which represent the same partition) are modified to $\mathbf{x} = [1,2,1,1,2,2,2,3,3,3]$.

Another important aspect of three of the proposed methods (namely, WCA, FA and ESA) is how the similarity between the different individuals is computed, which lies at the core of the constituent operators of such methods. To this end, the well-known Hamming distance has been selected. Some studies of the literature have previously used this function for similar purposes in other combinatorial problems [22], verifying its good performance for this purpose. Specifically, this function calculates the number of non-corresponding elements of both solutions. For example, assuming two individuals $\mathbf{x} = [1, 1, 2, \mathbf{3}, \mathbf{3}, \mathbf{2}, 2, 2, 3, 3, 1]$ and $\mathbf{x}' = [1, 1, 2, \mathbf{2}, 3, \mathbf{3}, 2, 3, 3, \mathbf{2}]$, their Hamming Distance $D_H(\mathbf{x}, \mathbf{x}')$ would equal 3.

Furthermore, four different movement operators have been developed for evolving individuals along the search process. These functions are applied depending on the distance between two individuals (in the case of FA and WCA), or depending on the nature of the solution (in the case of ESA and PVNS). These operators are called CE_1, CE_3, CC_1 and CC_3. For each of these functions, the subscript represents the number of randomly selected nodes, which are extracted from its corresponding cluster. In CE_* operators, the taken nodes are re-inserted in already existing clusters, while in CC_* nodes can be inserted also in newly generated clusters.

WCA: This solver was first conceived for solving continuous optimization problems. For this reason, an adaptation has been made in order to address combinatorial optimization problems such as the one addressed in this paper. Along with the encoding and the measurement of distance method, the most critical aspect is how rivers and streams flow to their corresponding leading raindrop. Following the philosophy of the original WCA, the movement of each stream $p_{str} \in \mathcal{P}_{str}$ towards its river $\lambda(p_{str})$ at each generation $t \in \{1, \ldots, T\}$ is set to:

$$\mathbf{x}^{p_{str}}(t + 1) = \Psi\left(\mathbf{x}^{p_{str}}(t), \min\left\{V, \lfloor rand \cdot \theta \cdot D_H(\mathbf{x}^{p_{str}}(t), \mathbf{x}^{\lambda(p_{str})}(t))\rfloor\right\}\right), \quad (4)$$

where θ is a heuristic parameter, $rand$ is a continuous random variable uniformly distributed in $\mathbb{R}[0, 1]$, and $\Psi(\mathbf{x}, Z) \in \{CE_1, CE_3, CC_1, CC_3\}$, each parametrized by the number of times Z this operator is applied to the raindrop $\mathbf{x}$. The best position resulting from the Z movements performed on $\mathbf{x}$ is chosen as the output of the operator. The same philosophy is followed for the movement of a stream or a river towards the sea, simply by replacing $\mathbf{x}^{\lambda(p_{str})}(t)$ by $\mathbf{x}^{p_{sea}}(t)$.

Furthermore, in order to further enhance the exploration capacity of the technique, the *inclination* mechanism recently introduced in [22] is also used in the WCA developed in this work. Thanks to this mechanism, the method intelligently selects the proper movement function to use at each iteration for each raindrop, depending on its specific situation. Particularly, each time a raindrop is ready to perform a movement, the so-called *inclination* $\xi(\mathbf{x}, \mathbf{x}')$ is calculated, using as reference the $D_H(\mathbf{x}, \mathbf{x}')$ to its designated river/sea $\mathbf{x}'$. Specifically, $\xi(\cdot, \cdot)$ is set equal to $V/D_H(\cdot, \cdot)$. Taking into account that the bigger $D_H(\cdot, \cdot)$ is, the higher $\xi(\cdot, \cdot)$ should be, a *fast move* should be enforced with a higher probability if the inclination is high. On the other hand, if $D_H(\cdot, \cdot)$ is small the inclination decreases, suggesting that the search is in a promising area of the solution space, and performing a *slow move* with higher probability. In this research, instead of having only two different functions (as occurs in [22]), four different operators are available: CC_* functions are considered *fast moves*, whereas

CE_* are deemed *slow moves*. Finally, the philosophy of both evaporation and raining concepts remain in the same way as in the basic WCA, acting similarly to a mutation operator in Genetic Algorithms. Concretely, the raining process comprises a number R of consecutive CC_3 movements.

FA: As in the case of WCA, the classic FA cannot be either applied directly to address a discrete problem. For this reason, some modifications have been included. First, each firefly in the swarm represents a solution for the problem. The concept of light absorption is also considered for this adaptation, which is crucial for the adjustment of fireflies' attractiveness. The distance between two different fireflies is represented by the Hamming Distance. Finally, the movement of a firefly attracted to another brighter firefly is determined similarly to Expression (4). At last, and emulating the concept of *inclination* introduced for the WCA, when a firefly is prepared to perform a movement to another firefly, it examines its distance. If it is higher than $V/2$, it can be assumed that it is far from its counterpart. Therefore, it carries out a *wide move*, using a CC_* operator. Otherwise, a *short move* is performed by a CE_* function.

ESA: For the sake of fairness w.r.t. the rest of considered solvers, a population-based evolutionary version of the naïve SA has been used [32]. ESA counts with the same four CC_* and CE_* movement functions, meaning that each individual has its own randomly assigned operator. Furthermore, each population element has its own temperature value drawn at random from $\mathbb{R}^+[0.7, 1.0]$ and kept fixed over the entire search process. A step forward in the adaptation of this method is done by also exploiting Eq. (4) for the individual's movement, using the distance to the best individual of the population as the reference. This way, each individual performs a number of $D_H(\cdot, \cdot)$ separated movements, from which the best one is selected. Finally, the best individual performs a random number of movements between 1 and Z.

PVNS: A population-based approach of the original VNS has been designed for this problem. Following the same philosophy considered for ESA, each individual of the population has its own main movement operator, randomly selected among CE_1, CE_3, CC_1 and CC_3. At each generation, every individual of the population performs a movement using its main operator, but it may choose a different one with probability 0.25.

4   Experimentation and Results

In order to assess the performance of the above 4 heuristic solvers, computer experiments are run over a heterogeneous set of synthetically generated network instances. Specifically, the benchmark is composed by networks with $V \in \{35, 50, 75\}$ nodes. For each network, a different number of *ground of truth* communities is modeled by first creating a partition of the network (with random sizes for its constituent groups $\{\mathcal{V}_m\}_{m=1}^M$), and then by connecting nodes within every group with probability p_{in} and nodes of different groups with probability p_{out}. Therefore, the *ground of truth* partition should be more discriminable if p_{in} is high and p_{out} is low. Weights $w_{v,v'}$ for every edge (v, v') have been drawn uniformly at random from ranges $\mathbb{R}[0.0, 10.0]$ (inter-community edges) and $\mathbb{R}[10.0, 20.0]$ (intra-community edges). This network construction process permits to assess the performance of the proposed solvers over *noisy* versions of a graph characterized by a controlled underlying community distribution, as opposed to the common practice by which such a comparison is made based on the attained fitness value of every technique. Finally, 15 independent runs have been executed for each dataset, with the main goal of providing statistically reliable insights on the performance of every method. In relation to the ending criterion, each run finishes when there are $V + \sum_{v=1}^V v = V(V+3)/2$ iterations without improvements in the best solution found. The population size is set to 50 individuals for all cases. In the specific case of WCA, the number of rivers has been set to 9 raindrops (approximately 20% of the whole population), leading to a number of streams equal to 40. On the other hand, the maximum distance for evaporation) and R have been respectively set to 5% and a uniform random value from $\mathbb{N}[0, \lfloor 0.5V \rfloor]$.

Table 1. Obtained NMI results (average/best/standard deviation) using WCA, ESA, FA and PVNS. Best average results have been highlighted in bold.

(V, M, p_{in}, p_{out})	WCA Avg	Best	Std	t_{conv}	ESA Avg	Best	Std	t_{conv}	FA Avg	Best	Std	t_{conv}	PVNS Avg	Best	Std	t_{conv}
$(35, 4, 0.6, 0.1)$	**0.526**	0.526	0.000	0.89	0.515	0.526	0.010	1.33	0.521	0.547	0.010	1.40	**0.526**	0.526	0.000	3.6
$(35, 4, 0.9, 0.4)$	**0.876**	0.876	0.000	1.34	0.860	0.876	0.010	1.23	0.745	0.768	0.010	1.40	**0.876**	0.876	0.000	4.43
$(35, 7, 0.6, 0.1)$	**1.000**	1.000	0.000	0.98	0.972	1.000	0.010	1.30	0.900	0.929	0.010	1.91	**1.000**	1.000	0.000	3.19
$(35, 7, 0.6, 0.4)$	0.807	0.807	0.000	1.85	**0.827**	0.863	0.010	1.32	0.800	0.828	0.010	1.59	0.806	0.807	0.010	3.95
$(35, 7, 0.8, 0.1)$	**1.000**	1.000	0.000	0.75	0.997	1.000	0.010	1.28	0.927	0.949	0.010	1.66	**1.000**	1.000	0.000	3.67
$(35, 7, 0.9, 0.4)$	**1.000**	1.000	0.000	0.80	0.997	1.000	0.010	1.29	0.914	0.935	0.010	1.95	**1.000**	1.000	0.000	4.34
$(35, 18, 0.6, 0.1)$	**0.960**	0.969	0.010	2.57	0.931	0.962	0.010	1.30	0.952	0.973	0.010	1.47	0.954	0.969	0.010	3.29
$(35, 18, 0.9, 0.4)$	**0.998**	1.000	0.010	1.21	0.971	0.974	0.010	1.34	0.974	0.990	0.010	1.97	**0.998**	1.000	0.010	3.52
$(50, 5, 0.6, 0.1)$	**1.000**	1.000	0.000	1.59	0.998	1.000	0.010	2.36	0.821	0.851	0.010	3.73	**1.000**	1.000	0.000	9.13
$(50, 5, 0.6, 0.4)$	**0.694**	0.699	0.010	4.80	0.680	0.699	0.010	2.71	0.640	0.658	0.010	4.66	0.689	0.699	0.010	9.12
$(50, 5, 0.9, 0.1)$	**1.000**	1.000	0.000	1.45	0.996	1.000	0.010	2.65	0.825	0.905	0.030	3.02	**1.000**	1.000	0.000	7.51
$(50, 10, 0.7, 0.4)$	**0.972**	0.972	0.000	1.80	0.971	1.000	0.010	2.63	0.893	0.908	0.010	4.20	**0.972**	0.972	0.010	8.72
$(50, 10, 0.9, 0.4)$	**1.000**	1.000	0.000	1.35	0.989	1.000	0.010	2.67	0.941	0.962	0.010	3.74	**1.000**	1.000	0.000	8.04
$(50, 25, 0.6, 0.1)$	**0.979**	0.989	0.010	6.31	0.952	0.965	0.010	2.74	0.955	0.969	0.010	3.14	0.967	0.977	0.010	9.27
$(50, 25, 0.6, 0.4)$	**0.955**	0.968	0.010	4.94	0.942	0.961	0.010	2.77	0.944	0.961	0.010	3.27	0.947	0.968	0.010	8.55
$(50, 25, 0.9, 0.4)$	**0.990**	0.991	0.010	3.73	0.971	0.987	0.010	2.79	0.970	0.980	0.010	3.19	0.982	0.991	0.010	8.99
$(75, 8, 0.6, 0.1)$	**0.987**	1.000	0.010	6.65	0.959	1.000	0.010	4.71	0.828	0.844	0.010	8.03	0.971	1.000	0.010	22.57
$(75, 8, 0.8, 0.3)$	**1.000**	1.000	0.000	2.69	**1.000**	1.000	0.000	4.55	0.865	0.888	0.010	7.85	**1.000**	1.000	0.000	21.83
$(75, 8, 0.9, 0.4)$	**1.000**	1.000	0.000	2.25	**1.000**	1.000	0.000	4.27	0.896	0.919	0.010	9.99	**1.000**	1.000	0.000	22.72
$(75, 15, 0.6, 0.2)$	**0.986**	0.987	0.010	5.98	0.982	0.989	0.010	5.48	0.892	0.917	0.010	8.27	0.984	0.989	0.010	22.98
$(75, 30, 0.6, 0.1)$	**0.971**	0.976	0.010	12.60	0.949	0.973	0.010	5.48	0.943	0.956	0.010	8.53	0.956	0.966	0.010	18.50
$(75, 30, 0.8, 0.4)$	**0.966**	0.970	0.010	16.53	0.951	0.971	0.010	5.77	0.939	0.955	0.010	8.64	0.958	0.979	0.010	23.30
$(75, 38, 0.9, 0.1)$	**0.984**	0.993	0.010	13.27	0.972	0.981	0.010	6.28	0.972	0.979	0.010	8.42	0.973	0.981	0.010	20.78
$(75, 38, 0.9, 0.4)$	**0.985**	0.993	0.010	18.35	0.968	0.981	0.010	6.06	0.970	0.982	0.010	9.56	0.973	0.994	0.010	22.57
Friedman's non-parametric test (mean ranking)																
Rank	1.3333				3.1042				3.7292				1.8333			

In Table 1, the results (average/best/standard deviation) obtained by the four methods are displayed in terms of the Normalized Mutual Information (NMI) with respect to the *ground of truth* partition of every network. The NMI score quantifies the level of agreement between both community partitions ignoring label permutations: if $NMI(\widetilde{\mathcal{V}}, \widetilde{\mathcal{V}'}) = 1$ both community distributions $\widetilde{\mathcal{V}}$ and $\widetilde{\mathcal{V}'}$ are equal to each other, whereas lower values of this score denote that there are differences between them. A first inspection reveals that WCA outperforms the other methods in almost all the instances, with consistently superior average NMI scores for most networks. As expected, results degrade when the values of (p_{in}, p_{out}) imprint topological noise on the *ground of truth* partition of every network, as exemplified by instances $(V, M, p_{in}, p_{out}) = (50, 5, 0.6, 0.4)$ (for which the best partition found attains NMI = 0.699) and $(50, 5, 0.9, 0.1)$ (which is perfectly resolved by WCA and PVNS in all runs). The average iteration index at which every solver converges as per the aforementioned convergence criterion is also included in the table (column t_{conv}). Although ESA and FA converge faster on average (specially for large networks), their lower NMI scores when compared to PVNS and WCA belittle this computational advantage.

A Friedman's non-parametric test for multiple comparison has been carried out to resolve the statistical significance of the results (last row of the Table). The mean ranking returned by this test is displayed for each of the compared algorithms. Furthermore, the Friedman statistic obtained is 53.0125. The confidence interval has been set in 99%, being 11.34 the critical point in a χ^2 distribution with 3 degrees of freedom. Since $53.0125 > 11.34$, it can be concluded that there are significant differences among the results. Furthermore, to prove the significance between the best two techniques – namely, WCA and PVNS, a Wilcoxon Signed-Rank test has been applied. The confidence interval has been established at 99% also for this test. Regarding the difference in the obtained results, the obtained Z-value is -3.0594, with a p-value equal to 0.00222. These results support the significance of the difference at 99% confidence level. Besides that, the obtained Wilcoxon test statistic is 0. The critical value of this statistic for the datasets in which results are not at $p \leq 0.010$ is 7. Therefore, the result is significant at this confidence level, thereby concluding that the WCA approach is the best performing alternative in the designed benchmark.

5   Conclusions and Future Research Lines

In this work community detection in weighted directed graphs has been approached by using nature-inspired heuristics. To this end, the discovery of optimal partitions is formulated as an optimization problem driven by a measure of modularity adapted to accommodate the directional and weighted nature of the edges of the network. To efficiently undertake this optimization problem, different heuristic techniques have been designed and adapted to deal with the particularities of the solution space, such as the potential representational ambiguity of label encoding and the definition of distance between solutions to the problem. In addition, each designed heuristic is also modified to account for a better evolution of the individuals (partitions) found during the search process. Their performance has been compared over 24 network instances composed by

35, 50 and 75 nodes, using NMI with respect to their *ground of truth* partition as the comparison criterion. The obtained results reveal that WCA dominates the benchmark with statistically significance, specially for large networks.

In light of the promising results obtained in this research work, we plan to conduct further efforts in different directions. Additional nature-inspired, evolutionary and swarm intelligent methods will be included in the benchmark, which will also consider network instances of higher scales than the ones used in this work. Moreover, we will explore how to hybridize the aforementioned heuristic solvers with local search techniques mimicking the operation of other heuristics found in the literature, such as recently contributed message passing procedures [26] and other techniques renowned for their good scalability [19].

Acknowledgements. E. Osaba and J. Del Ser would like to thank the Basque Government for its funding support through the EMAITEK program. I. Fister Jr. and I. Fister acknowledge the financial support from the Slovenian Research Agency (Research Core Fundings No. P2-0041 and P2-0057). A. Iglesias and A. Galvez acknowledge the financial support from the projects TIN2017-89275-R (AEI/FEDER, UE), PDE-GIR (H2020, MSCA program, ref. 778035), and JU12 (SODERCAN/FEDER UE).

References

1. Aldecoa, R., Marín, I.: Deciphering network community structure by surprise. PloS ONE **6**(9), e24195 (2011)
2. Bello-Orgaz, G., Jung, J.J., Camacho, D.: Social big data: recent achievements and new challenges. Inf. Fusion **28**, 45–59 (2016)
3. Blondel, V.D., Guillaume, J.L., Lambiotte, R., Lefebvre, E.: Fast unfolding of communities in large networks. J. Stat. Mech.: Theory Exp. **2008**(10), P10008 (2008)
4. Chakraborty, T., Dalmia, A., Mukherjee, A., Ganguly, N.: Metrics for community analysis: a survey. ACM Comput. Surv. (CSUR) **50**(4), 54 (2017)
5. Chakraborty, T., Srinivasan, S., Ganguly, N., Mukherjee, A., Bhowmick, S.: On the permanence of vertices in network communities. In: ACM SIGKDD International Conference on Knowledge Discovery and Data Mining, pp. 1396–1405. ACM (2014)
6. Cockbain, E., Brayley, H., Laycock, G.: Exploring internal child sex trafficking networks using social network analysis. Policing: J. Policy Pract. **5**(2), 144–157 (2011)
7. Del Ser, J., Lobo, J.L., Villar-Rodriguez, E., Bilbao, M.N., Perfecto, C.: Community detection in graphs based on surprise maximization using firefly heuristics. In: IEEE Congress on Evolutionary Computation (CEC), pp. 2233–2239. IEEE (2016)
8. Eskandar, H., Sadollah, A., Bahreininejad, A., Hamdi, M.: Water cycle algorithm - a novel metaheuristic optimization method for solving constrained engineering optimization problems. Appl. Soft Comput. **110**(111), 151–166 (2012)
9. Falkenauer, E.: Genetic Algorithms and Grouping Problems. Wiley, New York (1998)
10. Guerrero, M., Montoya, F.G., Baños, R., Alcayde, A., Gil, C.: Adaptive community detection in complex networks using genetic algorithms. Neurocomputing **266**, 101–113 (2017)

11. Hafez, A.I., Zawbaa, H.M., Hassanien, A.E., Fahmy, A.A.: Networks community detection using artificial bee colony swarm optimization. In: Kömer, P., Abraham, A., Snášel, V. (eds.) Proceedings of the Fifth International Conference on Innovations in Bio-Inspired Computing and Applications IBICA 2014. AISC, vol. 303, pp. 229–239. Springer, Cham (2014). https://doi.org/10.1007/978-3-319-08156-4_23

12. Harris, J.M., Hirst, J.L., Mossinghoff, M.J.: Combinatorics and Graph Theory, vol. 2. Springer, New York (2008). https://doi.org/10.1007/978-0-387-79711-3

13. Hassan, E.A., Hafez, A.I., Hassanien, A.E., Fahmy, A.A.: A discrete bat algorithm for the community detection problem. In: Onieva, E., Santos, I., Osaba, E., Quintián, H., Corchado, E. (eds.) HAIS 2015. LNCS (LNAI), vol. 9121, pp. 188–199. Springer, Cham (2015). https://doi.org/10.1007/978-3-319-19644-2_16

14. Honghao, C., Zuren, F., Zhigang, R.: Community detection using ant colony optimization. In: IEEE Congress on Evolutionary Computation (CEC), pp. 3072–3078. IEEE (2013)

15. Hruschka, E.R., Campello, R.J., Freitas, A.A.: A survey of evolutionary algorithms for clustering. IEEE Trans. Syst. Man Cybern. Part C (Appl. Rev.) **39**(2), 133–155 (2009)

16. Jia, G., et al.: Community detection in social and biological networks using differential evolution. In: Hamadi, Y., Schoenauer, M. (eds.) LION 2012. LNCS, pp. 71–85. Springer, Heidelberg (2012). https://doi.org/10.1007/978-3-642-34413-8_6

17. Lara-Cabrera, R., Pardo, A.G., Benouaret, K., Faci, N., Benslimane, D., Camacho, D.: Measuring the radicalisation risk in social networks. IEEE Access **5**, 10892–10900 (2017)

18. Leicht, E.A., Newman, M.E.: Community structure in directed networks. Phys. Rev. Lett. **100**(11), 118703 (2008)

19. Lu, H., Halappanavar, M., Kalyanaraman, A.: Parallel heuristics for scalable community detection. Parallel Comput. **47**, 19–37 (2015)

20. Newman, M.E.: Analysis of weighted networks. Phys. Rev. E **70**(5), 056131 (2004)

21. Newman, M.E., Girvan, M.: Finding and evaluating community structure in networks. Phys. Rev. E **69**(2), 026113 (2004)

22. Osaba, E., Del Ser, J., Sadollah, A., Bilbao, M.N., Camacho, D.: A discrete water cycle algorithm for solving the symmetric and asymmetric traveling salesman problem. Appl. Soft Comput. **71**, 277–290 (2018)

23. Pizzuti, C.: GA-Net: a genetic algorithm for community detection in social networks. In: Rudolph, G., Jansen, T., Beume, N., Lucas, S., Poloni, C. (eds.) PPSN 2008. LNCS, vol. 5199, pp. 1081–1090. Springer, Heidelberg (2008). https://doi.org/10.1007/978-3-540-87700-4_107

24. Pizzuti, C.: Evolutionary computation for community detection in networks: a review. IEEE Trans. Evol. Comput. **22**(3), 464–483 (2018)

25. Rahimi, S., Abdollahpouri, A., Moradi, P.: A multi-objective particle swarm optimization algorithm for community detection in complex networks. Swarm Evol. Comput. **39**, 297–309 (2018)

26. Shi, C., Liu, Y., Zhang, P.: Weighted community detection and data clustering using message passing. J. Stat. Mech.: Theory Exp. **2018**(3), 033405 (2018)

27. Tasgin, M., Herdagdelen, A., Bingol, H.: Community detection in complex networks using genetic algorithms. arXiv preprint arXiv:0711.0491 (2007)

28. Villar-Rodríguez, E., Del Ser, J., Torre-Bastida, A.I., Bilbao, M.N., Salcedo-Sanz, S.: A novel machine learning approach to the detection of identity theft in social networks based on emulated attack instances and support vector machines. Concurr. Comput.: Pract. Exp. **28**(4), 1385–1395 (2016)

29. Wang, X., Tang, L.: A population-based variable neighborhood search for the single machine total weighted tardiness problem. Comput. Oper. Res. **36**(6), 2105–2110 (2009)
30. Westlake, B.G., Bouchard, M.: Liking and hyperlinking: community detection in online child sexual exploitation networks. Soc. Sci. Res. **59**, 23–36 (2016)
31. Yang, X.S.: Firefly algorithm, stochastic test functions and design optimisation. Int. J. Bio-Inspir. Comput. **2**(2), 78–84 (2010)
32. Yip, P.P., Pao, Y.H.: Combinatorial optimization with use of guided evolutionary simulated annealing. IEEE Trans. Neural Netw. **6**(2), 290–295 (1995)

A Metaheuristic Approach
for the α-separator Problem

Sergio Pérez-Peló⬤, Jesús Sánchez-Oro$^{(\boxtimes)}$⬤, and Abraham Duarte⬤

Department of Computer Sciences, Universidad Rey Juan Carlos, Madrid, Spain
{sergio.perez.pelo,jesus.sanchezoro,abraham.duarte}@urjc.es

Abstract. Most of the critical infrastructures can be easily modeled as a network of nodes interconnected among them. If one or more nodes of the network fail, the connectivity of the network can be compromised, to the point of completely disconnecting the network. Additionally, disconnecting the network can result in cascade failures, because the remaining nodes may be overloaded because of heavy traffic in the network. One of the main objectives of an attacker is to isolate the nodes whose removal disconnect the network in minimum size subnetworks. On the contrary, a defender must identify those weak points in order to maintain the network integrity. This work is focused on solving the α separator problem, whose main objective is to find a minimum set of nodes that disconnect a network in isolated subnetworks of size smaller than a given value. The problem is tackled from a metaheuristic point of view, analyzing the solutions given by a Greedy Randomized Adaptive Search Procedure over different network topologies. The results obtained are compared with the best algorithm found in the literature.

Keywords: Alpha-separator · GRASP · Networks · Critical nodes

1 Introduction

From large companies and institutions to individual users, cybersecurity is becoming one of the main concerns of the last years. Most of the Internet users nowadays have their personal and professional information stored in the cloud, including sensitive and private information that cannot be revealed. Therefore, analyzing the risks of the underlying network is one of the main tasks to be accomplished by the owner of the network, in order to guarantee the maximum security to their users. However, network attacks are continuously increasing since attackers obtain benefits if they are able to access to that private information. If a company suffers a cyber attack, the loss of information and privacy usually result in the loss of a vast number of clients, thus implying important economic and social damage [1]. Two of the most common cyber attacks are Denial of Service (DoS) and Distributed Denial of Service (DDoS), which are rather dangerous since they can disable the complete service of an Internet provider, or even produce a cascade failure that affects to a large number of clients [3].

© Springer Nature Switzerland AG 2018
H. Yin et al. (Eds.): IDEAL 2018, LNCS 11315, pp. 336–343, 2018.
https://doi.org/10.1007/978-3-030-03496-2_37

The analysis of the security of the network usually starts identifying its most relevant nodes. The responsible of the security of the network is interested of knowing which nodes should be reinforced with modern and robust security measures in order to reduce the impact of a cyber attack in the network. Meanwhile, the attacker needs to know which are the most vulnerable nodes of the network in order to perform an efficient attack to a small number of nodes, maximizing the damage caused. Therefore, both actors are competing for finding the most relevant nodes of a network in the shortest time.

Networks can be easily modeled as graphs. Let $G = (V, E)$ be an undirected graph that represents a network where the set of nodes is modeled as the set of vertices V, with $|V| = n$, and the set of connections among nodes is modeled as the set of edges E, with $|E| = m$. A separator S for a network G is defined as a set of vertices $S \subseteq V$ which is able to split the network into two or more connected components $C_1, C_2, \ldots, C_n$ if nodes in S are removed from the network. In mathematical terms,

$$V \setminus S = C_1 \cup C_2 \ldots \cup C_p \forall\, (u, v) \in E^\star \; \exists\, C_i \; : \; u, v \in C_i \tag{1}$$

where $E^\star$ represents the set of edges with no endpoint in the separator S, i.e., $E^\star = \{(u, v) \in E \; : \; u, v \notin S\}$.

The α-separator problem (α-SP) tackled in this work consists in finding a separator $S^\star$ that divides a network G into connected components satisfying just one constraint: the size of each connected component must be smaller than a predefined threshold $\alpha \cdot n$, being α a parameter of the problem. More formally,

$$S^\star \leftarrow \underset{S \in \mathbb{S}}{\arg\min} |S| \; : \; \max_{1 \leq i \leq p} |C_i| \leq \alpha \cdot n \tag{2}$$

considering that the network is divided into p connected components after removing nodes in S.

The number of resulting connected components is not an indicator of the solution quality for the α-separator problem, since the objective function value is evaluated as the number of vertices included in the separator (i.e., $f(S) = |S|$). The α-SP has been proven $\mathcal{NP}$-hard for general network topologies for $\alpha \geq \frac{2}{3}$ [7]. Additionally, some special network topologies can be solved in polynomial time, including trees and cycles [11]. The polynomial-time algorithms for special networks are not considered in real-life problems since the topology of the network is not known a priori and the network does not usually follow a special structure.

Figure 1(a) shows a network with 9 vertices and 10 edges, while Figs. 1(b) and (c) we present two different solutions for the α-SP. Notice that we consider $\alpha = \frac{2}{3}$, and therefore, the connected components remaining after selecting a separator must contain, at most, $\lceil \frac{2}{3} \cdot 9 \rceil = 6$ vertices. Figure 1(b) presents a feasible solution S_1 where vertices B, C, and I are included in the separator, dividing the network into the following connected components: $C_1 = \{\text{A}, \text{D}\}$ (2 vertices), $C_2 = \{\text{F}, \text{G}, \text{H}\}$ (3 vertices, and $C_3 = \{\text{E}\}$ (1 vertex). The objective function of S_1 is 3, since there are three vertices in the separator. Solution S_2, depicted in Fig. 1(c), includes vertices A and B in the separator, dividing the network in $C_1 = \{D\}$, $C_2 = \{E\}$, and $C_3 = \{C, F, G, H, I\}$, all of them satisfying the size constraint. The objective function value of S_2 is $f(S_2) = 2$, which is smaller than $f(S_1)$, thus being S_2 a better solution than S_1, since it requires from a small number of nodes to be removed in order to disconnect the network.

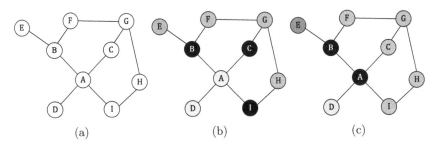

(a)   (b)   (c)

Fig. 1. 1(a) Example of a graph derived from a network, 1(b) a feasible solution with 3 nodes in the separator (B,C, and I), and 1(c) a better solution with 2 nodes in the separator (A and B)

The literature presents several exact and heuristic approaches for solving the α-SP. In particular, trees an cycles can be solved in polynomial time, and also a greedy algorithm with approximation ratio of $\alpha \cdot n + 1$ was presented [11]. The problem of detecting node separators in the Internet Autonomous Systems was tackled by means of a heuristic algorithm [14]. As fast as we know, the best heuristic approach is a random walk algorithm based on a Markov Chain Monte Carlo method [10].

The α-SP is equivalent to some well-known problems when considering specific values of α. In particular, if $\alpha = \frac{1}{n}$, then it is equivalent to the minimum vertex cover problem, while if $\alpha = \frac{2}{n}$, the α-SP becomes equivalent to the minimum dissociation set problem. As a consequence, α-SP is a generalization of these problems, which are also $\mathcal{NP}$-hard [9]. Moreover, the research on α-SP is interesting for different problems related to network theory and resilience assurance [4].

The remaining of the paper is structured as follows: Sect. 2 presents the algorithmic proposal for solving α-SP, Sect. 3 shows the experiments performed to test the proposal and, finally, Sect. 4 draws some conclusions on the research.

2   Algorithmic Proposal

This paper tackles the α-SP from a metaheuristic point of view following the Greedy Randomized Adaptive Search Procedure (GRASP) methodology. GRASP is a metaheuristic formally defined by Feo et al. [8]. We refer the reader to Resende and Ribeiro [13] for a complete survey on this methodology.

GRASP is a trajectory-based metaheuristic that follows a multi-start scheme conformed with two different phases: construction and improvement. The former consist of a greedy, randomized, and adaptive construction of a solution, while the latter is designed for improving the quality of the generated solution through a local search method, or a more complex improvement strategy like Tabu Search, for example. These two stages are applied iteratively until reaching a certain stopping criterion, which is usually a number of construction or a predefined computing time.

2.1 Constructive Method

The α-SP tries to find the minimum number of nodes that must be removed from the network in order to disconnect it, satisfying a constraint with respect to the size of the remaining subgraphs. With this in mind, we can use structural features of a node in a network, such as its degree, or its position in the network, among others, as a selection criterion for the separator.

Following GRASP methodology, the initial solution is constructed by considering a greedy constructive procedure which requires a greedy function value to evaluate the relevance of adding a node to the separator. In this work, we propose a novel approach that leverages the centrality measures traditionally used in the field of social network in order to evaluate the relevance of a node. In particular, we propose closeness centrality [2] as the greedy function for evaluating the nodes of a network. Given a vertex $v \in V$, its closeness is evaluated as:

$$C(v) = \frac{1}{\sum_{u \in V \setminus \{v\}} d(v, u)} \tag{3}$$

where $d(v, u)$ is evaluated as the distance between nodes v and u. In the context of α-SP, we evaluate $d(v, u)$ as the length of the shortest path that connects v and u in the network.

Analyzing the definition of the closeness metric, the larger the closeness, the shorter the distances to other nodes and, therefore, the closer it is to the remaining nodes of the network. Thus, the node with the largest value of closeness in the network is the most reachable one, which usually coincides with the most relevant one. Therefore, removing those nodes with large closeness values will eventually disconnect the network faster.

A completely greedy algorithm always produces the same solution, but it has been shown that considering some randomness in the construction usually leads to better initial solutions [5,12], which corresponds to the randomized part of the GRASP methodology.

We propose a greedy randomized adaptive constructive procedure which starts by selecting the first vertex v_f at random from the graph. Next up, a Candidate List CL is created with all the vertices in V except v_f, i.e., $v \in V \setminus \{v\}$. Then, the method iterates until reaching a feasible solution which satisfies the constraint of the α-SP. Specifically, in each iteration a Restricted Candidate List, RCL, is created with all the vertices whose closeness value is larger than a predefined threshold μ. The threshold is evaluated as:

$$\mu = g_{\max} - \beta \cdot (g_{\max} - g_{\min}) \tag{4}$$

where $g_{\min}$ and $g_{\max}$ are the smallest and largest closeness values, respectively, among all the vertices in the CL. The next vertex to be included in the solution is then randomly selected from the RCL. These steps are repeated until reaching a feasible solution. It is worth mentioning that $\beta \in [0, 1]$ is an input parameter of the constructive procedure which controls the randomness of the method. On the one hand, if $\beta = 0$, the RCL is constructed with those vertices with maximum closeness value, resulting in a totally greedy algorithm. On the other hand, when $\beta = 1$, all vertices are included in the RCL, which corresponds to a completely random procedure. Experiments performed in Sect. 3 will test different values for this parameter, analyzing how it affects to the quality of the generated solutions.

2.2   Local Improvement

The second stage of a GRASP algorithm consists of improving the constructed solution in order to find a local optimum. To achieve this, it is possible to consider a simple local search method or more complex search procedures, even hybridizing GRASP with other complete metaheuristics as Variable Neighborhood Search, or Tabu Search, among others. Notice that the more complex the improvement, the more computationally demanding.

Bearing in mind that the α-SP requires from obtaining solutions in the shortest possible computing time, we have considered a local search method in order to find a local optimum for the solution constructed with the aforementioned constructive procedure.

The local search proposed for the α-SP considers removing two vertices from the solution and include a single one to replace them. Notice that any feasible move performed with this local search will lead us to a better solution, since the objective function will be reduced in one unit. However, the main drawback of this local search is that it is not easy to reach a feasible solution, since the interchange of two vertices by a single one usually violates the constraint regarding the size of the resulting connected components.

For this reason, the proposed local search traverses all available pairs of vertices in the solution, interchanging them for every vertex not considered in the incumbent solution. This exhaustive exploration of the search space lead us to maximize the probability of finding a feasible solution thus improving the results.

Traditionally, local search methods follows two different schemes: best or first improvement. The former explores the complete neighborhood, performing the best move found, while the latter performs the first move that leads to a better solution. Notice that, in the context of the proposed local search, all the feasible moves leads to the same improvement of the objective function value. Because of this, there is no reason for considering a best improvement approach and, therefore, the proposed local search method follows a first improvement scheme.

3 Computational Results

This Section analyzes and discusses the results obtained by the proposed algorithms, comparing them with the best method found in the state of the art. All the algorithms have been implemented in Java 9, and the experiments have been conducted on an Intel Core 2 Duo 2.66 GHz with 4 GB RAM.

In order to have a fair comparison, we have considered the same type of instances that the ones proposed in the best previous method found in the literature [10]. However, due to the impossibility to contact with the previous authors, the instances are not exactly the same. We have generated the set of instances following the Erdös-Rényi model [6], as stated in the previous work. In particular, we have considered a set of 50 instances with nodes ranging from 100 to 200 and edges ranging from 200 to 2000.

The experiments have been divided into two different sections. The preliminary experiments are devoted to find the best value for the β parameter of the GRASP algorithm, while the final experimentation compares the results obtained by the GRASP algorithm with the ones obtained by the best previous method.

In all the experiments we report the same metrics: Avg., the average objective function value; Time (s), the average computing time in seconds; Dev (%), the average deviation with respect to the best solution found in the experiment; and # Best, the number of times that an algorithm reaches the best solution of the experiment.

The preliminary experiment, designed for tuning the β parameter, considers a subset of 20 out of 50 representative instances to avoid overfitting. We have tested the following values for $\beta = \{0.25, 0.50, 0.75, RND\}$, where RND value indicates that a random value in the range 0–1 is selected for each constructed solution. Table 1 shows the results obtained when constructing and improving 100 independent solutions, returning the best solution found.

Table 1. Performance of GRASP constructive procedure with different values for the β parameter.

β	Avg.	Time (s)	Dev(%)	#Best
0.25	54.20	298.34	1.58	12
0.5	54.55	289.95	1.93	10
0.75	55.95	308.25	4.31	7
−1.00	54.45	294.75	1.88	10

Analyzing Table 1 we can clearly see that the best results are obtained when considering $\beta = 0.25$, closely followed by $\beta = RND$. Specifically, $\beta = 0.25$ is able to find 12 out of 20 best solutions, and the average deviation of 1.58% indicates that in those instances in which it is not able to reach the best value, the constructive method obtains a high quality solution really close to the best one. The value of $\beta = 0.25$ indicates that the best results are obtained when introducing a small random part in the constructive procedure, and increasing the randomness of the method results in worse solutions. The worst results are obtained with the largest β value, 0.75, obtaining just 7 out of 20 best solutions with an average deviation of 4.31%. This behavior also confirms that the closeness metric is a good selection as a greedy function value for the constructive procedure.

The final experiment is intended to compare the best variant of GRASP with the best previous method found in the state of the art [10]. Specifically, it consists of a random walk (RW) algorithm with Markov Chain Monte Carlo method. It is worth mentioning that we have not been able to contact to the authors of the previous work neither to obtain the set of instances nor an executable file of the algorithm. Therefore, we have reimplemented the previous algorithm following, in detail, all the steps described in the manuscript. Table 2 shows the results obtained by the proposed algorithm (GRASP with $\beta = 0.25$) and the best previous method found (RW).

Table 2. Comparison of the best variant of RVNS with the best previous method found in the state of the art.

	Avg.	Time (s)	Dev (%)	#Best
RW	71.78	1070.35	27.26	4
GRASP	55.1	299.19	0.00	50

The results obtained shows that our proposal reaches better results than previous work. In particular, GRASP is able to obtain 50 out of 50 best solutions, while RW only reaches the best solution in 4 out of 50 instances. Furthermore, the computing time for GRASP is less than half of the time required by RW. Finally, the average deviation of GRASP is zero, which indicates that in all instances GRASP is able to match the best solution. However, the deviation of RW is higher, indicating that its results are not close to the best solution obtained by the GRASP.

4   Conclusions

This work has proposed a Greedy Randomized Adaptive Search Procedure algorithm for detecting critical nodes in networks. The greedy criterion for this GRASP algorithm is a criterion adapted from the social network field of research, which is named closeness. The GRASP proposal is able to obtain better results than the best previous method found in the literature, which consists of a random walk algorithm in both quality and computing time. This results, supported by non-parametric statistical tests, confirms the superiority of the proposal. The adaptation of a social network metric to the problem under consideration in this work has led us to obtain high quality solutions, which reveals the relevance of the synergy among different fields of research.

Acknowledgements. This work has been partially founded by Ministerio de Economía y Competitividad with grant ref. TIN2015-65460-C2-2-P.

References

1. Andersson, G., et al.: Causes of the 2003 major grid blackouts in North America and Europe, and recommended means to improve system dynamic performance. IEEE Trans. Pow. Syst. **20**(4), 1922–1928 (2005)
2. Bavelas, A.: Communication patterns in task-oriented groups. Acoust. Soc. Am. J. **22**, 725 (1950)
3. Crucitti, P., Latora, V., Marchiori, M.: Model for cascading failures in complex networks. Phys. Rev. E **69**, 045104 (2004)
4. Cuadra, L., Salcedo-Sanz, S., Del Ser, J., Jiménez-Fernández, S., Geem, Z.W.: A critical review of robustness in power grids using complex networks concepts. Energies **8**(9), 9211–9265 (2015)
5. Duarte, A., Sánchez-Oro, J., Resende, M.G., Glover, F., Martí, R.: Greedy randomized adaptive search procedure with exterior path relinking for differential dispersion minimization. Inf. Sci. **296**, 46–60 (2015)
6. Erdős, P., Rényi, A.: On random graphs. Publications Mathematicae **6**, 290 (1959)
7. Feige, U., Mahdian, M.: Finding small balanced separators. In: Kleinberg, J.M. (ed.) STOC, pp. 375–384. ACM (2006)
8. Feo, T.A., Resende, M.G., Smith, S.H.: Greedy randomized adaptive search procedure for maximum independent set. Oper. Res. **42**(5), 860–878 (1994)
9. Garey, M., Johnson, D.: Computers and Intractability - A Guide to the Theory of NP-Completeness. Freeman, San Fransisco (1979)
10. Lee, J., Kwak, J., Lee, H.W., Shroff, N.B.: Finding minimum node separators: a Markov chain Monte Carlo method. In: 13th International Conference on DRCN 2017 - Design of Reliable Communication Networks, pp. 1–8, March 2017
11. Mohamed-Sidi, M.: K-Separator Problem (Problème de k-Sèparateur). Ph.D. thesis, Telecom & Management SudParis, Èvry, Essonne, France (2014)
12. Quintana, J.D., Sánchez-Oro, J., Duarte, A.: Efficient greedy randomized adaptive search procedure for the generalized regenerator location problem. Int. J. Comput. Intell. Syst. **9**(6), 1016–1027 (2016)
13. Resende, M.G.C., Ribeiro, C.C.: GRASP: greedy randomized adaptive search procedures. In: Burke, E., Kendall, G. (eds.) Search Methodologies, pp. 287–312. Springer, Boston (2014). https://doi.org/10.1007/978-1-4614-6940-7_11
14. Wachs, M., Grothoff, C., Thurimella, R.: Partitioning the internet. In: Martinelli, F., Lanet, J.L., Fitzgerald, W.M., Foley, S.N. (eds.) CRiSIS, pp. 1–8. IEEE Computer Society (2012)

Author Index

Printed in the United States
By Bookmasters